M000209283

Matroid Theory

D. J. A. Welsh

*Merton College
and
The Mathematical Institute,
University of Oxford*

DOVER PUBLICATIONS, INC.
Mineola, New York

Copyright

Copyright © 1976 by D. J. A. Welsh
All rights reserved.

Bibliographical Note

This Dover edition, first published in 2010, is an unabridged republication of the work originally published in 1976 by Academic Press, Inc., New York.

Library of Congress Cataloging-in-Publication Data

Welsh, D. J. A.
   Matroid theory / D. J. A. Welsh. -- Dover ed.
      p. cm.
   Originally published: New York : Academic Press, 1976.
   Includes bibliographical references and index.
   ISBN-13: 978-0-486-47439-7
   ISBN-10: 0-486-47439-9
    1. Matroids. I. Title.

QA166.6.W44 2010
511'.6—dc22

2009048800

Manufactured in the United States by Courier Corporation
47439901
www.doverpublications.com

Preface

This book is an attempt to show the unifying and central role which matroids have played in combinatorial theory over the past decade. This is not to say that all aspects of combinatorial theory can be covered by the matroid umbrella; however, many parts of graph theory, transversal theory, block designs and combinatorial lattice theory can be more clearly understood by the use of matroids. Furthermore, since matroids are closely related to classical linear algebra and geometry they serve as a link between combinatorics and the more mainstream areas of mathematics.

The first half of this book can be regarded as a basic introduction to matroid theory. Most theorems are proved or an exact reference is given. The second half is an attempt to place the reader at the frontier of the subject. At this level I have found it impossible to prove every result. However I have, I hope, treated in some detail the more central and important topics, others I have set as exercises. Those exercises which are followed by a reference to some paper will usually be non-routine.

I have lectured on most of this book at various universities; the first half is the core of a course on combinatorics that I have given to third year undergraduates at Oxford. The other chapters have been covered at various times in M.Sc. level courses at Oxford and Waterloo.

I take this opportunity to acknowledge a deep sense of gratitude to a number of friends who in different ways helped in the production of this book.

C. St. J. A. Nash-Williams first introduced me to the subject with a most stimulating seminar on the applications of matroids in 1966. I am also very grateful to him for making it possible for me to visit the University of Waterloo where an early draft of the first half was prepared. F. Harary encouraged me to write this book and enabled me to try out a very preliminary version at the University of Michigan, Ann Arbor. The bulk of the book in its present form was drafted in the very happy and stimulating atmosphere of the mathematics department of the University of Calgary. I am deeply grateful to the department there and in particular to E. C. Milner for his unfailing kindness and patience in listening to my problems and queries.

I began to make a list of the people who had helped me over the last seven years either in discussion or by correspondence—it came to almost half the names listed in the bibliography. However, I do need to say that special thanks are due to D. W. T. Bean, A. W. Ingleton, C. J. H. McDiarmid, L. R. Mathews and J. G. Oxley, each of whom read substantial parts of the manuscript and had many helpful discussions on various points that have arisen.

Finally I would like to thank Clare Bass, Sheila Robinson and my wife Bridget for their help in preparing the final manuscript.

Merton College   Dominic Welsh
and
The Mathematical Institute, Oxford
July 1975

Contents

Chapter 20. Infinite Structures

Preliminaries

1. BASIC NOTATION

A reference to "item k" refers to item k of the same section; $(j.k)$ refers to the same chapter; $(i.j.k)$ refers to item k of section j of chapter i.

A reference $[k]$ refers to item k of the bibliography. References are by author's name and the last two digits of the year of publication, with additional letters to distinguish publications of the same author in the same year.

There are exercises at the end of most sections: the open problems are marked $^\circ$.

2. SET THEORY NOTATION

Throughout we adopt the usual set theoretic conventions: set-union, set-intersection, set inclusion and proper inclusion are denoted by the familiar symbols $\cup, \cap, \subseteq, \subset$, respectively.

For the most part we shall be considering structures on a finite set S. Elements of S will be denoted by lower case italic letters and subsets of S by italic capital letters. The *empty* set is denoted by $\varnothing$; $A\backslash B$ denotes the *set difference* of A and B; $A \triangle B$ denotes the *symmetric difference* of A and B,

$$A \triangle B = (A\backslash B) \cup (B\backslash A).$$

Two sets A, B are *incomparable* if neither is a subset of the other. As usual $\{x_1, \ldots, x_k\}$ denotes the set with elements $x_1, \ldots, x_k$ and when it is clear from the context that we are referring to a set rather than an element we abbreviate $\{x\}$ to x. For example $X \cup x$ means $X \cup \{x\}$, $X\backslash x$ means $X\backslash\{x\}$.

$|A|$ denotes the *cardinality* of the set A, and we write X is a *k-set* (*k-subset*) if X is a set (subset) of cardinality k. Suppose that we have a set A_i for each i in a non-empty *indexing set* I. We use $A(I)$ to denote the union of the A_i; $i \in I$. That is

$$A(I) = \{x : x \in A_i \text{ for some } i \in I\}.$$

1

A *function* or *map* from S to T is denoted by $f: S \rightarrow T$. If x belongs to the domain of f, $f(x)$ is called the image of x, and for $A \subseteq S$, the image of A, denoted by $f(A) = \{y: y \in T, f(x) = y \text{ for some } y \in A\}$.

If U is a subset of S we denote the *restriction* of f to U by $f|_U$.

The *power set* 2^S of S is the collection of subsets of S and a map $\phi: S \rightarrow T$ defines in the obvious way a map

$$2^\phi: 2^S \rightarrow 2^T$$

where for $X \subseteq S$,

$$2^\phi X = \{y: y = \phi x \quad \text{for some} \quad x \in X\}.$$

As usual we normally write ϕX rather than $2^\phi X$.

If S and I are sets and $\phi: I \rightarrow S$ is a map with $\phi(i) = x_i$ for all $i \in I$ we will often denote this map ϕ by the symbol $(x_i : i \in I)$ and call it a *family* of elements of S *indexed* by I or with *index set* I. A family is thus a map not a set, though loosely speaking it can often be thought of as a collection of labelled objects of S. A subset X of S is a *maximal* (*minimal*) subset of S possessing a given property P if X possesses property P and no set properly containing X (contained in X) possesses P.

The set of integers is denoted by $\mathbb{Z}$. The set of real numbers by $\mathbb{R}$. The sets of non-negative integers and real numbers are denoted respectively by $\mathbb{Z}^+$ and $\mathbb{R}^+$.

3. ALGEBRAIC STRUCTURES

I shall assume familiarity with the basic algebraic structures such as a group, field, or vector space.

The finite field with q elements will be denoted by $GF(q)$ and I use $F(x_1, \ldots, x_n)$ to denote the minimal extension field of the field F generated by $x_1, \ldots, x_n$. $V_n(F)$ denotes the vector space of dimension n over the field F. $V_n(q)$ denotes the vector space of dimension n over the field $GF(q)$.

A typical element v of a vector space will be denoted by v or $(v_1, \ldots, v_n)$, where n is the dimension of the space. The zero vector is denoted by 0.

Now for any field F consider the vector space V of all vectors $(a_0, \ldots, a_n)$, $a_i \in F$. If u, v are two members of $V \backslash 0$ we write $u \sim v$ if there exists some non-zero member b of F such that $u = bv$. It is easy to check that $\sim$ is an equivalence relation on $V \backslash 0$. The equivalence classes under this relation are the points of the *projective geometry of dimension n* over F. When F is the finite field $GF(q)$ we denote this projective geometry by $PG(n, q)$.

For a further discussion of projective geometries we refer to Chapter 12 (where we study them in more detail), or to the recent book of Bumcrot [69].

Unless specified a vector will mean a row vector and if x is a vector, x' is its transpose and xy' will denote the *scalar product* of x and y. Similarly A' will denote the *transpose* of the matrix A.

4. GRAPH THEORY

We assume familiarity with the concepts of a *graph* and a directed or oriented graph which we call a *digraph*. We denote a graph G by a pair $(V(G), E(G))$ where $V = V(G)$ is the *vertex* set and $E = E(G)$ is the set of *edges*.

The edge $e = (u, v)$ is said to *join* the vertices u and v. If $e = (u, v)$ is an edge of G, then u and v are *adjacent* vertices while u and v are called the *endpoints* of the edge $e = (u, v)$. If two edges e_1, e_2 have a common endpoint they are said to be *incident*. We often denote the edge $e = (u, v)$ by uv or vu.

A *loop* of a graph is an edge of the type (x, x). Two edges are *parallel* if they have common endpoints and are not loops.

A graph is *simple* if it has no loops or parallel edges.

The *degree* of a vertex v is the number of edges having G as an endpoint, and is denoted by $\deg(v)$. A graph is *regular* if all its vertices have the same degree.

Two graphs G_1, G_2 are *isomorphic* if there is a bijection $\phi : V(G_1) \to V(G_2)$ such that if $u, v \in V(G)$ the number of edges joining u, v in G_1 equals the number of edges joining $\phi(u)$ and $\phi(v)$ in G_2.

A vertex u of G is an *isolated* vertex if $\deg(u) = 0$. An edge uv of G is a *pendant* edge if either u or v, but not both, has degree 1 in G.

If G is a graph with vertex set $\{v_1, \ldots, v_n\}$ the *adjacency matrix* of G corresponding to the given labelling of the vertex set is the $n \times n$ matrix $A = (a_{ij})$, in which a_{ij} is the number of edges of G joining v_i and v_j.

A *complete graph* is a simple graph in which an edge joins each pair of vertices. The complete graph on n vertices is denoted by K_n.

If the vertex set of a graph can be divided into two disjoint sets V_1, V_2 in such a way that every edge of the graph joins a vertex of V_1 to a vertex of V_2 the graph is said to be a *bipartite graph*. We often denote such a graph by $\Delta(V_1, V_2; E)$ if we wish to specify the two sets involved, and where Δ signifies that the graph in question is bipartite.

If a bipartite graph $\Delta(V_1, V_2; E)$ has the property that every vertex of V_1 is joined to every vertex of V_2, and it is simple then it is called a *complete bipartite graph* and is usually denoted by $K_{m,n}$ where $m = |V_1|$ and $n = |V_2|$.

Figure 1 shows the complete graph K_6 and the complete bipartite graph $K_{3,2}$.

A *path* in G is a finite sequence of distinct edges of the form $(v_0, v_1), (v_1, v_2), \ldots, (v_{m-1}, v_m)$. The *length* of this path is m and it is said to *connect* v_0 and v_m.

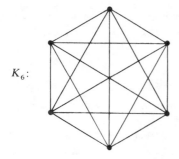

K_6:

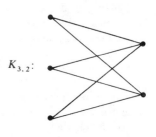

$K_{3,2}$:

Figure 1

The vertices v_0 and v_m are called respectively the *initial* and *terminal* vertices of this path. The vertices v_i, $i \neq 0$, $i \neq m$ are *interior* vertices of this path. A *cycle* is a path in which $v_i \neq v_j$ for $i \neq j$ except that $v_0 = v_m$.

If we define a relation $\sim$ on $V(G)$ by $x \sim y$ if $x = y$ or there is a path in G joining x and y then it is easy to verify that $\sim$ is an equivalence relation on $V(G)$. The distinct equivalence classes are called the *connected components* of G. If there is one component G is *connected*.

A *subgraph* of a graph $G = (V, E)$ is a pair (U, F) where $U \subseteq V$ and $F \subseteq E$, with the proviso that if $e = (u, v) \in F$ then u and v are members of U.

If A is any set of edges of G we let $V(A)$ denote those vertices of G which are endpoints of some edge of A and then call $(V(A), A)$ the *subgraph generated* by A. We will use $G | A$ to denote the subgraph generated by A, though often we abbreviate this to A when it is clear from the context.

Thus for example if P is the set of edges of some path in G and $e \notin P$ is an edge of G, the statement "$e \cup P$ is a path" will mean e is incident with either the initial or terminal vertex of P and is not incident with any interior vertex of P. Similarly if $A \subseteq E(G)$ the statement "A is isomorphic to the complete graph K_4" means that the subgraph generated by A is isomorphic to K_4. If $X \subseteq V(G)$, we let $G \backslash X$ denote the subgraph obtained by deleting X and all edges incident with X from G.

A subgraph of G is a *spanning subgraph* if it has vertex set $V(G)$.

A *tree* is a graph which has no cycles and is connected. A *forest* is a graph which has no cycles. A *spanning forest* of G is union of spanning trees of its connected components. A *spanning tree* is a spanning subgraph which is a tree. It is easy to prove that G is connected if and only if it has a spanning tree, and that such a spanning tree must have exactly $|V(G)| - 1$ edges.

If $U \subseteq V(G)$ we let $G \backslash U$ denote the subgraph of G obtained by removing U and all its incident edges. A connected graph is *n-connected* for some positive integer n if there exists no set $U \subseteq V(G)$ with $|U| < n$ such that $G \backslash U$ is

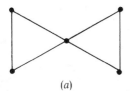

(a)

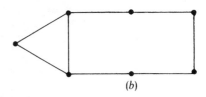

(b)

Figure 2

disconnected or a single vertex. Figures 2(a) and (b) show respectively a 1-connected and 2-connected graph.

A vertex u of a connected graph G such that $G\backslash\{u\}$ is disconnected is called a *cut vertex* of G.

More technical terms such as planarity, homeomorphism and so on are defined as they arise in the text. Moreover we often redefine graph theory terms as they are used to save the reader not too familar with graphs having to look back.

In a digraph $G = (V, E)$ we often denote an edge directed from the vertex u to the vertex v by (u, v) and call $u(v)$ respectively the *initial (terminal)* vertices of the edge. A *path* in a digraph is a finite sequence of distinct edges of the form (v_0, v_1), (v_1, v_2), ..., (v_{m-1}, v_m) where the v_i are distinct vertices. Such a path is said to *pass through* the vertices v_i. Two paths are *vertex disjoint* if they do not pass through a common vertex. Other terms for digraphs are defined by obvious analogy with the corresponding terms for graphs.

CHAPTER 1

Fundamental Concepts and Examples

1. INTRODUCTION

Matroid theory dates from the 1930's when van der Waerden in his "Moderne Algebra" first approached linear and algebraic dependence axiomatically and Whitney in his basic paper [35] first used the term matroid. As the word suggests Whitney conceived a matroid as an abstract generalization of a matrix, and much of the language of the theory is based on that of linear algebra. However Whitney's approach was also to some extent motivated by his earlier work in graph theory and as a result some of the matroid terminology has a distinct graphical flavour.

Apart from isolated papers by Birkhoff [35], MacLane [36], [38], and Dilworth [41], [41a], [44], on the lattice theoretic and geometric aspects of matroid theory, and two important papers by Rado on the combinatorial applications of matroids [42] and infinite matroids [49], the subject lay virtually dormant until Tutte [58], [59], published his fundamental papers on matroids and graphs and Rado [57] studied the representability problem for matroids. Since then interest in matroids and their applications in combinatorial theory has accelerated rapidly. This is probably due to the discovery independently by Edmonds and Fulkerson [65] and Mirsky and Perfect [67] of a new, important class of matroids called transversal matroids. It is in the field of transversal theory that matroids seem so far to have had the most effect (measured in terms of new results obtained or easier proofs found of known results). In graph theory the main benefit of a matroid treatment seems to be a much more natural understanding of dual concepts such as the structure of the set of cocycles or the effect of contraction of a set of edges of a graph. The beauty and importance of matroids is perhaps best appreciated by the study of two covering and packing theorems of Edmonds [65], [65a]. These results give as easy corollaries, earlier very difficult, and intricate theorems of graph theory due to Tutte [61], and Nash-Williams [61], [64] a theorem about vector spaces due to Horn [55] and several results in transversal theory proved earlier by Higgins [59], Ore [55]

6

and others. They illustrate perfectly the principle that mathematical generalization often lays bare the important bits of information about the problem at hand.

As is often the case with young subjects, the matroid terminology varies considerably from author to author. Even the term matroid is rejected by many. Mirsky and Perfect [67] use "independence space", Crapo and Rota in their monograph [70] on combinatorial geometries use "pregeometry" for "matroid"; Rado's work is in terms of "independence functions"; Cohn [65] uses the term "transitive dependence relation".

2. AXIOM SYSTEMS FOR A MATROID

As will be seen, a matroid may be defined in many different but equivalent ways, several of which were described in Whitney's original paper. Deciding which set of axioms would be the most natural to start with was difficult. I have eventually settled on "independence axioms" because I think they will be the most natural to the average reader.

Matroid theory has exactly the same relationship to linear algebra as does point set topology to the theory of real variable. That is, it *postulates* certain sets to be "independent" ($=$linearly independent) and develops a fruitful theory from certain axioms which it demands hold for this collection of independent sets.

A *matroid* M is a finite set S and a collection $\mathscr{I}$ of subsets of S (called *independent* sets) such that (I1)–(I3) are satisfied.

(I1) $\varnothing \in \mathscr{I}$.
(I2)   If $X \in \mathscr{I}$ and $Y \subseteq X$ then $Y \in \mathscr{I}$.
(I3)   If U, V are members of $\mathscr{I}$ with $|U| = |V| + 1$ there exists $x \in U \backslash V$ such that $V \cup x \in \mathscr{I}$.

A subset of S not belonging to $\mathscr{I}$ is called *dependent*.

Example. Let V be a finite vector space and let $\mathscr{I}$ be the collection of linearly independent subsets of vectors of V. Then $(V, \mathscr{I})$ is a matroid.

Following the analogy with vector spaces we make the following definitions.

A *base* of M is a maximal independent subset of S, the collection of bases is denoted by $\mathscr{B}$ or $\mathscr{B}(M)$.

The *rank function* of a matroid is a function $\rho: 2^S \to \mathbb{Z}$ defined by

$$\rho A = \max(|X| : X \subseteq A, X \in \mathscr{I}) \qquad (A \subseteq S).$$

The *rank of the matroid*, M sometimes denoted by ρM, is the rank of the set S.

A subset $A \subseteq S$ is *closed* or a *flat* or a *subspace* of the matroid M if for all $x \in S \backslash A$

$$\rho(A \cup x) = \rho A + 1.$$

In other words no element can be added to A without increasing its rank. If for $x \in S$, $A \subseteq S$, $\rho(A \cup x) = \rho A$ we say that x *depends* on A and denote it by $x \sim A$. We define the *closure operator* of the matroid to be a function $\sigma : 2^S \to 2^S$ such that σA is the set of elements which depend on A. (It will turn out that σA is the smallest closed set containing A.) A subset X is *spanning* in M if and only if it contains a base.

All the above concepts are very familiar from vector space theory. Motivated by analogies in graph theory we define a *circuit* of M to be a minimal dependent set, and denote the collection of circuits by $\mathscr{C}$ or $\mathscr{C}(M)$.

Now it is fairly obvious that knowledge of the bases, or circuits, or rank function, or closure operator is sufficient to uniquely determine the matroid. Hence it is not surprising that there exist axiom systems for a matroid in terms of each of these concepts. We list some of these axioms in the following theorems.

THEOREM 1 (Base axioms). *A non-empty collection $\mathscr{B}$ of subsets of S is the set of bases of a matroid on S if and only if it satisfies the following condition:*

(B1)   *If $B_1, B_2 \in \mathscr{B}$ and $x \in B_1 \backslash B_2$, $\exists y \in B_2 \backslash B_1$ such that $(B_1 \cup y) \backslash x \in \mathscr{B}$.*

THEOREM 2 (Rank axioms). *A function $\rho : 2^S \to \mathbb{Z}$ is the rank function of a matroid on S if and only if for $X \subseteq S$, $y, z \in S$:*

(R1) $\rho \varnothing = 0$;
(R2) $\rho X \leqslant \rho(X \cup y) \leqslant \rho X + 1$;
(R3)   if $\rho(X \cup y) = \rho(X \cup z) = \rho X$ then $\rho(X \cup y \cup z) = \rho X$.

THEOREM 3 (Rank axioms). *A function $\rho : 2^S \to \mathbb{Z}$ is the rank function of a matroid on S if and only if for any subsets X, Y of S:*

(R1′) $0 \leqslant \rho X \leqslant |X|$;
(R2′) $X \subseteq Y \Rightarrow \rho X \leqslant \rho Y$;
(R3′) $\rho(X \cup Y) + \rho(X \cap Y) \leqslant \rho X + \rho Y$.

THEOREM 4 (Closure axioms). *A function $\sigma : 2^S \to 2^S$ is the closure operator of a matroid on S if and only if for X, Y subsets of S, and $x, y \in S$;*

(S1) $X \subseteq \sigma X$;
(S2) $Y \subseteq X \Rightarrow \sigma Y \subseteq \sigma X$;
(S3) $\sigma X = \sigma \sigma X$;

(S4)   if $y \notin \sigma X, y \in \sigma(X \cup x)$, then $x \in \sigma(X \cup y)$.

THEOREM 5 (Circuit axioms). *A collection $\mathscr{C}$ of subsets of S is the set of circuits of a matroid on S if and only if conditions* (C1) *and* (C2) *are satisfied.*

(C1)   If $X \neq Y \in \mathscr{C}$, then $X \nsubseteq Y$.
(C2)   If C_1, C_2 are distinct members of $\mathscr{C}$ and $z \in C_1 \cap C_2$ there exists $C_3 \in \mathscr{C}$ such that $C_3 \subseteq (C_1 \cup C_2)\backslash z$.

The proofs of Theorems 1–5 are routine, though sometimes laborious, and we defer them to the following sections. (For example we shall prove Theorem 5 in Section 9 when we are dealing with circuits.)

A note on terminology

We shall call a set S with any of the above systems ($\mathscr{B}, \rho, \sigma, \mathscr{C}$) a matroid and will usually just denote it by M. Unless otherwise specified S will always be the set supporting M and throughout ρ and σ will denote a rank function and closure operator of a general matroid.

Two matroids M_1 and M_2 on S_1 and S_2 respectively are *isomorphic* if there is a bijection $\phi: S_1 \to S_2$ which preserves independence. It is clear that equivalently ϕ is an isomorphism if and only if it preserves the rank function, circuits and so on. We write $M_1 \simeq M_2$ if M_1 and M_2 are isomorphic.

EXERCISES 2

1. Prove that for $n = 1, 2, 3$ there are exactly 2^n non-isomorphic matroids on a set of n elements.

2. If an element belongs to every base of M show that it can belong to no circuit of M.

3. A function f on the set of finite sequences of elements of S is an *independence function* (*I-function*) if it takes the values 0 and 1 and satisfies for any positive integer n and any elements $x_1, \ldots, x_n, y_1, \ldots, y_{n+1}$ of S the following axioms.

   (IF1) f is *decreasing*, that is $f(x_1, \ldots, x_n) \geqslant f(x_1, \ldots, x_{n+1})$.
   (IF2) f is *commutative*, that is for any permutation $i \to \alpha i$ of $\{1, \ldots, n\}$
   $$f(x_{\alpha 1}, \ldots, x_{\alpha n}) = f(x_1, \ldots, x_n).$$
   (IF3) f is *non-reflexive*, that is for all $x \in S$ $f(x, x) = 0$.
   (IF4) f is *distributive*, that is
   $$f(x_1, \ldots, x_n)f(y_1, \ldots, y_{n+1}) \leqslant \sum_{i=1}^{n+1} f(x_1, \ldots, x_n, y_i).$$
   (IF5) $f(\varnothing) = 1$.

   Show that a set together with an independence function is essentially a matroid, and conversely. (These independence function axioms were introduced by Rado [42].)

3. EXAMPLES OF MATROIDS

The uniform matroid $U_{k,n}$

Let S be a set of cardinality n and let $\mathscr{I}$ be all subsets of S of cardinality $\leqslant k$. This is a matroid on S, called the uniform matroid of rank k and denoted by $U_{k,n}$. We list its bases, circuits and so on:

$$\mathscr{B}(U_{k,n}) = \{X : |X| = k\},$$

$$\mathscr{C}(U_{k,n}) = \{X : |X| = k + 1\},$$

$$\rho A = \begin{cases} |A| & |A| \leqslant k \\ k & |A| \geqslant k, \end{cases} \qquad (A \subseteq S)$$

$$\sigma A = \begin{cases} A & |A| < k, \\ S & |A| \geqslant k. \end{cases} \qquad (A \subseteq S).$$

We call the matroid with every set independent the *free matroid*.

Vectorial matroids

Let S be any finite subset of a vector space V. Let a set $X = \{x_1, \ldots, x_k\} \in \mathscr{I}$ if and only if the vectors $x_1, x_2, \ldots, x_k$ are linearly independent in V. Then $\mathscr{I}$ is the collection of independent sets of a matroid M. The rank function ρ is just the rank (or dimension) function of V restricted to the set S. Any matroid isomorphic to one obtained this way is called *vectorial*.

Cycle matroids of graphs

Let G be a graph, let S be its set of edges $E(G)$ and let $X \in \mathscr{I}$ if and only if X does not contain a cycle of G. Then $\mathscr{I}$ is the collection of independent sets of a matroid on S, called the *cycle matroid* of the graph G and denoted by $M(G)$. We prove this and discuss this class of matroids later.

Example. When G is the complete graph K_3 its cycle matroid

is the uniform matroid $U_{2,3}$.

Algebraic matroids

Let F be a field and let K be an extension of F. Now a set $\{x_1, x_2, \ldots, x_k\}$ of elements of K is *algebraically dependent* if the elements $x_1, \ldots, x_k$ satisfy a polynomial equation of the form $f(x_1, \ldots, x_k) = 0$ where the coefficients of f are members of F. Otherwise $x_1, \ldots, x_k$ are *algebraically independent* over F.

THEOREM 1. *Let S be a finite subset of K and let $X \in \mathscr{I}$ if and only if $X \subseteq S$ and the elements of X are algebraically independent over F. Then $\mathscr{I}$ is the collection of independent sets of a matroid on S.*

The proof of this result is not difficult, but for those unfamiliar with field extension theory we sketch a proof in Chapter 11.

Affine dependence

Let S be a finite set of points in the d-dimensional Euclidean space $\mathbb{R}^d$. Recall that an element $x \in \mathbb{R}^d$ is *affinely dependent* on the set $\{x_1, \ldots, x_r\}$ of $\mathbb{R}^d$ if there exist real numbers $\lambda_i (1 \leqslant i \leqslant r)$ such that

$$x = \sum_{i=1}^{r} \lambda_i x_i$$

and

$$\sum_{i=1}^{r} \lambda_i = 1.$$

A subset $X \subseteq \mathbb{R}^d$ is *affinely independent* if no element $x \in X$ is affinely dependent on $X \setminus x$. It is easy to prove that if $\mathscr{I}$ is the collection of subsets of S which are affinely independent then $\mathscr{I}$ is a matroid on S.

Abelian groups

Let J be an abelian group with no elements of finite order (except 0). Call an element g of J *dependent on* a set X of J if in X there are elements $x_1, \ldots, x_n$ and integers $m, k_1, \ldots, k_n$ $(m \neq 0)$ such that

$$mg = k_1 x_1 + \ldots + k_n x_n.$$

Call a subset Y of J *independent* if no element y of Y is dependent on $Y \setminus y$. Then if S is any finite subset of J, the independent subsets of S form a matroid M. The closed sets are simply the subgroups H of J for which the factor group J/H has no elements of finite order (except 0).

However every torsion free abelian group can be embedded in a vector space. Hence the matroids obtained in this way are all isomorphic to vectorial matroids.

EXERCISES 3

1. Find a matroid which is not vectorial.

2. Consider the matroid $U_{2,4}$. Prove that it is not isomorphic to any matroid which is vectorial over $GF(2)$ but that it is isomorphic to a vectorial matroid over $GF(3)$.

3. Prove that in a vectorial matroid if $x_1, \ldots, x_n$ form a circuit there exist scalars a_i such that $a_i \neq 0$ for all i and

$$\sum_{i=1}^{n} a_i x_i = 0.$$

4. LOOPS AND PARALLEL ELEMENTS

We define a *loop* of a matroid M to be an element x of S such that $\{x\}$ is a dependent set; and define two elements x, y of S to be *parallel* if they are not loops but $\{x, y\}$ is a dependent set.

Example. If $M = M(G_0)$ is the cycle matroid of the graph G_0 of Figure 1 then the edges x_1, x_2 are loops and the pairs of edges y_1, y_2 and z_1, z_2 are parallel in $M(G_0)$.

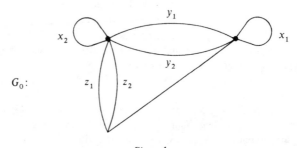

Figure 1

Thus when the matroid is the cycle matroid of a graph the concepts of loop and parallelism coincide with the usual graphical ones.

However when M is a vectorial matroid the only possible loop is the zero element (if it belongs to the ground set S).

The following statements about loops are obvious.

(1) x is a loop if and only if $x \in \sigma(\varnothing)$.
(2) x is a loop if and only if $\rho\{x\} = 0$.
(3) If x is a loop and $x \in A$ then A is dependent.
(4) If x is a loop $x \in \sigma A$ for all subsets $A \subseteq S$.
(5) x is a loop if and only if $\{x\}$ is a circuit.
(6) x is a loop if and only if it is not contained in any base.

Similarly there are many equivalent definitions of being parallel. The following statements are easy to check.

(7)   Distinct elements x and y are parallel if and only if $\{x, y\}$ is a circuit.

(8)   If x is parallel to y and y is parallel to z, and $x \neq z$, then x is parallel to z.

(9)   If $x \neq y$, then x is parallel to y if and only if $x \in \sigma\{y\}$, and $y \in \sigma\{x\}$ and x and y are not loops.

(10) If $x \in \sigma A$ for some subset A of S, and y is parallel to x, then $y \in \sigma A$.

(11) If A contains two parallel elements then A must be dependent.

For many purposes one can "forget about" loops and parallel elements in matroid theory, in the same way as many authors in graph theory only consider graphs with no loops or parallel edges. However as we will see, if we adopt this course the intrinsic simplicity of duality theory (to be studied in the next chapter) vanishes.

As we have used the adjective "simple" to describe a graph which has no loops or parallel edges we define a *simple matroid* to be a matroid with no loops or parallel elements.

EXERCISES 4

1. Prove that for any matroid M on S

   (a)   a loop belongs to every flat,
   (b)   if $x \sim A$ and y is parallel to x then $y \sim A$,
   (c) $\varnothing$ is a flat if and only if M has no loops.

5. PROPERTIES OF INDEPENDENT SETS AND BASES

In this section we obtain some elementary properties of independent sets and bases of a matroid M on S. It is clear that if A is independent there exists a base B such that $A \subseteq B$.

The following stronger result is used extensively.

THEOREM 1 (The Augmentation Theorem). *Suppose that X, Y, are independent in M and that $|X| < |Y|$. Then there exists $Z \subseteq Y \backslash X$ such that $|X \cup Z| = |Y|$ and $X \cup Z$ is independent in M.*

Proof. Let Z_0 be a set such that over all $Z \subseteq Y \backslash X$ such that $X \cup Z$ is independent, $|X \cup Z_0|$ is a maximum. If $|X \cup Z_0| < |Y|$ then $\exists Y_0 \subseteq Y$, $|Y_0| = |X \cup Z_0| + 1$ and since Y_0 is independent by (I3) $\exists y \in Y_0 \backslash (X \cup Z_0)$ such that $X \cup Z_0 \cup y$ is independent. The set $Z_0 \cup y$ contradicts the choice of Z_0.∎

An immediate consequence of this is the following result, which extends the well known property of bases of a vector space.

COROLLARY. *All bases of a matroid on S have the same cardinality, which is the rank of S.*

Proof. If not let B_1, B_2 be bases with $|B_1| < |B_2|$. Then there exists $Z \subseteq B_2 \backslash B_1$ such that $B_1 \cup Z \in \mathscr{I}$. This contradicts the maximality of B_1 in $\mathscr{I}$.

Now we can prove Theorem 2.1 that a non-empty collection $\mathscr{B}$ of subsets of S is the set of bases of a matroid on S if and only if all members of $\mathscr{B}$ satisfy

(B1)   If $B_1, B_2 \in \mathscr{B}$ and $x \in B_1 \backslash B_2$ $\exists y \in B_2 \backslash B_1$ such that $(B_1 \cup y) \backslash x \in \mathscr{B}$.

Proof of Theorem 2.1. Clearly since $\varnothing \in \mathscr{I}(M)$, $\mathscr{B}(M) \neq \varnothing$. Similarly if we apply the augmentation theorem to $B_1 \backslash x$ and B_2 there must exist $y \in B_2$ such that $|(B_1 \backslash x) \cup y| = |B_2|$. Thus $y \in B_2 \backslash B_1$.

Conversely let $\mathscr{B}$ be a non-empty collection of subsets of S satisfying (B1). Define $\mathscr{I}$ to be the collection of subsets X of S such that X is contained in some member of $\mathscr{B}$.

Trivially $\mathscr{I}$ satisfies (I1) and (I2). Hence suppose that X, Y are distinct members of $\mathscr{I}$ with $X \subseteq B_1$, $Y \subseteq B_2$, and B_1, $B_2 \in \mathscr{B}$. Let

$$X = \{x_1, \ldots, x_k\}$$
$$B_1 = \{x_1, \ldots, x_k, b_1, \ldots, b_q\}$$
$$Y = \{y_1, \ldots, y_k, y_{k+1}\}$$
$$B_2 = \{y_1, \ldots, y_k, y_{k+1}, c_1, \ldots, c_{q-1}\}.$$

Consider $B_1 \backslash b_q$; there exists $z \in B_2$ such that $(B_1 \backslash b_q) \cup z \in \mathscr{B}$. If $z \in Y$ then $X \cup z \in \mathscr{I}$ and the exchange axiom (I3) is satisfied.

If $z \notin Y$ consider $((B_1 \backslash b_q) \cup z) \backslash b_{q-1}$. Again $\exists z_1 \in B_2$ such that $((B_1 \backslash b_q \cup z) \backslash b_{q-1}) \cup z_1$ belongs to $\mathscr{B}$. If $z_1 \in Y$ we are finished, if not remove b_{q-2} and so on. Since $|\{b_1, \ldots, b_q\}| > |\{c_1, \ldots, c_{q-1}\}|$, after at most q steps we arrive at a situation where we replace b_i by a member of Y. Hence $\mathscr{I}$ satisfies the exchange axiom (I3) and is the collection of independent sets of a matroid. Clearly the bases of this matroid is the collection of sets $\mathscr{B}$.∎

We can also easily obtain a new axiomatization of a matroid by its independent sets.

THEOREM 2. *A collection $\mathscr{I}$ of subsets of S is the set of independent sets of a* *matroid on S if and only if $\mathscr{I}$ satisfies conditions* (I1) (I2) *and the following statement.*

(I3′)   *If A is any subset of S all maximal subsets Y of A with $Y \in \mathscr{I}$ have the same cardinality.*

Proof. Suppose that $\mathscr{I}$ is the collection of independent sets of a matroid M. Then if (I3′) does not hold there exist Y_1, Y_2 maximal independent with $|Y_1| \geqslant |Y_2| + 1$. Apply (I3) and we get a set $T_2 \cup y$ with $Y_2 \cup y \subseteq A$, $Y_2 \cup y \in \mathscr{I}$, $y \in Y_1 \backslash Y_2$, and this contradicts the maximality of Y_2.

Conversely if $\mathscr{I}$ satisfies (I1) (I2) and (I3′) let U, V be members of $\mathscr{I}$, $|U| = |V| + 1$. Let $A = U \cup V$. Then since all maximal subsets of $U \cup V$ will have the same cardinal, there exists $x \in U \cup V$ such that $|x \cup U| = |V|$ and the result follows.■

EXERCISES 5

1. If M is a matroid on S with $A \subseteq S$ let $X \in \mathscr{I}'$ if X is independent in M and $X \cap A = \varnothing$. Prove $\mathscr{I}'$ is the collection of independent sets of a matroid on S.

2. Let M be a matroid on S and let B_1, B_2 be distinct bases of M. Prove that

   (a) there exists a bijection $\pi: B_1 \to B_2$ such that for all $e \in B_1, (B_2 \backslash \pi(e)) \cup e$ is a base of M.
   (b) for any $e \in B_1 \, \exists f \in B_2$ such that $(B_1 \backslash e) \cup f$ and $(B_2 \backslash f) \cup e$ are bases of M.
   (c) there exists a bijection $\pi': B_1 \to B_2$ such that $(B_1 \backslash e) \cup \pi'(e)$ is a base of M for all $e \in B_1$.

3. If B_1, B_2 are bases of the matroid M and $X_1 \subseteq B_1$ prove that there exists $X_2 \subseteq B_2$ such that $(B_1 \backslash X_1) \cup X_2$ and $(B_2 \backslash X_2) \cup X_1$ are both bases of M. (Greene [73] see also Woodall [74], Brylawski [73a]).

6. PROPERTIES OF THE RANK FUNCTION

As we have already remarked a matroid is uniquely determined once we know the rank of every set. Indeed Whitney in his fundamental paper [35] defined a matroid in terms of its rank function. We have already stated two sets of rank axioms in Section 2. In this section we shall obtain other elementary properties of the rank function and also prove Theorems 2.2 and 2.3.

First it is obvious that if M is a matroid on S its rank function ρ is well defined and satisfies:

(R1) $\rho\varnothing = 0;$

(R2') $A \subseteq B \Rightarrow \rho A \leqslant \rho B$ $(A, B \subseteq S)$;

(R1') $0 \leqslant \rho A \leqslant |A|$ $(A \subseteq S)$.

Also since the addition of an element y to a set X can increase the cardinal of the maximal independent subset contained in X by at most 1,

(R2) $\rho X \leqslant \rho(X \cup y) \leqslant \rho X + 1$ $(X \subseteq S, y \in S)$.

Now to prove that ρ satisfies for any $X \subseteq S$, $x, y \in S$,

(R3)   if $\rho(X \cup y) = \rho(X \cup x) = \rho x$ then $\rho(X \cup x \cup y) = \rho x$.

Let $A = X \cup x \cup y$ and suppose $\rho A > \rho X$ and let Y be a maximal independent subset of X. By the augmentation theorem we must be able to augment Y to a maximal independent subset of A. Hence $Y \cup x$ or $Y \cup y \in \mathscr{I}(M)$. This contradicts the equality of the ranks of $X \cup y$, $X \cup x$, and X, and hence ρ satisfies R3.

Proof of Theorem 2.2. We now have only to prove that a function ρ satisfying (R1), (R2) and (R3) is the rank function of some matroid. Define $\mathscr{I}(\rho)$ to be the following collection of subsets of S;

$$X \in \mathscr{I}(\rho) \Leftrightarrow \rho X = |X|.$$

Clearly $\varnothing \in \mathscr{I}(\rho)$. Let $A \in \mathscr{I}(\rho)$ and let $B \subseteq A$. Suppose $B \notin \mathscr{I}(\rho)$, then $\rho B < |B|$ so that if $\{c_1, c_2, \ldots, c_k\} = A \backslash B$, we have $\rho(B \cup c_1) \leqslant \rho B + 1 < |B| + 1$. By repeated application of (R2) we arrive at

$$\rho A = \rho(B \cup c_1 \cup c_2 \cup \ldots \cup c_k) < |B| + k = |A|$$

which is a contradiction. Hence $\mathscr{I}(\rho)$ satisfies (I2). Now suppose that $X, Y,$ are members of $\mathscr{I}(\rho)$ and let

$$X = \{x_1, x_2, \ldots, x_q, y_{q+1}, \ldots, y_k\}$$
$$Y = \{x_1, x_2, \ldots, x_q, z_{q+1}, \ldots, z_k, z_{k+1}\}$$

where $y_i \neq z_j$ for any i and j.

Suppose that $X \cup z_i \notin \mathscr{I}(\rho)$ for $(q + 1 \leqslant i \leqslant k + 1)$. Then

$$\rho X = \rho(X \cup z_i) = |X| (q + 1 \leqslant i \leqslant k + 1).$$

Hence by (R3)

$$\rho(X \cup z_i \cup z_j) = \rho X = |X| (q + 1 \leqslant i, j \leqslant k + 1).$$

Applying (R3) repeatedly we get eventually

$$\rho Y \leqslant \rho(X \cup z_{q+1} \cup \ldots \cup z_{k+1}) = |X| < |Y|$$

which contradicts $Y \in \mathcal{I}(\rho)$. Hence $X \cup z_i \in \mathcal{I}(\rho)$ for some i and $\mathcal{I}(\rho)$ satisfies the exchange axiom (I3). Thus $\mathcal{I}(\rho)$ is the collection of independent sets of a matroid and it is easy to show that the rank function of this matroid is in fact ρ.∎

Similar and easily proved properties of ρ are listed below.

(1) For $A \subseteq S$ and $x, y \in S$,

$$\rho(A \cup x \cup y) - \rho(A \cup x) \leqslant \rho(A \cup y) - \rho A.$$

(2) For $A, B \subseteq S$ and $x \in S$,

$$\rho(A \cup B \cup x) - \rho(A \cup B) \leqslant \rho(A \cup x) - \rho A.$$

(3) For $A, B, C \subseteq S$,

$$\rho(A \cup B \cup C) - \rho(A \cup B) \leqslant \rho(A \cup C) - \rho A.$$

These are all essentially special cases of the stated Theorem 2.3 which says that ρ is the rank function of a matroid if and only if it satisfies (R1′), (R2′) and for any subsets $A, B \subseteq S$, ρ satisfies the *submodular inequality*.

(R3′) $$\rho(A \cup B) + \rho(A \cap B) \leqslant \rho A + \rho B.$$

Proof of Theorem 2.3. That the rank function of a matroid M satisfies (R1′) and (R2′) is obvious. Suppose that $\rho(A \cup B) = p$ and $\rho(A \cap B) = q$. Let X be an independent subset of $A \cap B$ with $|X| = q$. By the augmentation theorem there exists Y with $Y \supseteq X$, $Y \subseteq A \cup B$, $|Y| = p$, and Y independent in M. Write $Y = X \cup V \cup W$ where X, V, W are pairwise disjoint and where $V \subseteq A \backslash B$ and $W \subseteq B \backslash A$. Then $X \cup V$ is an independent subset of A and $X \cup W$ is an independent subset of B. Hence

$$\rho A + \rho B \geqslant |X \cup V| + |X \cup W| = 2|X| + |V| + |W|$$
$$= |Y| + |X| = \rho(A \cup B) + \rho(A \cap B).$$

Conversely let $\rho : 2^S \to \mathbb{Z}$ be a function satisfying (R1′), (R2′) and (R3′). It is easy to see that ρ must satisfy (R1) and (R2). Now let $A \subseteq S$ and $x, y \in S \backslash A$. Then if $\rho A = \rho(A \cup x) = \rho(A \cup y)$, the submodularity of ρ implies that

$$\rho(A \cup x \cup y) + \rho A \leqslant \rho(A \cup x) + \rho(A \cup y)$$

and hence $\rho(A \cup x \cup y) = \rho A$. Hence by Theorem 2.2 ρ must be the rank function of a matroid.∎

7. THE CLOSURE OPERATOR

MacLane [36] first noticed the intimate relationship between matroids and incidence geometries, or in his terminology, "schematic configurations".

More recently Rota, in his fundamental paper on combinatorial theory [64] has used this relationship in his approach to matroid theory via geometric lattices. Here we examine the more elementary ideas of this relationship.

If A is any subset of S and x is any element of S recall that x depends on A in the matroid M if

$$\rho(A) = \rho(A \cup x)$$

and write this as $x \sim A$. The reader can verify the following elementary properties of $\sim$.

(1) x depends on $\varnothing$ if and only if x is a loop.
(2) x depends on the singleton set $\{y\}$ if and only if x is a loop or x is parallel to y, or $x = y$.
(3) If $x \in A$ then $x \sim A$.
(4) If B is a base of M then $x \sim B$ for all $x \in S$.
(5) If x is a loop then x depends on every subset of S.
(6) If $x \sim A$ and $A \subseteq B$, then $x \sim B$.

It is clear that in vectorial matroids $x \sim A$ if and only if x can be written as a linear combination of some members of A.

Now since the closure operator $\sigma : 2^S \to 2^S$ is defined by

$$x \in \sigma A \Leftrightarrow x \sim A \qquad (A \subseteq S)$$

we have immediately that

(S1) $A \subseteq \sigma A$.

From (6) we get

(S2) if A, B, are subsets of S, and $A \subseteq B$, then $\sigma A \subseteq \sigma B$.

We now prove a series of elementary lemmas concerning σ.

(7) If X is a maximal independent subset of $A \subseteq S$, then $\sigma X = \sigma A$.

Proof. As $X \subseteq A$, $\sigma X \subseteq \sigma A$. Let $y \in \sigma A \backslash \sigma X$. Then

$$\rho(A \cup y) = \rho A, \qquad \rho(X \cup y) = \rho X + 1 = \rho A + 1.$$

Since $\rho(X \cup y) \leqslant \rho(A \cup y)$ this is a contradiction.

As a corollary we have:

(8) If $X \subseteq A$ and $\rho X = \rho A$ then $\sigma X = \sigma A$.
(9) For $A \subseteq S$, $\rho A = \rho(\sigma A)$.

Proof. Suppose $\rho(\sigma A) \geqslant \rho A + 1$. Let Y be a maximal independent subset of A. By the augmentation theorem there exists $Z \neq \varnothing$, $Z \subseteq (\sigma A) \backslash A$ such that $Z \cup Y$ is independent. Let $x \in Z$. Then $x \cup Y$ is independent so that $x \notin \sigma Y$. But by (7) $\sigma Y = \sigma A$, contradiction.

We now show that σ has the basic property of closure operators:

(S3)   For $X \subseteq S$, $\sigma X = \sigma\sigma X$.

Proof. Let A be a maximal independent subset of X so that by (9) A is maximal independent in σX and hence by (7) $\sigma(\sigma X) = \sigma A = \sigma X$.

We can now prove Theorem 2.4, that a function $\sigma : 2^S \to 2^S$ is the closure operator of a matroid on S if and only if it satisfies (S1)–(S4).

Proof of Theorem 2.4. Let σ be the closure operator of M. We have already shown (S1)–(S3) are satisfied.

Now suppose that $y \notin \sigma X$, $y \in \sigma(X \cup x)$, but $x \notin \sigma(X \cup y)$. Then

$$\rho(X \cup y) = \rho X + 1$$
$$\rho(X \cup x \cup y) = \rho(X \cup x)$$
$$\rho(X \cup y \cup x) = \rho(X \cup y) + 1$$
$$= \rho X + 2.$$

Hence we have the impossibility that $\rho(X \cup x) = \rho X + 2$ and σ must satisfy (S1)–(S4).

Conversely let $\sigma : 2^S \to 2^S$ be a function which satisfies the axioms (S1)–(S4). Define the collection $\mathscr{I}(\sigma)$ of subsets of S by

$$A \in \mathscr{I}(\sigma) \Leftrightarrow x \in A \Rightarrow x \notin \sigma(A \backslash x).$$

We show that $\mathscr{I}(\sigma)$ is the collection of independent subsets of a matroid. The null set belongs to $\mathscr{I}(\sigma)$. If $A \in \mathscr{I}(\sigma)$ suppose $B \subseteq A$ but $B \notin \mathscr{I}(\sigma)$. Then $\exists x \in B$ such that

$$x \in \sigma(B \backslash x).$$

Hence by (S2) $x \in \sigma(A \backslash x)$, so $A \notin \mathscr{I}(\sigma)$, contradiction. Thus $\mathscr{I}(\sigma)$ satisfies (I2). We now show that $\mathscr{I}(\sigma)$ satisfies (I3′) and hence by Theorem 5.2, $\mathscr{I}(\sigma)$ will be the collection of independent sets of matroid. We do this by three lemmas.

Call a set $X \subseteq S$, σ-*independent* if $X \in \mathscr{I}(\sigma)$.

LEMMA 1. *Let $A \subseteq S$ and let X be a maximal σ-independent subset of A. Then $\sigma X = \sigma A$.*

Proof. We first show that $A \subseteq \sigma X$. Suppose not. Then there exists $a \in A \backslash \sigma X$ and we assert $X \cup a$ is σ-independent.

Clearly $a \notin \sigma X$ and suppose $\exists x \in X$ such that

$$x \in \sigma((X \cup a)\backslash x) = \sigma((X\backslash x) \cup a).$$

Then since $x \notin \sigma(X\backslash x)$, by (S4)

$$a \in \sigma((X\backslash x) \cup x) = \sigma X$$

which is a contradiction. Hence $X \cup a$ is σ-independent subset of A which contradicts the maximality of X. Thus $A \subseteq \sigma X$.

Now since $X \subseteq A$, by (S2) $\sigma X \subseteq \sigma A$ and thus

$$A \subseteq \sigma X \subseteq \sigma A$$

and by (S2) again

$$\sigma A \subseteq \sigma\sigma X \subseteq \sigma\sigma A.$$

by (S3) this means

$$\sigma A \subseteq \sigma X \subseteq \sigma A$$

which proves the lemma.

LEMMA 2. *If X is σ-independent and Y is a proper subset of X then σY is a proper subset of σX.*

Proof. If $x \in X\backslash Y$, $x \notin \sigma(X\backslash x)$. By (S2) $\sigma Y \subseteq \sigma(X\backslash x)$. Thus $x \in \sigma X\backslash\sigma Y$.

LEMMA 3. *If $A \subseteq S$ and X, Y are maximal, σ-independent subsets of A then $|X| = |Y|$.*

Proof. Suppose not. Let X, Y be chosen so that $|X| < |Y|$ and $|X \cap Y|$ is maximum. Clearly $X \nsubseteq Y$ so take $y \in Y\backslash X$ and let $D = \sigma(X\backslash Y)$. By Lemmas 1 and 2 D is a proper subset of σA. Add elements $x_1, x_2, \ldots, x_k$ of X to $Y\backslash y$ until

$$\sigma\{(Y\backslash y) \cup x_1 \cup \ldots \cup x_k\} = \sigma A.$$

This addition process must stop after at most $k = |X|$ steps since by Lemma 1, $\sigma X = \sigma A$.

Now consider $Y' = (Y\backslash y) \cup x_k$. We assert that Y' is σ-independent. If not then either

$$x_k \in \sigma(Y\backslash y)$$

or for some $y_i \in Y\backslash y = Z$

$$y_i \in \sigma\big((Z\backslash y_i) \cup x\big)$$

Now by choice of x_k since $\sigma(Y\backslash y)$ is a proper subset of $\sigma A_1 x_k \notin \sigma(Y\backslash y)$. Hence $y_i \in \sigma((Z\backslash y_i) \cup x)$. But $y_i \notin \sigma(Z\backslash y_i)$ so that by (S4) $x \in \sigma((Z\backslash y_i) \cup y_i) = \sigma Z = \sigma(Y\backslash y)$ which we have already shown to be impossible. Hence Y' is σ-independent and $|Y' \cap X| > |Y \cap X|$ contradicting the choice of X and Y. Thus $|X| = |Y|$.

From Lemmas 1–3 we see by Theorem 5.2 that $\mathscr{I}(\sigma)$ is the collection of independent sets of a matroid and it is easy to see that the closure operator of this matroid is in fact σ.∎

A characterization of the closure operator which is particularly useful when dealing with the cycle matroid of a graph is the following.

THEOREM 1. *An element x of S belongs to the closure in M of a set A if and only if $x \in A$ or there exists a circuit C of M for which $C\backslash A = \{x\}$.*

Proof. Clearly when $x \in A$, $x \in \sigma A$. Suppose C is such a circuit. Since $\rho(C\backslash x) = \rho C$, $x \sim C\backslash x$ and hence by (6) $x \sim A$, thus $x \in \sigma A$. Conversely let A be a minimal set such that there exists $x \in (\sigma A)\backslash A$ but there is no circuit C with $x \in C \subseteq A \cup x$. By (7), A must be independent or it would not be minimal. But $x \in (\sigma A)\backslash A$ implies that $A \cup x$ is dependent and therefore there must exist a circuit D such that $x \in D \subseteq A \cup x$. This is a contradiction.∎

EXERCISES 7

1. Let $\sim$ be a relation between elements of S and subsets of S which satisfies for any $x, y_1, \ldots, y_n, z_1, \ldots, z_m$ of S the following axioms:

   (D1) $y_i \sim \{y_1, y_2, \ldots, y_n\}$ $(1 \leqslant i \leqslant n)$;
   (D2) if $n \geqslant 1$ and $x \sim \{y_1, \ldots, y_n\}$ and $x \not\sim \{y_1, \ldots, y_{n-1}\}$ then $y_n \sim \{y_1, \ldots, y_{n-1}, x\}$;
   (D3) If $x \sim \{y_1, \ldots, y_n\}$ and for each i, $1 \leqslant i \leqslant n$,
   $y_i \sim \{z_1, \ldots, z_m\}$ then $x \sim \{z_1, \ldots, z_m\}$.

Interpreting $x \sim A$ as "x depends on A" prove that (D1–D3) are in an obvious sense dependence axioms for a matroid. (They are essentially the original axioms of van der Waerden [37]).

8. CLOSED SETS = FLATS = SUBSPACES

We have defined a subset A to be *closed* or a *flat* or a *subspace* in the matroid M if for all $x \in S\backslash A$, $\rho(A \cup x) = \rho A + 1$. As suggested by the heading of this section authors seem to be very divided about whether to call a set closed or a "flat" or a "subspace". We shall use mainly the terminology "flat" but because of its association with the closure operator reserve the right to occasionally revert to the "closed set", or "subspace" terminology when the context seems more natural to do so.

The following theorem is very easy to check.

THEOREM 1. *The following statements about a subset A and a matroid M are equivalent*:

(1) A *is a flat;*
(2) $\sigma A = A$;
(3) $x \in S \backslash A \Rightarrow x \nsim A.$

THEOREM 2. *If X and Y are flats of M then so is $X \cap Y$.*

Proof. Applying (S1) and (S2)

$$X \cap Y \subseteq \sigma(X \cap Y) \subseteq \sigma X \cap \sigma Y = X \cap Y,$$

hence $X \cap Y = \sigma(X \cap Y)$ so by Theorem 1, $X \cap Y$ is a flat.∎

THEOREM 3. *For any set $A \subseteq S$, σA is the intersection of all flats containing A, and is therefore the minimal flat containing A.*

Proof. If σA is not a flat $\exists y \notin \sigma A$ such that

$$\rho(\sigma(A) \cup y) = \rho(\sigma A) = \rho A.$$

But $y \notin \sigma A \Rightarrow \rho(A \cup y) > \rho A$. Since $\rho(A \cup y) < \rho((\sigma A) \cup y)$ this is a contradiction. Hence σA is a flat and must therefore contain the flat $F = \bigcap_i \{F_i : F_i$ a flat of M which contains $A\}$. Suppose $x \in (\sigma A) \backslash F$. Then

$$\rho(x \cup A) = \rho A$$

$$\rho(x \cup F) = \rho F + 1 \geqslant \rho A + 1.$$

But $A \subseteq F$, hence by (7.6) $x \sim A \Rightarrow x \sim F$, thus we have a contradiction. ∎

EXERCISES 8

1. Give an example to show that the union of two flats of a matroid is not in general a flat.

2. Prove that a matroid of rank r has at least 2^r flats. (Lazarson [57]).

9. CIRCUITS

For the graph theorist the most natural way to define a matroid is by its circuit axioms. This is the approach used by Tutte [65].

Obviously every dependent set contains a circuit and other easy observations are the following statements.

(1) If C is a circuit then $\rho C = |C| - 1$.
(2) If C is a circuit then $|C| \leqslant \rho S + 1$.
(3) The only matroid with no circuits has a single base S.
(4) Every proper subset of a circuit is independent.

(C1) If C_1, C_2 are distinct circuits then $C_1 \not\subseteq C_2$.
Slightly less obvious is the following result.
(C2) If C_1, C_2 are circuits and $z \in C_1 \cap C_2$ there exists a circuit C_3 such that $C_3 \subseteq (C_1 \cup C_2)\backslash z$.

Proof. Suppose C_1, C_2 are distinct circuits of M but no such C_3 exists. Then $(C_1 \cup C_2)\backslash z$ must be independent. Thus

$$\rho((C_1 \cup C_2)\backslash z) = \rho((C_1\backslash z) \cup (C_2\backslash z))$$
$$= |C_1 \cup C_2| - 1.$$

But $\rho(C_1) = |C_1| - 1$, $\rho(C_2) = |C_2| - 1$ so that by the submodularity of the rank function

$$\rho(C_1 \cup C_2) + \rho(C_1 \cap C_2) \leqslant \rho(C_1) + \rho(C_2)$$
$$= |C_1| + |C_2| - 2$$
$$= |C_1 \cup C_2| + |C_1 \cap C_2| - 2.$$

But $\rho(C_1 \cup C_2) \geqslant \rho((C_1 \cup C_2)\backslash z)) = |C_1 \cup C_2| - 1$ and since $\rho(C_1 \cap C_2) = |C_1 \cap C_2|$ we have a contradiction.

LEMMA. *If A is independent in M then for $x \in S$, $A \cup x$ contains at most one circuit.*

Proof. Suppose A is an independent set such that there exists $x \in S$ with two distinct circuits C_1, C_2 satisfying

$$C_1 \cup C_2 \subseteq A \cup x.$$

Then $x \in C_1 \cap C_2$ and hence by (C2) there exists C_3, a circuit of M, such that

$$C_3 \subseteq (C_1 \cup C_2)\backslash x \subseteq A$$

contradicting the independence of A.

COROLLARY. *If B is a base of M and $x \in S \backslash B$ then there exists a unique circuit $C = C(x, B)$ such that*

$$x \in C \subseteq B \cup x.$$

This circuit $C(x, B)$ is called the fundamental circuit of x in the base B.

In fact we have the following stronger result.

THEOREM 1. *If M is a matroid on S, and B is a base of M, then for any $x \in S \backslash B$, $(B \backslash y) \cup x$ is a base of M if and only if $y \in C(x, B)$ or $y = x$.*

Proof. Let $(B \backslash y) \cup x$ be a base. Then if $y \notin C(x, B)$ we have $C(x, B) \subseteq (B \backslash y) \cup x$ which cannot be a base.

Conversely suppose that $y \in C(x, B)$ but $(B \backslash y) \cup x$ is not a base. Then $B \cup x$ must contain a circuit C' which clearly cannot be $C(x, B)$. Thus $B \cup x$ contains two distinct circuits contradicting the previous lemma. ∎

We now prove that a much stronger statement than (C2) can be made about the circuits of a matroid.

THEOREM 2. *If C_1, C_2 are distinct circuits of a matroid M and $x \in C_1 \cap C_2$, then for any element y of $C_1 \backslash C_2$ there exists a circuit C such that*

$$y \in C \subseteq (C_1 \cup C_2) \backslash x.$$

Proof. Suppose C_1, C_2, x, y, are such that the theorem is false, and that $|C_1 \cup C_2|$ is minimal with this property. By (C2) there exists a circuit $C_3 \subseteq (C_1 \cup C_2) \backslash x$, but $y \notin C_3$. Now $C_3 \cap (C_2 \backslash C_1)$ cannot be null, or C_3 would be contained in C_1. Let $z \in C_3 \cap (C_2 \backslash C_1)$. Consider now C_2, C_3; $z \in C_2 \cap C_3$, $x \in C_2 \backslash C_3$ and since $y \notin C_2 \cup C_3$, $C_2 \cup C_3$ is a proper subset of $C_1 \cup C_2$. By the minimality of $C_1 \cup C_2$, therefore, there exists a circuit C_4 such that

$$x \in C_4 \subseteq (C_2 \cup C_3) \backslash z.$$

Now consider C_1, C_4; $x \in C_1 \cap C_4$, $y \notin C_2 \cup C_3$ and hence $y \in C_1 \backslash C_4$. Also $C_1 \cup C_4 \subseteq C_1 \cup C_2$. Hence by the minimality argument again there must exist a circuit C_5 such that

$$y \in C_5 \subseteq (C_1 \cup C_4) \backslash x.$$

Since $C_1 \cup C_4 \subseteq C_1 \cup C_2$ we have found a circuit C_5 such that

$$y \in C_5 \subseteq (C_1 \cup C_2) \backslash x.$$

This is a contradiction. ∎

Notice that this proof of Theorem 2 only uses the conditions (C1) and (C2). Hence in proving Theorem 2.5 (see below) that (C1) and (C2) axiomatise a matroid by its circuits, we can assume that the conditions (C1) and (C2) are equivalent to (C1) and (C3) where (C3) is what is sometimes known as the *strong circuit axiom*.

(C3)   *If C_1, C_2 are distinct members of $\mathscr{C}$ and $y \in C_1 \backslash C_2$ then for each $x \in C_1 \cap C_2$, there exists $C_3 \in \mathscr{C}$ such that $y \in C_3 \subseteq (C_1 \cup C_2) \backslash x$.*

Note: Whitney [35] took (C1) and (C3) as his circuit axioms, the equivalence of these with the apparently weaker (C1) and (C2) was proved by Lehman [64].

Proof of Theorem 2.5. That (C1) and (C2) hold for the circuits of a matroid is true from Theorem 2. Suppose $\mathscr{C}$ satisfies (C1) and (C2). We shall define a function r associated with $\mathscr{C}$ and show that r is the rank function of a matroid M which has $\mathscr{C}$ as its set of circuits. Let $(x_1, \ldots, x_q)$ be any (ordered) family of elements of S. Set $\theta_i = 0$ if $\{x_1, \ldots, x_i\}$ contains a member C of $\mathscr{C}$ such that $x_i \in C$ and set $\theta_i = 1$ otherwise. Let t be a function on the ordered subsets of S defined by

$$t\{x_1, \ldots, x_q\} = \sum_{i=1}^{q} \theta_i$$

LEMMA. *For any permutation π of $(1, \ldots, q)$,*

$$t\{x_1, \ldots, x_q\} = t\{x_{\pi(1)}, \ldots, x_{\pi(q)}\}.$$

Proof. It is sufficient to show that

$$t\{x_1, \ldots, x_{q-2}, x_{q-1}, x_q\} = t\{x_1, \ldots, x_{q-2}, x_q, x_{q-1}\}.$$

Let Y be the ordered set $\{x_1, \ldots, x_{q-2}\}$ and then in the obvious notation let

$$t(Y) = a, \qquad t(Y, x_{q-1}) = a_1, \qquad t(Y, x_q) = a_2$$
$$t(Y, x_{q-1}, x_q) = a_{12}, \qquad t(Y, x_q, x_{q-1}) = a_{21}.$$

Case 1

There is no member of $\mathscr{C}$ in $Y \cup x_{q-1}$ containing x_{q-1} and none in $Y \cup x_q$ containing x_q. Then

$$a_1 = a_2 = a + 1.$$

If there is a member of $\mathscr{C}$ in $Y \cup x_{q-1} \cup x_q$ containing x_{q-1} and x_q then

$$a_{12} = a_1 = a_2 = a_{21};$$

otherwise

$$a_{12} = a_1 + 1 = a_2 + 1 = a_{21}.$$

Case 2

There is a member of $\mathscr{C}$ say C_2 contained in $Y \cup x_{q-1}$ and $x_{q-1} \in C_2$, and there is a member C_1 of $\mathscr{C}$ in $Y \cup x_{q-1} \cup x_q$ containing x_{q-1} and x_q. Then by (C3) there is a member C_3 of $\mathscr{C}$ such that

$$x_q \in C_3 \subseteq Y \cup x_q.$$

Hence

(α)
$$a_{12} = a_1 = a = a_2 = a_{21}.$$

Case 3

There is a C_2 as in case 2 but no C_1 as in case 2. If there is a C_3 as above then (α) holds. Otherwise

$$a_{12} = a_1 + 1 = a + 1 = a_2 = a_{21}.$$

Case 4

There is a member C of $\mathscr{C}$ such that

$$x_q \in C \subseteq Y \cup x_q.$$

The proof of cases 2 and 3 applies here also.

If A is any subset of S we now let

$$r(A) = t\{x_1, \ldots, x_p\}$$

where $x_1, \ldots, x_p$ is any ordering of the elements of A. By the lemma, r is well defined and it is obvious that the rank axioms (R1) and (R2) hold for r.

Now let $A \subseteq S$, and suppose

$$r(A \cup x) = r(A \cup y) = r(A).$$

Then there exist members C_1 and C_2 of $\mathscr{C}$ such that

$$x \in C_1 \subseteq A \cup x,$$
$$y \in C_2 \subseteq A \cup y.$$

Hence $r(A \cup x \cup y) = r(A)$ and (R3) is satisfied so that r is the rank function of a matroid M which is easily seen to have as its circuits the family $\mathscr{C}$.∎

A useful result in connection with the above is the following theorem of Tutte [65].

THEOREM 3. *Let $\mathscr{D}$ be a collection of non-null subsets of S such that for any two distinct members X, Y, of $\mathscr{D}$ such that $x \in X \cap Y$, $y \in X \backslash Y$, there exists $Z \in \mathscr{D}$*

such that

$$y \in Z \subseteq (X \cup Y) \backslash x.$$

Then the collection $\mathscr{D}'$ of minimal members of $\mathscr{D}$ is the set of circuits of a matroid.

The proof of this follows easily from the following lemma.

LEMMA. *Let $\mathscr{D}$ be a family of subsets satisfying the hypotheses of Theorem 3. Then if $a \in W \in \mathscr{D}$ there exists $V \in \mathscr{D}'$ such that*

$$a \in V \subseteq W$$

Proof. There exists $V \in \mathscr{D}$ such that

$$a \in V \subseteq W.$$

Choose such a V so that $|V|$ is a minimum. Suppose $V \notin \mathscr{D}'$. Then there exists $Z \in \mathscr{D}$ such that $Z \subseteq V \backslash a$. Take b to be any element of Z. Then by hypothesis there exists $Z' \in \mathscr{D}$ such that

$$a \in Z' \subseteq (V \cup Z) \backslash b.$$

But then $a \in Z' \subseteq W$, contradicting the choice of V. Hence $V \in \mathscr{D}'$ and the proof of the lemma is complete.

Proof of Theorem 3. Let X, Y be two distinct members of $\mathscr{D}'$ and let $x \in X \cap Y$. Then by hypothesis there exists $z \in \mathscr{D}$ such that

$$Z \subseteq (X \cup Y) \backslash x.$$

Let $a \in Z$. Then by the above lemma there exists $W \in \mathscr{D}'$ such that

$$a \in W \subseteq Z \subseteq (X \cup Y) \backslash x.$$

Hence $\mathscr{D}'$ satisfies (C2) of Theorem 2.5. Clearly no member of $\mathscr{D}'$ properly contains another and hence $\mathscr{D}'$ is the set of circuits of a matroid.∎

EXERCISES 9

1. If C is a circuit of the matroid M and $a \in C$ prove that there exists a base B such that $C = C(a, B)$.

2. Given an arbitrary collection $\mathscr{D}$ of incomparable subsets of S does there exist a matroid M which has circuit set $\mathscr{C}(M) \supseteq \mathscr{D}$?

3. If $C_1, \ldots, C_m$ are distinct circuits of a matroid M on S with

$$C_i \nsubseteq \bigcup_{j \neq i} C_j \qquad (1 \leqslant i \leqslant m)$$

and $D \subseteq S, |D| < m$, prove that there exists a circuit C of M such that

$$C \subseteq \bigcup_{i=1}^{m} C_i \backslash D.$$

4. The circuits $C_1, \ldots, C_q$ of the matroid M on $S = \{e_1, \ldots, e_n\}$ form a *fundamental set of circuits* if it is possible to order $\{e_1, \ldots, e_n\}$ in such a way that C_i contains e_{r+i} but no e_j for $j > r + i$ where $r = \rho(M)$ and $q = |S| - r$.

   Prove that every matroid has a fundamental set of circuits. Show that there exist non-isomorphic matroids M_1, M_2 with a common fundamental set of circuits. (Whitney [35])

5. A *refinement* of a matroid M on S is a matroid N on S such that each circuit of M is a circuit of N. Prove that a matroid has a refinement if and only if no circuit of M has cardinality $\rho(M) + 1$. (Bean [72]).

6. If M is a matroid on $S = \{e_1, \ldots, e_n\}$ let $\{A_i : 1 \leqslant i \leqslant n\}$ be a collection of mutually disjoint sets. Let $\mathscr{C}(M')$ be all sets of the form $A_{i_1} \cup \ldots \cup A_{i_k}$ such that $\{x_{i_1}, \ldots, x_{i_k}\}$ is a circuit of M. Prove that $\mathscr{C}(M')$ is the set of circuits of a matroid on $\bigcup A_i$ (called the *enlargement* of M).

10. THE CYCLE MATROID OF A GRAPH

In this section we examine in more detail the cycle matroid of a graph, briefly defined in Section 3.

THEOREM 1. *If G is a graph the cycles of G are the circuits of a matroid $M(G)$ on the edge set $E(G)$.*

This matroid $M(G)$ is called the *cycle matroid* of G (or in Tutte's work [65] the *polygon matroid* of G).

Proof. Clearly no cycle properly contains another and it is an elementary piece of graph theory to show that if C_1, C_2 are distinct cycles of G with a common edge e, there exists a third cycle $C_3 \subseteq (C_1 \cup C_2) \backslash e$. Thus the set of cycles of G satisfy the axioms (C1) and (C2) so that by Theorem 2.5 they are the circuits of a matroid.∎

Now by definition a maximal subgraph of G which contains no cycle is a spanning forest. Hence we can list the following basic properties of $M(G)$:

(1) If G is a connected graph the bases of $M(G)$ are the spanning trees of G.

(2) If G is a disconnected graph the bases of $M(G)$ are the spanning forests of G.

(3) A set X of edges of G is independent in $M(G)$ if and only if X contains no cycle, that is X is a forest.

Now if G is connected any spanning tree is well known to have $|V(G)| - 1$ edges. Thus counting up the connected components we get:

(4) The rank of the matroid $M(G)$ is $|V(G)| - k(G)$ where $k(G)$ is the number of connected components of G.

More generally if A is any subset of edges of G let $k(A)$ denote the number of components in the subgraph generated by A.

(5) For any subset $A \subseteq E(G)$ the rank of A in $M(G)$ is given by

$$\rho_G A = |V(A)| - k(A).$$

We can also now justify our choice of terminology in Section 4 since it is easy to prove the following statements.

(6) The edge e of G is a loop in the matroid $M(G)$ if and only if it is a loop of the graph.

(7) Two edges e, f of G are parallel in $M(G)$ if and only if they are parallel edges in G.

Using Theorem 7.1 it is clear how to characterize the closure operator and hence implicitly the flats of $M(G)$.

(8) Let A be a set of edges of G, let $e \in E(G) \backslash A$. Then e belongs to the closure of A in $M(G)$ if and only if there is a cycle C of G with

$$e \in C \subseteq A \cup e.$$

However despite the power of matroid theory as a tool in the clarification of certain graphical ideas (which we hope to show later) we warn that many problems of graph theory cannot even be posed in matroid language. Crudely this is because there is no simple exact counterpart of a vertex in a matroid. Also, non-isomorphic graphs may have isomorphic cycle matroids.

Example 1. Let G_1, G_1' be the trees of Fig. 1. Then $M(G_1) \simeq M(G_1')$ and in fact $M(G_1) \simeq M(H)$ for any forest H with the same number of edges as G_1.

G_1: 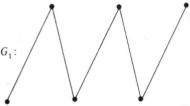 G_1':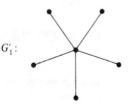

Figure 1

Example 2. Let G_2 and G_2' be graphs of Fig. 2, they are clearly non-isomorphic. However it is easy to check that the map $x \to x'$ is an isomorphism between $M(G_2)$ and $M(G_2')$.

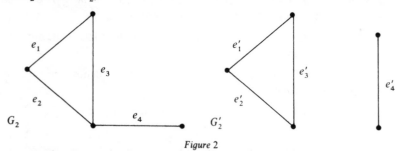

Figure 2

The characterization of those graphs uniquely determined (up to isomorphism) by their cycle matroids will be given in Chapter 6. We close this section by mentioning that "most" matroids are not the cycle matroids of graphs.

A matroid M is *graphic* if there exists some graph G such that M is isomorphic to the cycle matroid $M(G)$. The smallest non-graphic matroid is $U_{2,4}$ the uniform matroid of rank 2 on a set of 4 elements. (Try to draw a graph with 4 edges such that its cycles are all edge sets of cardinal 3?) As we shall show in Chapter 10 Tutte has now characterized graphic matroids in what is certainly the deepest theorem on this subject, if not in the whole of combinatorics.

EXERCISES 10

1. For what values of k is the uniform matroid of rank k on a set of n elements the cycle matroid of some graph?

2. Prove that in a graphic matroid every pair of circuits C_1, C_2 has the property that $C_1 \triangle C_2$ is the union of disjoint circuits. Find a non-graphic matroid which also has this property and also find a matroid which does not have this property.

11. A EUCLIDEAN REPRESENTATION OF MATROIDS OF RANK $\leqslant 3$

Matroids of rank $\leqslant 3$ have the following very useful geometric description in the plane.

Let M be a simple matroid of rank 3 on the set $S = \{x_1, \ldots, x_n\}$. Place the n points $x_1, \ldots, x_n$ on the plane and draw a line through each closed set A such that

$$|A| \geqslant 3, \qquad \rho A = 2.$$

Then the bases of M are all subsets of S of cardinal 3 which are not collinear in this diagram.

Example. Let M be the matroid on $\{x_1, x_2, \ldots, x_6\}$ whose bases are all 3-sets except

$$\{x_1, x_3, x_6\}, \qquad \{x_1, x_4, x_5\}, \qquad \{x_2, x_5, x_6\}, \qquad \{x_2, x_3, x_4\}.$$

Then the Euclidean representation of M is Figure 1.

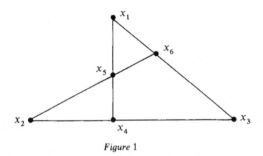

Figure 1

Conversely any diagram of points and lines in the plane in which a pair of lines meet in at most one point represents a unique simple matroid whose bases are those 3-sets of points which are not collinear in the diagram.

Proof. Let D be any such diagram and let S be the set of points of D. Let $B(D)$ consist of those subsets X of S such that $|X| = 3$, and the members of X are not collinear in D. We show that $B(D)$ satisfies the base axioms of a matroid.

Clearly distinct members X, Y of $B(D)$ satisfy $X \nsubseteq Y$. Also if $x \in X$, $Y = \{y_1, y_2, y_3\}$ and $(X \backslash x) \cup y_i \notin B(D)$ for $1 \leqslant i \leqslant 3$, since $X \neq Y$ this contradicts the assumption that no two distinct lines of D meet in at most one point. Hence $(X \backslash x) \cup y_i \in B(D)$ for some i. Thus $B(D)$ satisfies the base axioms of a matroid.

Example. The diagram D of Fig. 2 is the Euclidean representation of the matroid on $\{x_1, \ldots, x_7\}$ with bases all 3-sets except $\{x_1, x_2, x_6\}, \{x_1, x_4, x_7\}, \{x_1, x_3, x_5\}, \{x_2, x_3, x_4\}, \{x_2, x_5, x_7\}, \{x_3, x_6, x_7\}, \{x_4, x_5, x_6\}$.
The reader familiar with projective geometry will recognize this matroid

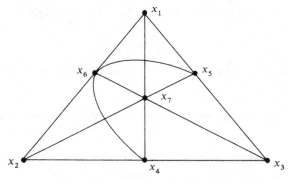

Figure 2

as the well-known Fano plane coordinatizable over the field $GF(2)$ under the map

$$x_1 \to (1, 0, 0), \qquad x_2 \to (0, 1, 0), \qquad x_3 \to (0, 0, 1)$$
$$x_4 \to (0, 1, 1), \qquad x_5 \to (1, 0, 1), \qquad x_6 \to (1, 1, 0)$$
$$x_7 \to (1, 1, 1).$$

This matroid, called the *Fano matroid*, is more well known as the projective geometry $PG(2, 2)$.

It is easy to see how to deal with non-simple matroids of rank $\leqslant 3$. We illustrate with examples.

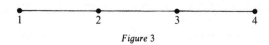

Figure 3

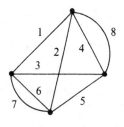

Figure 4

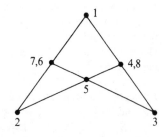

Figure 5

Figure 3 is the Euclidean representation of the uniform matroid of rank 2 on 4 elements.

Figure 5 shows the Euclidean representation of the cycle matroid of the graph of Figure 4. Notice that a pair of parallel elements is represented by a single point.

We close by emphasising that this Euclidean representation is in no way connected with the problem of whether or not a matroid can be represented (or coordinatized) in some vector space. We discuss this problem more fully in Chapter 9.

EXERCISE 11

1. Prove that the Fano matroid has no "straight line" Euclidean representation.

NOTES ON CHAPTER 1

This chapter is based mainly on Whitney [35].

The approach is not similar to that of Tutte [65], for example Tutte's rank function is a *decreasing* set function. Useful early papers on matroids which take alternative approaches to this basic material are those of Lehman [64], Rota [64], Minty [66], Mirsky and Perfect [67]. More recently we have the books of Crapo and Rota [70], Tutte [70], Bruter [70], Mirsky [71], the papers of Harary and Welsh [69], Lovasz [71], Wilson [73], Brylawski, Kelly and Lucas [74], and the forthcoming book on combinatorial optimization by Lawler [75]. The independence axioms given by Theorem 5.2 are due to Edmonds [65]. The closure axioms are those of Rota [64].

Section 11 has an extension to matroids of higher rank, see Mason [71].

CHAPTER 2

Duality

1. THE DUAL MATROID

Most of the concepts discussed in Chapter 1 were familiar in linear algebra. The concept of matroid duality is less familiar, but is of fundamental importance in the applications of matroids to combinatorial theory.

The basic result due to Whitney [35] is the following theorem.

THEOREM 1. *If $\{B_i: i \in I\}$ is the set of bases of a matroid M on S then $\{S \backslash B_i: i \in I\}$ is the set of bases of a matroid M^* on S.*

We call M^* the *dual matroid* of M. Obviously the relation between M and M^* is symmetrical, that is

$$(1) \qquad\qquad (M^*)^* = M.$$

In order to prove Theorem 1 we need the following easy lemma.

LEMMA. *Let B_1, B_2 be bases of a matroid and let $x \in B_1 \backslash B_2$ then there exists $y \in B_2 \backslash B_1$ such that $(B_2 \backslash y) \cup x$ is a base.*

Proof. Consider $B_2 \cup x$, it is dependent and of rank $r = |B_2|$. Since $\{x\}$ is independent, by the augmentation theorem there exists a set A, such that A is a base of M and

$$\{x\} \subseteq A \subseteq B_2 \cup x.$$

Let $\{y\} = B_2 \backslash A$, then $(B_2 \backslash y) \cup x$ is our required base.

Proof of Theorem 1. Let $B_1^* = S \backslash B_1, B_2^* = S \backslash B_2$. Let $b \in B_1^* \backslash B_2^*$ so that $b \in B_2 \backslash B_1$. By the preceding lemma there exists $y \in B_1 \backslash B_2$ such that $(B_1 \backslash y) \cup b$ is a base B_3 of M. Now $y \in B_1 \backslash B_2 \Rightarrow y \in B_2^* \backslash B_1^*$. Also

$$S \backslash ((B_1 \backslash y) \cup b) = (B_1^* \cup y) \backslash b$$
$$= (B_1^* \backslash b) \cup y = S \backslash B_3,$$

where B_3 is a base of M. Thus we have shown that the collection of sets $\{S \backslash B_i : B_i \text{ a base of } M\}$ satisfies the base axiom (B1) and is therefore the set of bases of a matroid. ∎

It should be emphasized immediately that this idea of duality has no connection with vector space duality (the algebraic interpretation of M^* in the case of a vectorial matroid will be seen in Chapter 9).

The following properties of M^* follow almost immediately from its definition.

(2)   A subset $X \subseteq S$ is independent in M^* if and only if $S \backslash X$ is spanning in M.

(3)   An element $x \in S$ is a loop of M if and only if x belongs to every base of M^*.

(4)   The rank of M^* is $|S| - \rho(M)$.

This last result is a special case of the very useful

THEOREM 2. *The rank functions* ρ, ρ^* *of* M, M^* *respectively are related by:*

(5) $\rho^*(S \backslash A) = |S| - \rho S - |A| + \rho A$ $(A \subseteq S)$.

Proof. Let $\lambda : 2^S \to \mathbb{Z}$ be defined by

$$\lambda A = |A| + \rho(S \backslash A) - \rho S   (A \subseteq S).$$

Since ρ satisfied (R3′)

$$\rho(S \backslash A) + \rho(S \backslash B) \geqslant \rho\big((S \backslash (A \cup B)\big) + \rho\big(S \backslash (A \cap B)\big)$$

and hence λ also satisfies (R3′); and (R1′), (R2′) trivially. Hence λ is the rank function of a matroid (say $M(\lambda)$) on S. However A is independent in $M(\lambda)$ if and only if

$$|A| + \rho(S \backslash A) - \rho S \geqslant |A|;$$

that is if and only if $S \backslash A$ is spanning in M. By (2) this means $M(\lambda) = M^*$. ∎

The function $\rho^* : 2^S \to \mathbb{Z}$ we call the *corank* function of M. A *cobase* of M is a base of M^*, a *cocircuit* of M is a circuit of M^*, if x is a loop of M^* it is called a *coloop* of M and so on.

Now since M^* determines M uniquely we have the obvious result that a matroid is uniquely determined by its cobases, cocircuits, corank function and so on. Moreover each of the Theorems 1.2.1–5 is capable of dualization to give an axiomatisation of a matroid in terms of these dual concepts. Consider for example the dualization of Theorem 1.2.1.

THEOREM 1.2.1*. *A nonempty collection* $\mathscr{B}$ *of subsets of* S *is the collection of cobases of a matroid on* S *if and only if the condition* (B1) *is satisfied.*

Another example is the dualization of the following statement.

(α) An element x is a loop of M if and only if x does not belong to any base of M.

(α)* An element x is a coloop of M if and only if x does not belong to any cobase of M.

These examples illustrate the very simple but useful *duality principle*—that to every statement about a matroid there is a dual statement.

We shall denote this dual statement by an asterisk, for example (1.3.8)* will denote the dual of the statement (1.3.8).

We now show how the basic independent set extension Theorem 1.5.1 has a stronger form. We call it the *dual augmentation theorem*.

THEOREM 3. *Let M be a matroid on S, let A, A^* be subsets of S with $A \cap A^* = \emptyset$, A independent in M, A^* independent in M^*. Then there exists bases B, B^* of M, M^* respectively with $A \subseteq B$, $A^* \subseteq B^*$, $B \cap B^* = \emptyset$.*

Proof. Letting ρ, ρ^* be the rank functions of M, M^* respectively we have, since A^* is independent in M^*,

(6) $\rho(S \backslash A^*) = |S| - |A^*| - \rho^* S + \rho^*(A^*)$

$\qquad\qquad = |S| - \rho^* S = \rho S.$

Since A is an independent subset of $S \backslash A^*$ we can augment it to a base B of $S \backslash A^*$ which by (6) must be a base of M. Then $S \backslash B$ is the required base of M^* which contains A^* and completes the proof. ■

We close this section by proving some easy linking theorems between a matroid and its dual.

THEOREM 4. *A subset X of S is a base of a matroid M if and only if X has non-null intersection with every cocircuit of M, and is minimal with respect to this property.*

Proof. Let B be a base of M. Suppose there exists a cocircuit C^* of M such that $B \cap C^* = \emptyset$. Then $S \backslash B$ contains C^* and therefore is dependent in M^*, which is a contradiction. Now let X be any proper subset of B which has a non-null intersection with every circuit of M^*. Then $S \backslash X$ contains no circuit of M^* and is independent in M^*. Since $S \backslash X$ properly contains the base $S \backslash B$ of M^* this is a contradiction.

Conversely if X has non-null intersection with every circuit of M^*, $S \backslash X$ contains no cocircuit and therefore is independent in M^*. If X is minimal with respect to this property then $S\backslash X$ must be a maximal independent subset of M^*. Thus $S\backslash X$ is a cobase of M, completing the proof of the theorem. ∎

From this it is easy to verify the following result.

THEOREM 5. *A subset Y of S is a cocircuit of a matroid M if and only if Y has non-null intersection with every base of M, and is minimal with respect to this property.*

We next obtain a new characterization of the cocircuits.

THEOREM 6. *A set $\mathscr{C}$ of subsets of S is the set of circuits of a matroid M on S if and only if the members of $\mathscr{C}$ are the minimal non-null subsets C of S such that $|C \cap C^*| \neq 1$ for every cocircuit C^* of M.*

The proof is not difficult and is left to the reader.

EXERCISES 1

1. Show that the dual of a simple matroid need not be simple.

2. Can a set be both a circuit and a cocircuit in a matroid?

3. Prove that if B is a base of the matroid M and $e \in B$ there is exactly one cocircuit C^* of M such that $C^* \cap (B\backslash e) = \varnothing$.

4. If C is a circuit of M and x, y are distinct elements of C prove there exists a cocircuit C^* containing x and y but no other element of C. (Minty [66]).

5. A *painting* of S is a partition of S into disjoint sets R, G, B with $|G| = 1$. We regard the elements of R as red, of G as green, and of B as blue. A *graphoid* is a triple $(S, \mathscr{C}, \mathscr{D})$ where $\mathscr{C}, \mathscr{D}$ are collections of non comparable subsets of S such that:

   (i)   if $C \in \mathscr{C}$ and $D \in \mathscr{D}$ then $|C \cap D| \neq 1$,

  (ii) for any painting of S there is either

      (a) a member of $\mathscr{C}$ consisting of the green element and otherwise only red elements, or

      (b) a member of $\mathscr{D}$ consisting of the green element and otherwise only blue elements

Prove that if $(S, \mathscr{C}, \mathscr{D})$ is a graphoid $\mathscr{C}$ is the set of circuits of a matroid M on S and $\mathscr{D}$ is the set of cocircuits of M.

Conversely given any matroid M on S prove that $(S, \mathscr{C}(M), \mathscr{C}(M^*))$ is a graphoid. (A theory of matroids based on graphoids is developed by Minty [66].)

2. THE HYPERPLANES OF A MATROID

A *hyperplane* of a matroid M is a maximal proper closed subset of S. It is easy to prove:

LEMMA 1. *The following statements about a subset H of S are equivalent.*

(1) *H is a hyperplane.*
(2) *$\sigma H \neq S$ but $\sigma(H \cup x) = S \qquad \forall x \in S\backslash H$.*
(3) *For any base B, $B \nsubseteq H$ but if $x \in S\backslash H$ there exists a base $B' \subseteq H \cup x$.*
(4) *H is a maximal subset of S which is not spanning.*
(5) *H is a maximal set of rank $\rho S - 1$.*

Since the intersection of flats is a flat we know that two hyperplanes will intersect in a flat. Except in special cases (vector spaces for example) the rank of their intersection will be strictly less than $\rho S - 2$. However certain "intersection type" properties do carry over to general matroids.

LEMMA. 2. *If X, Y are flats of M, with $Y \subseteq X$ and $\rho Y = \rho X - 1$, there exists a hyperplane H such that $Y = X \cap H$.*

Proof. Let $A = \{a_1, \ldots, a_k\}$ be a maximal independent subset of Y, $(k = \rho(Y))$. Extend A to an independent set $A \cup a_{k+1} \subseteq X$, $a_{k+1} \in X\backslash Y$. Then if we now extend $A \cup a_{k+1}$ to a base B of M the flat $H = \sigma(B\backslash a_{k+1})$ is the required hyperplane of M since by (R3) $H \cap X$ has rank $\rho(Y)$, and contains Y.

THEOREM 1. *Let X be a flat of M with rank t. Then there exist distinct hyperplanes $H_i (1 \leqslant i \leqslant \rho(M) - t)$ such that*

$$X = \bigcap \{H_i : 1 \leqslant i \leqslant \rho(M) - t\}.$$

Proof. When $k = \rho(M) - t = 1$, take $H_1 = X$. We now use induction on k. Let $r = \rho(M)$, and let $\{b_1, b_2, \ldots, b_t\}$ be a maximal independent subset of X. By the augmentation theorem there exists a base $B = \{b_1, \ldots, b_t, \ldots, b_r\}$ of M. Let $X' = \sigma(\{b_1, \ldots, b_{t+1}\})$ so that $\rho X' = t + 1$ and by the induction hypothesis there exist distinct hyperplanes H_i such that

$$X' = \bigcap \{H_i : 1 \leqslant i \leqslant \rho(M) - t - 1\}.$$

But by the previous lemma there exists a hyperplane H such that

$$X = H \cap X'.$$

Clearly $H \neq H_i (1 \leqslant i \leqslant \rho(M) - t - 1)$ and the theorem is proved. ∎

We now link hyperplanes and cocircuits.

THEOREM 2. *A set H is a hyperplane of the matroid M on S if and only if S\H is a cocircuit of M.*

Proof. Let C be a cocircuit of M. Then

$$\rho(S\backslash C) = |S| - \rho^*S - |C| + \rho^*C = \rho S - 1.$$

Moreover if $x \notin S\backslash C$,

$$\rho((S\backslash C) \cup x) = |S| - \rho^*S - |C\backslash x| + \rho^*(C\backslash x)$$

and since $C\backslash x$ is independent in M^* we get

$$\rho((S\backslash C) \cup x) = \rho S.$$

Thus $S\backslash C$ is a maximal proper subset of S which is closed in M. In other words $S\backslash C$ is a hyperplane of M.

Conversely let H be a hyperplane of M. We have

$$\rho^*(S\backslash H) = |S| - \rho S - |H| + \rho H = |S\backslash H| - 1.$$

But if $x \in S\backslash H$,

$$\rho^*((S\backslash H)\backslash x) = |S| - \rho S - |H \cup x| + \rho(H \cup x)$$
$$= |S\backslash H| - 1.$$

Thus $S\backslash H$ is dependent in M^* but every proper subset of it is independent, in other words it is a circuit of M^* and therefore a cocircuit of M. ∎

Hence by the duality principle we can dualize the circuit axioms (C1) and (C2) to get the following very useful axiomatization of a matroid in terms of its hyperplanes.

THEOREM 3. *A collection $\mathcal{H}$ of subsets of S is the set of hyperplanes of a matroid on S if and only if the conditions (H1), (H2) hold.*

(H1)   If $H_1, H_2 \in \mathcal{H}$ with $H_1 \neq H_2$, then $H_1 \nsubseteq H_2$.

(H2)   If $H_1, H_2 \in \mathcal{H}$ and $x \notin H_1 \cup H_2$ there exists $H_3 \in \mathcal{H}$ such that $H_3 \supseteq (H_1 \cap H_2) \cup x$.

EXERCISES 2

1. Prove that X is a flat of the matroid M on S if and only if $S\backslash X$ is the union of co-circuits of M.

2. Prove that $\mathcal{H}$ is the collection of hyperplanes of a simple matroid on S if in addition

to the normal hyperplane axioms it satisfies:

(a)   no element belongs to each hyperplane
(b)   if $p \neq q$ are elements of S there exists $H \in \mathscr{H}$ with $p \in H$, $q \notin H$.

3. PAVING MATROIDS

In this section we shall show from the hyperplane axioms (H1), (H2) that the class of d-partitions of a set S, studied by Hartmanis [59] is essentially a class of relatively well behaved matroids.

A *cover* of the set S is any finite collection of distinct subsets $\mathscr{A} = \{A_i : i \in I\}$ whose union is S. For any integer $d \geqslant 1$ the cover is said to be a *d-partition* if it satisfies the conditions (1)–(3).

(1) $|I| \geqslant 2$.
(2)   Each member A_i of $\mathscr{A}$ has cardinal at least d.
(3)   Each d-element subset of S is contained in a unique A_i.

Thus a d-partition is a natural generalization of an ordinary partition in the sense that a 1-partition is just a partition of the set S.

The sets A_i are called the *blocks* of the d-partition.

THEOREM 1. *If $\mathscr{A} = \{A_i : i \in I\}$ is a d-partition then the collection of blocks is the set of hyperplanes of a matroid, which we denoted by $P(\mathscr{A})$.*

Proof. We show that $\mathscr{A}$ satisfies the hyperplane axioms (H1), (H2). If $A_i \supseteq A_j$, $i \neq j$, then since $|A_j| \geqslant d$ there will exist a d-subset belonging to more than one A_i contradicting (3). Now let $x \notin A_i \cup A_j$, $i \neq j$. Then by (3) $|A_i \cap A_j| < d$ so that $|(A_i \cap A_j) \cup x| \leqslant d$. Thus $(A_i \cap A_j) \cup x$ must be contained in some A_k. Thus (H2) holds. ∎

The following properties of the matroid $P(\mathscr{A})$ are easy to prove.

(4)   The rank of S in $P(\mathscr{A})$ is $d + 1$.
(5) · Every set of cardinal $\leqslant d$ is independent and every circuit of $P(\mathscr{A})$ has cardinality $d + 1$ or $d + 2$.
(6) X is spanning in $P(\mathscr{A})$ if and only if X is not contained in any block.

We call a matroid M a *paving matroid* if there exists some d-partition $\mathscr{A}$ of S such that M is isomorphic to the matroid $P(\mathscr{A})$. We now show that paving matroids can be characterised by properties (4) and (5) above.

THEOREM 2. *A matroid of rank $r \geqslant 2$ is a paving matroid if and only if it has no circuit of cardinal $\leqslant r - 1$.*

Proof. Because of (4) and (5) we have only to prove that if every circuit C of M has $|C| \geqslant r$ then M is the paving matroid obtained from a *d-partition*, $d = r - 1$.

Consider the collection $\{H_i : i \in I\}$ of all hyperplanes of M. Since $\rho(M) = r$, $\rho(H_i) = r - 1$ and hence $|H_i| \geqslant d \; \forall i$. No matroid of rank > 1 can possess only one hyperplane. It remains to prove that (3) holds. Suppose X exists, $|X| = r - 1$, with $X \subseteq H_i \cap H_j$ where H_i, H_j are distinct hyperplanes. Then $\rho(H_i \cap H_j) \leqslant r - 2$ so that

$$\rho X \leqslant r - 2 \leqslant |X|.$$

Thus X is dependent in M and contains a circuit of cardinal $\leqslant r - 2$. This is a contradiction so the hyperplanes of M must be the blocks of an $(r - 1)$-partition. ∎

EXERCISES 3

1. Prove that the Fano matroid is a paving matroid.

○2. In the enumeration of non-isomorphic simple matroids on a set of $\leqslant 9$ elements paving matroids predominate. Does this hold in general? (Blackburn, Crapo and Higgs [73].)

4. THE COCYCLE MATROID OF A GRAPH

Harary [69] calls a set X of edges of a connected graph G a *cutset* of G if the removal of X from G results in a disconnected graph, and then defines a cocycle of G to be a minimal cutset of G. We are more general here and define a *cocycle* of a graph G to be a set of edges whose removal from G increases the number of connected components of G and is minimal with respect to this property. The difference between our definition and that of Harary is minimal, for when G is a connected graph the definitions coincide. If G is disconnected with connected components say $G_1, G_2, \ldots, G_p$ then if $\mathscr{C}^*(H)$ denotes the set of cocycles of a graph H,

$$\mathscr{C}^*(G) = \mathscr{C}^*(G_1) \cup \mathscr{C}^*(G_2) \cup \ldots \cup \mathscr{C}^*(G_p)$$

and

$$\mathscr{C}^*(G_i) \cap \mathscr{C}^*(G_j) = \varnothing \quad \text{when} \quad i \neq j.$$

The main results of this section are the following two theorems.

THEOREM 1. *If G is a graph and $\mathscr{C}^*(G)$ denotes the set of cocycles of G, then $\mathscr{C}^*(G)$ is the set of circuits of a matroid $M^*(G)$ on $E(G)$.*

We call $M^*(G)$, the *cocycle matroid* of G.

THEOREM 2. *For any graph G, the cycle and cocycle matroids of G are dual matroids, that is*

(1) $M^*(G) = (M(G))^*$,
(2) $M(G) = (M^*(G))^*$.

Example 1. Let G be the graph of Fig. 1.

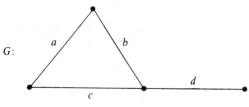

G:

Figure 1

$M(G)$ has the single circuit $\{a, b, c\}$.
The circuits of $M^*(G)$ are $\{d\}$, $\{bc\}$, $\{c, a\}$, $\{a,b\}$.

Example 2. Let H be the disconnected graph of Figure 2.

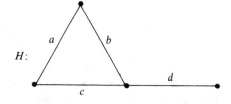

H:

Figure 2

The circuits of $M(H)$ are $\{a, b, c\}$ and $\{e, f\}$.
The circuits of $M^*(H)$ are $\{d\}$ $\{b, c\}$, $\{c, a\}$, $\{a, b\}$, $\{e, f\}$.

As an exercise the reader can verify that Theorem 2 is true for the graphs G and H.

The reader familiar with graph theory will not find it difficult to verify Theorem 1 directly, by showing that the cocycles of a graph satisfy the circuit axioms of a matroid. It is more efficient to prove Theorems 1 and 2 together. We use the following easy graphical lemma:

LEMMA. *If G is a graph X is a cocycle of G if and only if X is a minimal subset of E(G) which has non-null intersection with every spanning forest of G.*

Proof. Left as an exercise to the reader.

Proof of Theorems 1 and 2. Consider the cycle matroid $M(G)$. By Theorem 1.5 a subset X is a cocircuit of $M(G)$ if and only if X is a minimal set having non-null intersection with every base of $M(G)$. The bases of $M(G)$ are the spanning forests of G and hence by the above lemma, the cocircuits of $M(G)$ are the cocycles of G. Hence the cocycles of G are the circuits of the matroid $(M(G))^*$. This proves Theorem 1 and (1) of Theorem 2. The statement (2) follows immediately from (1) by duality. ∎

A consequence of Theorem 2 is:

COROLLARY. *A set X of edges of G is a base of the cocyle matroid $M^*(G)$ if and only if $E(G)\backslash X$ is a spanning forest of G.*

When G is a connected graph and T is a spanning tree of G, Harary [69] calls the set of edges in $E(G)\backslash T$ a *cotree* of G. Thus for a connected graph G, the bases of $M^*(G)$ are the set of cotrees of G. In Harary [69] the *cycle rank* of a graph G, written $m(G)$, is defined by

$$m(G) = |E(G)| - |V(G)| + k(G),$$

where $k(G)$ is the number of connected components of G.

A *bridge* or *isthmus* of a graph G is any edge whose removal increases the number of connected components of G. Thus an edge e of G is a bridge if and only if $\{e\}$ is a circuit of $M^*(G)$.

LEMMA. *The following statements about an edge e of a graph G are equivalent:*

(4)   *e is a bridge of G;*
(5)   *$\{e\}$ is a circuit of the cocycle matroid $M^*(G)$;*
(6)   *e belongs to no base of $M^*(G)$;*
(7)   *e belongs to every base of $M(G)$, that is e is a member of every spanning forest of G.*

Proof. Left to the reader.

Continuing our use of the "co-notation" we define a matroid M to be *cographic* if there exists a graph G such that M is isomorphic to the cocycle matroid of G.

We can easily check that the smallest matroid which is not cographic is the uniform matroid $U_{2,4}$. In Chapter 10 we characterize cographic matroids.

EXERCISES 4

1. Prove that the Fano matroid (see Figure 1.11.2) is not cographic.

2. Prove that $M(K_5)$ and $M(K_{3,3})$ are not cographic.

3. Prove that $M^*(K_5)$ and $M^*(K_{3,3})$ are cographic but not graphic.

○4. Characterize those graphs whose cycle matroid is isomorphic to their cocycle matroid.

NOTES ON CHAPTER 2

The concept of dual matroid was discovered by Whitney [35]. Theorems 2.4–6 are from Lehman [64], Minty [66]. An elegant simultaneous development of matroids and their duals through *graphoids* (see Exercise 1.5) is carried out by Minty [66].

Paving matroids were originally studied under the name of *generalized partition lattices* by Hartmanis [59], [61]. Their connection with matroids was pointed out by Crapo and Rota [70]. I have used the word "paving" (suggested in private communication by G. C. Rota) to avoid confusion with partition lattices introduced in the next chapter.

What we call the cocycle matroid of a graph is the *bond pregeometry* of Rota [64] or the *bond matroid* of Tutte [65].

CHAPTER 3

Lattice Theory and Matroids

1. BRIEF REVIEW OF LATTICE THEORY

In this chapter we look in detail at the collection of flats of a matroid, ordered by inclusion. We start by reviewing those aspects of lattice theory which we will need.

A *partially ordered set* or *poset* is a set P together with a binary relation $\leqslant$, defined for some pairs of elements x, y of P, which is reflexive, antisymmetric and transitive. That is, for any x, y and z of P.

(1) $x \leqslant x$,
(2) if $x \leqslant y$ and $y \leqslant x$, then $x = y$,
(3) if $x \leqslant y$ and $y \leqslant z$, then $x \leqslant z$.

Two elements x and y of P are *comparable* if $x \leqslant y$ or $y \leqslant x$, otherwise x and y are *incomparable*. The expression $x \geqslant y$ implies $y \leqslant x$, $x < y$ implies $x \leqslant y$ but $x \neq y$, and so on. A *maximal* element z of a poset $(P, \leqslant)$ has the property that if $x \geqslant z$ then $x = z$. *Minimal* elements are similarly defined. An element x *covers* an element y if $x > y$, and if $x \geqslant z \geqslant y$ then either $z = x$, or $z = y$. If $(P, \leqslant)$ is a poset with a unique minimal element, which we will always denote by o, then an *atom* is an element which covers o. If a, b are two elements of a poset P the *interval* $[a, b]$ is the set $\{z : a \leqslant z \leqslant b\}$, with the ordering induced by P. A *chain* of P is a totally ordered subset of P, that is $\{x_0, x_1, \ldots, x_k\}$ is a chain if and only if

$$x_0 < x_1 < x_2 < \ldots < x_k,$$

and the *length* of this chain is k. If P has a least element or universal lower bound o, the *height* $h(x)$ of an element $x \in P$ is the least upper bound of lengths of chains $o = x_0 < x_1 < \ldots < x_k = x$ between o and x.

A finite poset P satisfies the *Jordan–Dedekind chain condition* if all maximal chains between elements a, b have the same length, for all pairs of elements a and b.

Two posets $(P, \overset{1}{\leqslant})$ and $(Q, \overset{2}{\leqslant})$ are *isomorphic* if there is a bijection $\phi: P \to Q$ such that $x \overset{1}{\leqslant} y$ if and only if $\phi(x) \overset{2}{\leqslant} \phi(y)$.

A poset can be represented by a *Hasse diagram* in which distinct elements are represented by distinct points, in which x is placed above y whenever $x > y$, and x and y are joined by a straight line whenever x covers y. Thus $x > y$ if and only if one can move downward from x to y in the resulting diagram. Any finite poset is determined up to isomorphism by its diagram.

A *lattice* is a partially ordered set $\mathscr{L} = (\mathscr{L}, \leqslant)$ such that for any two elements x, y of $\mathscr{L}$:

(a)   the subset $\{z: z \geqslant x \text{ and } z \geqslant y\}$ of $\mathscr{L}$ with the induced ordering has a unique minimal element, denoted by $x \vee y$, and called the *least upper bound* or *join* of x and y,

(b)   the subset $\{z: z \leqslant x \text{ and } z \leqslant y\}$ of $\mathscr{L}$ with the induced ordering, has a unique maximal element denoted by $x \wedge y$ and called the *greatest lower bound* or *meet* of x and y. An element x is *irreducible* if $x = y \vee z \Rightarrow y$ or z is x.

It is clear that both $\wedge$ and $\vee$ are commutative, associative and idempotent, and

$$x \vee (y \wedge x) = x,$$

$$x \wedge (y \vee x) = x.$$

If $T = \{t_1, \ldots, t_k\}$ is a subset of elements of a lattice we let

$$\bigvee T = t_1 \vee t_2 \vee \ldots \vee t_k$$

and

$$\bigwedge T = t_1 \wedge t_2 \wedge \ldots \wedge t_k,$$

and call $\bigvee T$ and $\bigwedge T$ the *join* and *meet* of T respectively. A lattice is *finite* if $|\mathscr{L}|$ is finite and clearly any finite lattice has a *least element* (which we denote by o and a *greatest element* (which we denote by I).

A *sublattice* $\mathscr{L}'$ of $\mathscr{L}$ is a subset such that $x, y \in \mathscr{L}' \Rightarrow x \vee y \in \mathscr{L}'$ and $x \wedge y \in \mathscr{L}'$. It should be stressed that a subset of $\mathscr{L}$ may form a lattice without being a sublattice of $\mathscr{L}$.

Example. In the lattice $\mathscr{L}$ of Fig. 1 the subset o, x, y, I is a lattice but not a sublattice.

Two lattices $\mathscr{L}, \mathscr{L}'$ are *isomorphic* if there is a bijection $\phi: \mathscr{L} \to \mathscr{L}'$ such that $\phi(x \vee y) = \phi(x) \vee \phi(y)$ and $\phi(x \wedge y) = \phi(x) \wedge \phi(y)$ for all $x, y \in \mathscr{L}$.

A lattice $\mathscr{L}$ is *distributive* if for any $x, y, z \in \mathscr{L}$:

(4) $(x \wedge y) \vee (y \wedge z) \vee (z \wedge x) = (x \vee y) \wedge (y \vee z) \wedge (z \vee x).$

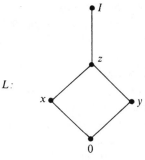

Figure 1.

A lattice $\mathscr{L}$ is *modular* if for $x, y \in \mathscr{L}$:

(5) $x \geqslant z \Rightarrow x \wedge (y \vee z) = (x \wedge y) \vee z.$

A finite lattice $\mathscr{L}$ is *semimodular* if for all $x, y \in \mathscr{L}$:

(6) x and y cover $x \wedge y \Rightarrow x \vee y$ covers x and $y.$

The following implications are well known.

(7) If $\mathscr{L}$ is distributive $\mathscr{L}$ is modular.

(8) If $\mathscr{L}$ is modular $\mathscr{L}$ is semimodular.

In neither (7) nor (8) is the converse true.

Useful characterizations of distributive and modular lattices are:

(9) $\mathscr{L}$ is distributive if and only if it has no sublattice isomorphic to the lattice $\mathscr{D}_5$ or $\mathscr{P}_5$.

(10) $\mathscr{L}$ is modular if and only if it has no sublattice isomorphic to the lattice $\mathscr{P}_5$.

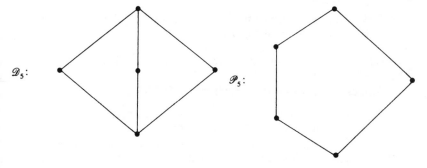

Figure 2.

Semimodular lattices are characterized by:

(11) $\mathscr{L}$ is semimodular if and only if it satisfies the Jordan–Dedekind chain condition and its height function h satisfies for all x, y,

(12) $h(x) + h(y) \geqslant h(x \wedge y) + h(x \vee y)$.

Other properties of semimodular lattices are:

(13)   For all $x, y \in \mathscr{L}$, x covers $x \wedge y \Rightarrow x \vee y$ covers y.

(14)   Any interval of a semimodular lattice is semimodular.

The *height* of a finite semimodular lattice is $h(I)$, and its elements of height equal to $h(I) - 1$ are called its *coatoms* or *copoints*.

For proofs of all these statements and a comprehensive study of lattice theory we refer the reader to Birkhoff's book [67].

EXERCISES 1

1. Prove that any sublattice of a modular (distributive) lattice is modular (distributive).

2. For any integer n prove that the set of divisors of n ordered by $a \leqslant b$ if $a|b$ forms a distributive lattice.

3. Prove that the set of subspaces of a vector space ordered by inclusion forms a modular lattice.

2. THE LATTICE OF FLATS OF A MATROID

If M is a matroid on S we can associate with M a partially ordered set $\mathscr{L}(M)$ whose elements are the flats of M ordered by inclusion.

Example 1. If M is the trivial matroid in which every subset is independent, $\mathscr{L}(M)$ is the distributive lattice of all subsets of S.

Example 2. If M is the uniform matroid of rank 3 on 5 elements $\mathscr{L}(M)$ is the lattice of Figure 1.

Obvious properties of $\mathscr{L} = \mathscr{L}(M)$ are the following.

(1) $\mathscr{L}$ is finite with a least element $o = \sigma(\varnothing)$ and a maximum element S.

(2)   The atoms of $\mathscr{L}$ are the flats of rank 1 in M.

(3)   An element of $\mathscr{L}$ is covered by S if and only if it is a hyperplane of M.

(4)   A flat X covers the flat Y in $\mathscr{L}$ if and only if $Y \subseteq X$ and $\rho X = \rho Y + 1$, where ρ is the rank function of M.

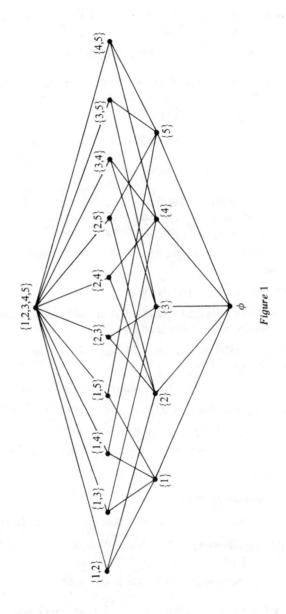

Figure 1

Using the properties of flats proved in Chapter 1 we can now derive the following properties of $\mathscr{L}(M)$.

(5)   Under the inclusion ordering $\mathscr{L}(M)$ is a lattice.

Proof. The intersection of any two flats is a flat and hence any two elements A, B of $\mathscr{L}$ have a meet $A \wedge B$ and join $A \vee B$ given by

$$A \wedge B = A \cap B$$

$$A \vee B = \bigcap \{X : X \in \mathscr{L}, A \cup B \subseteq X\} = \sigma(A \cup B).$$

Note that $A \vee B$ is well defined since $S \in \mathscr{L}$.

It follows from (4) that for any chain $o < A_1 < A_2 \ldots < \ldots < A_k = A$ of distinct elements of $\mathscr{L}$, the length k is at most ρA and that in fact ρA is the least upper bound of the lengths of all such chains. Thus we have:

(6) $\mathscr{L}(M)$ has as its height function the rank function of M restricted to the flats.

Another corollary of (4) is

(7) $\mathscr{L}(M)$ satisfies the Jordan Dedekind chain condition, and every maximal chain from o to A has length ρA.

But from (R3′) we know that for any pair of subsets A, B of S, if ρ is the rank function of M

$$\rho A + \rho B \geqslant \rho(A \cup B) + \rho(A \cap B).$$

Restricting ρ to flats we see that for any pair of flats A, B

$$\begin{aligned}
\rho(A \vee B) + \rho(A \wedge B) &= \rho(\sigma(A \cup B)) + \rho(A \cap B) \\
&= \rho(A \cup B) + \rho(A \cap B) \\
&\leqslant \rho A + \rho B
\end{aligned}$$

and thus we have proved

(8)   The lattice $\mathscr{L}(M)$ of flats of a matroid is semimodular.

However not every semimodular lattice is the lattice of flats of a matroid since we can prove:

(9)   Every element of $\mathscr{L}(M)$ is the join of atoms.

Proof. Let $A \in \mathscr{L} = \mathscr{L}(M)$ with $\rho A = k$. Then there exists $\{x_1, \ldots, x_k\} \in \mathscr{I}(M)$ with $\{x_1, \ldots, x_k\} \subseteq A$. Also since $\rho\{x_i, x_j\} = 2$, $i \neq j$, $x_i \notin \sigma\{x_j\}$ $i \neq j$. Thus the atoms $\{\sigma\{x_i\}; 1 \leqslant i \leqslant k\}$ are distinct and clearly

$$\sigma\{x_1\} \vee \sigma\{x_2\} \vee \ldots \vee \sigma\{x_k\} = A.$$

Example. Semimodular lattices which are not the lattices of flats of a matroid are shown in Figs 2(a), 2(b). In both cases the element g is not the join of atoms.

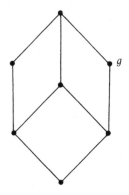 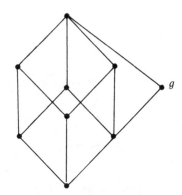

Figure 2.

EXERCISES 2

1. Find a characterization of paving matroids in terms of their lattices of flats.

2. Show that if $\mathcal{L}$ is the lattice of flats of either the free uniform matroid or a vector space then $\mathcal{L}$ is modular. Find a matroid whose lattice of flats is modular and which is neither free nor a vector space.

3. GEOMETRIC LATTICES AND SIMPLE MATROIDS

Birkhoff [67] calls a finite lattice *geometric* if it is semimodular and every point is the join of atoms. (In earlier editions of the book such lattices are called matroid lattices.) From the results of the previous section we know that $\mathcal{L}(M)$, the lattice of flats of a matroid is geometric. The main result of this chapter is the converse.

THEOREM 1. *A finite lattice $\mathcal{L}$ is isomorphic to the lattice of flats of a matroid if and only if it is geometric.*

Proof. We have shown in the last section that $\mathcal{L}(M)$ is geometric, so let $\mathcal{L}$

be an arbitrary geometric lattice with A as its set of atoms. Consider the collection $\mathscr{I}$ of subsets of A defined by

$$X \in \mathscr{I} \Leftrightarrow h(\vee X) = |X|$$

where h is the height function of $\mathscr{L}$.

By the submodularity of the height function, if $X = \{x_1, x_2, \ldots, x_k\}$,

$$h(\vee X) \leqslant \sum_{i=1}^{k} h(x_i) = |X|$$

and trivially for X, Y subsets of A,

$$X \subseteq Y \Rightarrow h(\vee X) \leqslant h(\vee Y).$$

Also if X, Y are two subsets of A

$$h(\vee X) + h(\vee Y) \geqslant h((\vee X) \vee (\vee Y)) + h((\vee X) \wedge (\vee Y))$$

$$\geqslant h(\vee(X \cup Y)) + h(\vee(X \cap Y))$$

since $(\vee X) \wedge (\vee Y) \geqslant \vee(X \cap Y)$. Thus the function

$$\rho X = h(\vee X)$$

satisfies the conditions (R1′)–(R3′) and is the rank function of a matroid on A. Denote this matroid by $M(\mathscr{L})$.

Now writing $A = \{x_1, \ldots, x_n\}$, consider the lattice $\mathscr{L}(M(\mathscr{L}))$ of flats of $M(\mathscr{L})$. If $M(\mathscr{L})$ has closure operator σ, it is easy to see that the map $\psi : \mathscr{L} \to \mathscr{L}(M(\mathscr{L}))$ defined for any element $X \in \mathscr{L}$, by

$$\psi(X) = \sigma\{a_1, \ldots, a_k\}$$

where $\{a_1, \ldots, a_k\}$ is the set of atoms below X, is a lattice isomorphism and hence $\mathscr{L}$ is the lattice of flats of $M(\mathscr{L})$. ∎

Notice that the proof of the existence of $M(\mathscr{L})$ depends only on the semimodular nature of $\mathscr{L}$ and therefore we have:

COROLLARY. *If $\mathscr{L}$ is a semimodular lattice, and $\mathscr{I}(\mathscr{L}) = \{X : X \text{ a set of atoms}$ of $\mathscr{L}$ such that $h(\vee X) = |X|\}$ then $\mathscr{I}(\mathscr{L})$ is the set of independent sets of a matroid $M(\mathscr{L})$ on the set of atoms of $\mathscr{L}$.*

It is not difficult to find an example to show that when $\mathscr{L}$ is semimodular but not geometric $\mathscr{L}(M(\mathscr{L})) \neq \mathscr{L}$.

Now unfortunately the structure of a matroid M is not completely specified by the geometric lattice of its flats.

Example. Let G_1, G_2 be the graphs of Fig. 1. Then $G_1 \neq G_2$ and $M(G_1) \neq M(G_2)$ but $\mathscr{L}(M(G_1)) \simeq \mathscr{L}(M(G_2))$.

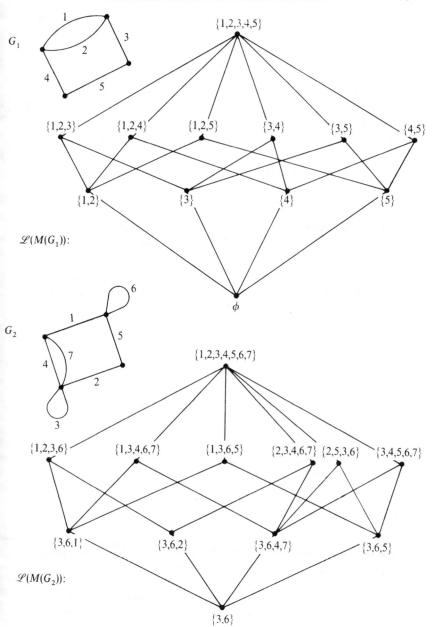

Figure 1

This indeterminacy of M from $\mathscr{L}(M)$ is due solely to the existence of loops and parallel elements; the loops are all absorbed at the o of $\mathscr{L}$ and parallel elements are indistinguishably merged in the same atom of $\mathscr{L}$.

The reader may recognize that something similar is going on to the process of constructing a projective geometry from a vector space, when a 1-dimensional subspace of the vector space simply becomes a single point in the projective space.

We note that with any matroid M we can in a canonical way associate a simple matroid $P(M)$ by imitating the procedure of getting a projective space from a vector space.

If M has ground set S and 1-flats $F_1, \ldots, F_m$ let $T = \{F_1, \ldots, F_m\}$ and define

$$\rho'\{F_i; i \in I\} = \rho(\bigcup_{i \in I} F_i)$$

where ρ is the rank function of M. It is easy to verify that ρ' is the rank function of a matroid $P(M)$ on T. We call $P(M)$ the *simple matroid determined by M*.

Example. When M is the cycle matroid of the graph G, $P(M)$ is the cycle matroid of the graph G' obtained by deleting all loops from G and replacing each pair of parallel elements by a single edge.

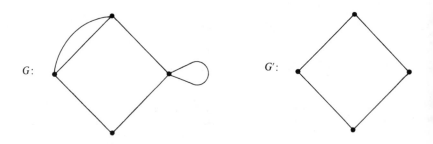

Figure 2.

The importance of simple matroids lies in the following theorem first noted by Birkhoff [36].

THEOREM 2. *The correspondence between a geometric lattice $\mathscr{L}$ and the matroid $M(\mathscr{L})$ on the set of atoms of $\mathscr{L}$ is a bijection between the set of finite geometric lattices and the set of simple matroids.*

Proof. Let x, y be distinct atoms of the geometric lattice $\mathscr{L}$. Then if ρ is the rank function of $M(\mathscr{L})$ clearly

$$\rho\{x\} = 1, \qquad \rho\{x, y\} = 2$$

and thus $M(\mathscr{L})$ is a simple matroid and clearly $\mathscr{L}(M(\mathscr{L}))$ is $\mathscr{L}$. Conversely if $\mathscr{L}$ is the geometric lattice $\mathscr{L}(M)$ for a simple matroid M then $M(\mathscr{L}(M)) \simeq M$. ∎

Thus the study of simple matroids is just the study of finite geometric lattices—this is the approach taken by Crapo and Rota in their book [70]. (They use the term *combinatorial geometry* for what we call simple matroid and then their *pregeometry* corresponds to matroid.) Many of the interesting properties of matroids are preserved if we just confine attention to simple matroids. However there is one important exception: the dual of a simple matroid need not be simple.

We will make free use of this close relationship between geometric lattices and matroids. We shall sometimes refer to a hyperplane of the underlying matroid as a *coatom* or *copoint* of the corresponding lattice, and we shall call the rank of a flat its height when we are in a lattice context.

It is also useful to "translate" some of the results about geometric lattices from Birkhoff [67] Chapter 4 to a matroid framework. For example a finite lattice $\mathscr{L}$ is said to be *complemented* if for every $x \in \mathscr{L}$, $\exists y \in \mathscr{L}$ such that $x \vee y = I, x \wedge y = 0$.

A lattice $\mathscr{L}$ is said to be *relatively complemented* if for any $a \leqslant b \in \mathscr{L}$, the interval $[a, b]$ is a complemented lattice.

THEOREM 3. *A geometric lattice is relatively complemented.*

In matroid language this says that given any flats $A \subseteq B$ of the matroid M and any flat X, $A \subseteq X \subseteq B$, there exists a flat Y of M such that

(1) $X \cap Y = A \qquad \sigma(X \cup Y) = B$.

A purely lattice-theoretic proof is given in Birkhoff [67, p. 88]. We sketch a set-theoretic matroid proof.

LEMMA. *Any interval of a geometric lattice is geometric.*

Proof. Deferred to next chapter but not difficult to prove directly.

Proof of Theorem 3. By the above lemma it is sufficient to show that a geometric lattice is complemented. Consider X, a flat of rank k, in the simple matroid M determined by the geometric lattice $\mathscr{L}$. Let $y_1 \notin X$ and let $X_1 = \sigma(X \cup y_1)$. Then $\rho X_1 = \rho(X) + 1$. Let $y_2 \notin X_1$ and take $X_2 = \sigma(X_1 \cup y_2)$.

Then $\rho X_2 = \rho X + 2$. Continuing in this way we arrive at a set X_t, $t + k = \rho S$ where $X_t = \sigma(X_{t-1} \cup y_t)$. Consider the set $Y = \sigma\{y_1, \ldots, y_t\}$.

$$\rho(\sigma(X \cup Y)) = \sigma X_t = \rho S,$$

and if $X \cap Y \neq \varnothing$, then since M is simple and has no loops $\rho(X \cap Y) \geqslant 1$ and we have the contradiction

$$\rho(X \cap Y) + \rho(X \cup Y) \geqslant \rho S + 1 \geqslant \rho X + \rho Y.$$

Thus Y is our required complement of X.∎

EXERCISES 3

1. Prove that any interval of a geometric lattice is a geometric lattice.

2. Prove that in a geometric lattice $\mathscr{L}$ a covers b if and only if there exists an atom $x \in \mathscr{L}$ with $b \vee x = a$.

3. Prove that every element of a geometric lattice can be expressed as the infimum of copoints.

4. Prove that a finite semimodular lattice is complemented if and only if its maximal element is the join of atoms, and is relatively complemented if and only if every element is the join of atoms, that is it is geometric.

°5. Given two geometric lattices $\mathscr{L}_1$, $\mathscr{L}_2$ is there any way of recognizing the fact that their associated simple matroids $M(\mathscr{L}_1)$, $M(\mathscr{L}_2)$ are dual apart from the obvious rather tedious exhaustive method of testing whether $M(\mathscr{L}_1) \simeq M(\mathscr{L}_2)^*$?

4. PARTITION LATTICES AND DILWORTH'S EMBEDDING THEOREM

Let S be a set of n elements. The *partition lattice* P_n of length $n - 1$ is the partially ordered set of equivalence relations $\{\pi\}$ on S, ordered by making $\pi_1 \leqslant \pi_2$ if and only if $x \pi_1 y \Rightarrow x \pi_2 y$, $x, y \in S$. In other words identifying the equivalence relation π with the equivalence classes (= partitions) it determines, $\pi_1 \leqslant \pi_2 \Leftrightarrow$ the partition determined by π_1 is a refinement of the partition determined by π_2.

Example. Figure 1 shows P_3, the ordered set of partitions of $\{1, 2, 3\}$

Thus the particularly ordered set P_3 is isomorphic with the lattice of flats of the cycle matroid $M(K_3)$. This is a special case of the following theorem.

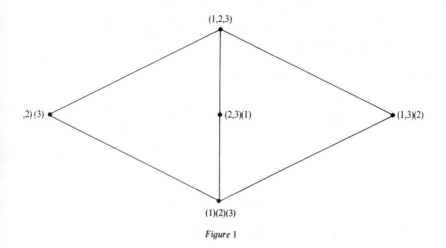

(1,2,3)

,2)(3)   (2,3)(1)   (1,3)(2)

(1)(2)(3)

Figure 1

THEOREM 1. *The partially ordered set of partitions of an n-set is a geometric lattice P_n which is isomorphic to the lattice of closed sets of the cycle matroid of the complete graph K_n.*

Proof. Let K_n have vertex set $\{1, 2, \ldots, n\}$. A typical closed set of $M(K_n)$ is of the form $E_1 \cup E_2 \cup \ldots \cup E_t$ where $E_1, E_2, \ldots, E_t$ are mutually disjoint sets of edges and for $i \neq j$ the vertex sets $\vee(E_i)$, $\vee(E_j)\ i \neq j$ have null intersection. Moreover E_i consists of all edges of the type (u, v) where u, v are members of $\vee(E_i)$. There is therefore an obvious 1–1 correspondence between the closed sets of $M(K_n)$ and partitions of $\{1, 2, \ldots, n\}$ and moreover this correspondence is order preserving. In other words P_n is isomorphic to $\mathscr{L}(M(K_n))$ which proves the theorem. ∎

We stress that these partition lattices have nothing to do with the paving matroids associated with d-partitions discussed in Chapter 2.

Associated with partition lattices is the following interesting embedding problem proposed by Birkhoff [67].

Given two lattices $\mathscr{L}_1$, $\mathscr{L}_2$, an *embedding* of $\mathscr{L}_1$ in $\mathscr{L}_2$ is a map $\phi: \mathscr{L}_1 \to \mathscr{L}_2$ such that for any $A \subseteq \mathscr{L}_1$

$$\phi(\bigvee_{x \in A} x) = \bigvee_{x \in A} \phi(x)$$

$$\phi(\bigwedge_{x \in A} x) = \bigwedge_{x \in A} \phi(x).$$

The question posed by Birkhoff was—can every finite lattice be embedded in a partition lattice P_n for some n?

As a major step towards proving this R. P. Dilworth proved the following very attractive theorem.

THEOREM 2. *Each finite lattice can be embedded in a finite geometric lattice.*

Dilworth originally proved this in 1941/42. For a very clear proof see Crawley and Dilworth [73], Chapter 14. It implies of course that the above embedding problem reduces to

Can each finite geometric lattice be embedded in a finite partition lattice?

EXERCISES 4

1. Find a geometric lattice which has a non-geometric sublattice.

2. Prove that any identity which is valid in every finite partition lattice is trivial in that it is valid in every lattice. (Sachs [61a]).

∘3. An *antichain* of a partially ordered set P is a collection $\{x_1, \ldots, x_t\}$ of elements of P such that $x_i \not< x_j$ for $i \neq j$. Find the maximum size of an antichain in the partition lattice P_n.

NOTES ON CHAPTER 3

The relation between matroids and finite geometric lattices was first pointed out by Birkhoff [41], and it is this approach which Rota [64] and Crapo and Rota [70] follow.

Partition lattices, on equivalently the lattice of equivalence relations were thoroughly studied by Ore [42], see also Sasaki and Fujiwara [52a], Sachs [61].

For alternative treatments of geometric lattices see the relevant chapters of Birkhoff [67], Szász [63] (where they are called Birkhoff lattices), or the recent books of Crawley and Dilworth [73], Maeda and Maeda [70].

For more on the combinatorial theory of lattices see Chapter 14 of Crawley and Dilworth [73].

CHAPTER 4

Submatroids

1. TRUNCATION

In this chapter we look at the different ways in which a matroid M on S induces 'smaller' matroids which in some sense preserve some of the structure of M. The simplest operation is the truncation of M, defined by the following theorem.

THEOREM 1. *Let M be a matroid on S and let $k \leq \rho S$ be a positive integer. Let*

$$\mathscr{I}_k = \{X : X \subseteq \mathscr{I}(M) : |X| \leq k\}.$$

Then $\mathscr{I}_k$ is the collection of independent sets of a matroid M_k on S.

We call M_k the *truncation* of M at k. The proof of Theorem 1 is very easy and left as an exercise. It is clear that the rank function ρ_k of M_k is related to the rank function ρ of M by

(1) $\quad \rho_k A = \min(k, \rho A) \qquad (A \subseteq S).$

The bases of M_k are the independent sets of M of cardinal k, and it is an easy exercise to verify:

(2) The geometric lattice $\mathscr{L}(M_k)$ is obtained from the lattice $\mathscr{L}(M)$ by removing from $\mathscr{L}$ those flats of height $\geq k$ and making $X \vee Y = S$ for any pair of flats X, Y whose sup in $\mathscr{L}$ was of height $\geq k$.

In other words, truncation at k corresponds to the lattice operation of 'knocking off' the lattice all elements of height larger than $k - 1$, except for the maximal element of $\mathscr{L}$.

Example. $U_{k, n}$, the uniform matroid of rank k on n elements is the truncation at k of the trivial matroid on n elements in which each subset is independent.

59

Now the bases of M_k^* are all subsets of S of the form $S \backslash X$, where X is independent in M and $|X| = k$. Recall from (2.1.2) that X is independent in M if and only if $S \backslash X$ is a spanning set of the dual matroid M^*. Then we see that the dual of M_k is a matroid on S which has as its bases all spanning sets of M^* which have cardinality $|S| - k$. Thus from duality we have the following two theorems.

THEOREM 2. *Let M be a matroid on S, let k be an integer ($\rho(S) \leqslant k \leqslant |S|$). Let $\mathscr{B}^k$ be the family of spanning sets of M which have cardinality k. Then $\mathscr{B}^k$ is the family of bases of a matroid M^k on S.*

We call M^k the *elongation* of M to height k, and from the above argument we have

THEOREM 3. *Let M be a matroid on S, let $k \leqslant \rho(S)$. Let $k' = |S| - k$. Then the truncation of M at k and the elongation of M^* to k' are dual matroids. That is*

$$(M_k)^* = (M^*)^{k'}.$$

Theorem 2 has the following interpretation when M is the cycle matroid of a graph G.

COROLLARY 1. *Let G be a graph. Let $k \geqslant \rho(M(G))$, and let $\mathscr{T}_k$ be the family of subsets of $E(G)$, defined by $T \in \mathscr{T}_k$ if and only if T is a spanning subgraph of G with exactly k edges. Then $\mathscr{T}_k$ is the set of bases of a matroid on $E(G)$.*

However, apart from the cycle matroid $M(G)$, the matroids on $E(G)$ defined by Corollary 1 do not seem to be either interesting or useful.

EXERCISES 1

1. Prove that if M is the cycle matroid of a graph a truncation of M is not in general the cycle matroid of a graph.

2. Let G be connected graph, describe the circuits of the elongation of $M(G)$ to height $|V(G)|$.

2. RESTRICTION

In this and the next section we show how a matroid M on S induces two matroids on a subset T of S which correspond in the natural way to subgraphs of a graph obtained by the operations of deletion and contracting of edges.

If $\mathscr{I}(M)$ is the set of independent sets of M on S and $T \subseteq S$, let

$$\mathscr{I}(M \,|\, T) = \{X : X \subseteq T,\ X \in \mathscr{I}(M)\}.$$

THEOREM 1. $\mathscr{I}(M \,|\, T)$ is the set of independent sets of a matroid on T.

We denote this matroid by $M \,|\, T$ and call it the *restriction* of M to T.

The proof of Theorem 1 is trivial. When S is the edge set $E(G)$ of a graph G and $T \subseteq E(G)$, recall that $G \,|\, T$ denotes the subgraph with edges, those edges of G not belonging to $E(G) \backslash T$. Then it is easy to verify

(1) $M(G \,|\, T) = M(G) \,|\, T.$

Because of this analogy between $M \,|\, T$ and the effect of removing the edges $S \backslash T$ the matroid $M \,|\, T$ is sometimes called the matroid obtained from M by *deleting* $S \backslash T$.

Obviously the rank function of $M \,|\, T$ is just the restriction of ρ to T. The following basic properties of $M \,|\, T$ are easily checked.

(2)   If X is dependent in M and $X \subseteq T$ then X is dependent in $M \,|\, T$.
(3) $M \,|\, T$ has rank $\rho(T)$.
(4) $M \,|\, T$ has as its circuits all circuits of M which are contained in T.
(5)   The closure operators σ, σ_T of M, $M \,|\, T$ respectively are linked by

$$\sigma_T A = (\sigma A) \cap T \qquad (A \subseteq T).$$

(6)   The hyperplanes of $M \,|\, T$ are the maximal proper subsets of T of the form $H \cap T$, H a hyperplane of M.

3. CONTRACTION

If G is a graph we say it is *contractible* to a graph H if H can be obtained from G by a finite sequence of elementary contractions, where an *elementary contraction* is either of the operations

(a)   deleting a loop

(b)   identifying the endpoints of an edge which is not a loop.

Figure 1 shows the effect of contracting the edges e, f of G.

Suppose we write $(G\phi e_1, e_2, \ldots, e_k)$ to denote the graph obtained from G by contracting the distinct edges $e_1, e_2, \ldots, e_k$ in the order specified. It is easy to prove that for any two edges e_1, e_2 of G, the graphs $(G\phi e_1, e_2)$ and $(G\phi e_2, e_1)$

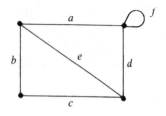
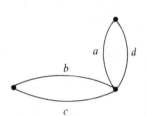

Figure 1

are isomorphic. Hence if X is any set of edges of G, we can write $G\phi X$ to denote the graph obtained from G by contracting out of G the edges of X in any order.

Write $G \cdot T$ for the graph obtained from G by contracting the edges of $E(G)\backslash T$. Then it is not difficult to prove:

LEMMA. *A set of edges B is a spanning forest of G. T if and only if there exists a spanning forest X of $G|(E(G)\backslash T)$ such that $X \cup B$ is a spanning forest of G.*

Proof. (Sketch). Take $E(G)\backslash T$ to be a single edge and use induction.

From the above lemma we see how to generalize the idea of graphical contraction to arbitrary matroids. Let M be a matroid on S and let $T \subseteq S$. Define $\mathscr{I}(M \cdot T)$ to be those subsets X of T such that there exists a maximal independent subset Y of $S\backslash T$ in M such that $X \cup Y \in \mathscr{I}(M)$.

THEOREM 1. *$\mathscr{I}(M \cdot T)$ is the set of independent sets of a matroid on T.*

We denote this matroid by $M \cdot T$ and call it the *contraction* of M to T.

Proof. Let $A \subseteq T$, let X_1, X_2 be maximal subsets of A which are members of $\mathscr{I}(M \cdot T)$. Then there exist Y_1, Y_2, both maximal independent subsets of $S\backslash T$ such that $X_1 \cup Y_1$ and $X_2 \cup Y_2$ are independent in M. Thus if $T_1 = (S\backslash T) \cup A$, $X_1 \cup Y_1$ and $X_2 \cup Y_2$ must be bases of $M|T_1$. Hence $|X_1 \cup Y_1| = |X_2 \cup Y_2|$ and since $|Y_1| = |Y_2|$ and $X_1 \cap Y_1 = X_2 \cap Y_2 = \varnothing$, $|X_1| = |X_2|$. Thus $\mathscr{I}(M \cdot T)$ satisfies axiom (I3'). The reader can verify that $\varnothing \in \mathscr{I}(M \cdot T)$ and that if $X \in \mathscr{I}(M \cdot T)$ then any subset of X belongs to $\mathscr{I}(M \cdot T)$. Hence $M \cdot T$ is a matroid on T ∎

From the definition of the matroid $M \cdot T$ we see that if ρ is the rank function of M and ρ^T is the rank function of $M \cdot T$, then for any $A \subseteq T$,

(1) $\quad \rho^T(A) = \rho(A \cup (S\backslash T)) - \rho(S\backslash T)$.

In particular the rank of the matroid $M \cdot T$ is $\rho(S) - \rho(S\backslash T)$.

Thus the rank of a matroid on S is preserved under contraction to a subset T of S, if and only if $S\backslash T$ consists of loops of the original matroid.

Using (1) we prove the fundamental result relating the contraction and restriction submatroids.

THEOREM 2. *If M is a matroid on S and $T \subseteq S$,*

(a) $(M\,|\,T)^* = M^* \cdot T,$
(b) $(M \cdot T)^* = M^*\,|\,T$

Proof. If $M\,|\,T$ has rank function λ, the rank function λ^* of $(M\,|\,T)^*$, satisfies for any $X \subseteq T$,

$$\lambda^*(T\backslash X) = |T| - \lambda T - |X| + \lambda X.$$

But $M^* \cdot T$ has by (1) rank function $(\rho^*)^T$ given by

$$\begin{aligned}
(\rho^*)^T(T\backslash X) &= \rho^*((T\backslash X) \cup (S\backslash T)) - \rho^*(S\backslash T) \\
&= \rho^*(S\backslash X) - \rho^*(S\backslash T) \\
&= |S| - |X| - \rho S + \rho X - (|S| - |T| - \rho S + \rho T) \\
&= \lambda^*(T\backslash X)
\end{aligned}$$

since λ is ρ restricted to the subsets of T. Since X is an arbitrary subset of T (a) is proved.

Putting M for M^* in (a) and taking duals gives (b). ∎

Remembering that C is a circuit of M^* if and only if its complement is a hyperplane of M we get an alternative characterization of $M \cdot T$.

(2)   The circuits of $M \cdot T$ are the minimal non-null sets of the form $C \cap T$, where C is a circuit of M.

From its definition we also have:

(3)   If G is a graph, and $T \subseteq E(G)$, then
$$M(G \cdot T) = M(G) \cdot T$$

Thus a corollary of (3) is:

(4)   The cycles of $G \cdot T$ are the minimal non-null sets of the form $C \cap T$, where C is a cycle of G.

The closed and spanning sets of the matroid $M \cdot T$ seem to have no nice characterization. For the sake of completeness we state the following results.

(5)   If $A \subseteq T$ and $x \in T\backslash A$, and σ, σ' are the closure operators of $M, M \cdot T$ respectively then $x \in \sigma'(A)$ if and only if $x \in \sigma(A \cup (S\backslash T))$.

Proof. From (1) $\rho^T(A \cup x) = \rho^T(A)$ if and only if $\rho(A \cup (S\backslash T) \cup \{x\}) = \rho(A \cup (S\backslash T))$.

(6)   A subset X of T is spanning in $M \cdot T$ if and only if $(S\backslash T) \cup X$ is a spanning set of M.

Proof. Find the subsets X such that $\rho^T(X) = \rho^T(T)$.

The contraction of a matroid M of rank 3, represented in the Euclidean form of Section 1.11 is also easy to visualize. Let M be a simple matroid on S, $T = S\backslash e$ for some $e \in S$. Then the matroid $M \cdot T$ has a Euclidean representation of points on a straight line and two points x, y of T are distinct in the Euclidean representation if and only if $\{x, y, e\}$ is not a collinear set.

Proof. $\{x, y\}$ will be independent in $M \cdot T$ if and only if $\{x, y, e\}$ is a base of M, that is if and only if the three points, x, y, e are not collinear in the Euclidean representation of M.

Example. Let M be the Fano matroid of Fig. 1.11.2. If T is any set of 6 points, $M \cdot T$ has the Euclidean representation

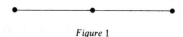

Figure 1

We close this section by emphasizing that it is not possible to have an exact matroid analogue of contraction in graphs. For example let G_1, G_2 be two non-isomorphic trees with the same number of edges. Then $M(G_1) \simeq M(G_2)$ so that $M(G_1)$ trivially contains $M(G_2)$ as a contraction but G_1 is not contractible to G_2.

We do have the one-way result;

(7)   If a graph G_1 is contractible to G_2 then $M(G_1)$ contains $M(G_2)$ as a contraction.

EXERCISES 3

1. Interpret the contraction of a subspace from a vector space as a vector space operation.

2. Prove that $M \cdot T$ has no loops if and only if $S\backslash T$ is a flat of M.

3. Prove that the contraction minor $M \cdot T$ may be defined as follows. Take any fixed base B of $M|(S\backslash T)$ and let $X \in \mathscr{F}$ if and only if $X \cup B \in \mathscr{I}(M)$ and $X \subseteq T$. Then $\mathscr{F} = \mathscr{I}(M \cdot T)$.

(This is a quicker way of finding the independent sets of $M \cdot T$)

4. Prove that for subsets A, B, of S,

$$\rho^A(A) + \rho^B(B) \leqslant \rho^{A \cup B}(A \cup B) + \rho^{A \cap B}(A \cap B).$$

(Brualdi [70a])

4. MINORS AND THEIR REPRESENTATION IN THE LATTICE

If M is a matroid on S and $T \subseteq S$, a matroid N on T is called a *minor* of M if N is obtained by any combination of restrictions and contractions of M.

THEOREM 1. *If M is a matroid on S and $T \subseteq U \subseteq S$,*

(a) $M|T = (M|U)|T,$

(b) $M \cdot T = (M \cdot U) \cdot T,$

(c) $(M|U) \cdot T = (M \cdot (S\backslash(U\backslash T)))|T,$

(d) $(M \cdot U)|T = (M|S\backslash(U\backslash T)) \cdot T.$

Proof. (a) is obvious. Let $N = (M \cdot U) \cdot T, M_1 = M \cdot U$.

Then

$$N^* = (M_1 \cdot T)^* = M_1{}^*|T.$$

But $M_1^* = M^*|U$, so by (a)

$$N^* = (M^*|U)|T = M^*|T = (M \cdot T)^*$$

and the uniqueness of the dual gives (b).

Proof of (c) is a routine exhibition of the equality of the rank functions and (d) can then be deduced by duality. ■

Now λ the rank function of $M|T$ is just ρ restricted to T. Hence for $A \subseteq T$, $x \in T$,

$$\lambda(A \cup x) = \rho(A \cup x) = \rho(A) = \lambda A$$

if and only if x is in the flat generated by A in M. Hence the flats of $M|T$ are precisely the sets of the form $X \cap T, X$ a flat of T.

THEOREM 2. *If $\mathscr{L}$ is the geometric lattice of the simple matroid M, the interval $[o, T]$ is the lattice of $M|T$, for any flat T of M.*

Proof. Any flat of $M|T$ is of the form $F \cap T$, F a flat of M and since T is a flat, $F \cap T \in \mathscr{L}$. The meet and intersection properties are easily checked. ∎

THEOREM 3. *If $\mathscr{L}$ is the geometric lattice of the simple matroid M and T is a flat of M, the interval $[T, S]$ of $\mathscr{L}$ is isomorphic to $\mathscr{L}(M.(S\backslash T))$.*

Proof. Let $U = S\backslash T$, and consider the map $\psi: [T, S]$ into subsets of U defined by

$$\psi X = X \cap U \qquad X \in [T, S].$$

It is routine (a) to verify that ψX is a flat of $M.T$ and hence ψ is a (1–1) map from $[T, S]$ to $\mathscr{L}(M.U)$, (b) to check that for all X, Y,

$$\psi(X \wedge Y) = \psi X \wedge \psi Y$$
$$\psi(X \vee Y) = \psi X \vee \psi Y.$$

Thus ψ is a lattice isomorphism which proves the result. ∎

Now suppose that U, T are flats of the simple matroid M on S with $T \subseteq U$. Consider the interval sublattice $[T, U]$ of the geometric lattice $\mathscr{L}(M)$. By the preceding lemmas

$$[T, U] \simeq \mathscr{L}((M|U).(U\backslash T))$$

but also

$$[T, U] \simeq \mathscr{L}(M.(S\backslash T)|(U\backslash T)).$$

Hence for simple matroids, we have the special case of Theorem 1(d) that for any flats $T \subseteq U \subseteq S$,

$$M.(S\backslash T)|(U\backslash T) = (M|U).(U\backslash T).$$

This also proves the assertion of Section 3.3. that any interval sublattice of a geometric lattice is geometric.

(1)   If $[A, B]$ is an interval of $\mathscr{L}(M)$ the closure operator σ_{AB} of the associated matroid is given in terms of the closure operator σ of M by

$$\sigma_{AB}X = \sigma[(X \cup A) \cap B]\backslash A.$$

    Hence

$$\sigma_{\varnothing A}X = \sigma(X \cap A)$$

is the closure operator of the matroid $M \mid A$ and

$$\sigma_{AS}X = \sigma(X \cup A)\backslash A$$

is the closure operator of $M \mid (S\backslash A)$.

When T is an arbitrary subset we leave the reader to verify.

(2)   For arbitrary $T \subseteq S$, the lattice of $M \mid T$ is obtained from $\mathscr{L}(M)$ by regarding the elements of T as atoms of $\mathscr{L}(M)$ and forming all suprema, in $\mathscr{L}(M)$, of all subsets of the set T.

(3)   For arbitrary $T, \mathscr{L}(M \cdot T)$ is isomorphic to the interval $[\sigma(S\backslash T), S]$ in $\mathscr{L}(M)$.

EXERCISES 4

1. Prove that no uniform matroid can contain either the Fano matroid or $M(K_4)$ as a minor.

2. Prove that a flat A is independent in a matroid M if and only if the interval $[o, A]$ of $\mathscr{L}(M)$ is isomorphic to the lattice of subsets of A ordered by inclusion.

3. Prove that if N is a minor of M then N^* is a minor of M^*.

NOTES ON CHAPTER 4

The basic matroid operations discussed in this chapter were first explicitly studied by Tutte [58] who developed the algebra of matroid minors—though again his notation and treatment differ from ours.

The lattice interpretation of minors as intervals in the corresponding geometric lattice is implicit in Rota [64], for more details of this approach see Crapo and Rota [70].

CHAPTER 5

Matroid Connection

1. TWO THEOREMS ABOUT CIRCUITS AND COCIRCUITS

Before discussing the theory of matroid separability or connection we prove two theorems about the circuits and cocircuits of a matroid.

It is well known that if C is the set of edges of a cycle of a graph G, and $e, f \in C$ there exists a cocycle C^* of G such that

$$C \cap C^* = \{e, f\}.$$

This turns out to be just a special case of a more general result.

THEOREM 1. *Let C be a circuit of a matroid M and let x, y be distinct elements of C. Then there exists a cocircuit C^* of M which contains x and y and no other elements of C.*

Proof. Augment the independent set $C\backslash x$ to a base B of M. Let $B^* = S\backslash B$ be the corresponding cobase. Clearly $x \in B^*$. Now let $C^*(y)$ be the fundamental cocircuit of y in the cobase B^* (Theorem 1.9.1*). Suppose $x \notin C^*(y)$. Then

$$C \cap C^*(y) = \{y\}$$

which contradicts Theorem 2.1.6. Hence $x \in C^*(y)$ and clearly $C^*(y) \cap C = \{x, y\}$. ∎

Another property of the set of circuits of a matroid which is crucial to the theory of connection is,

THEOREM 2. *Let M be a matroid on S and let x, y, z, be distinct elements of S. If there is a circuit C_1 containing x and y and a circuit C_2 containing y and z then there exists a circuit C_3 containing x and z.*

Proof. We use induction on $|S|$. The result is clearly true for $|S| \leqslant 3$. Let

$$\{x, y\} \subseteq C_1,$$
$$\{y, z\} \subseteq C_2.$$

If $C_1 \cup C_2 \neq S$, let $u \in S \backslash (C_1 \cup C_2)$ and then by the induction hypothesis, if $T = S \backslash u$ there is a circuit C_3 of $M | T$ such that $\{x, z\} \subseteq C_3$, and C_3 is a circuit of M. If $C_1 \cup C_2 = S$, by Theorem 1.9.2, there exists circuits C_3, C_4 such that

$$x \in C_3 \subseteq (C_1 \cup C_2) \backslash y,$$
$$z \in C_4 \subseteq (C_1 \cup C_2) \backslash y.$$

Clearly $C_3 \cap (C_2 \backslash C_1) \neq \emptyset$. Suppose $C_3 \cap C_1$ is a proper subset of $C_1 \backslash C_2$. Then we have the situation that $x \in C_3$, $z \in C_2$ and $C_3 \cap C_2 \neq \emptyset$ and since $|C_3 \cup C_2| < |S|$, by the induction hypothesis there is a circuit C_5 which contains x and z, and the proof is complete. Hence $C_3 \supseteq C_1 \backslash C_2$ and by a similar argument $C_4 \supseteq C_2 \backslash C_1$. Moreover $C_3 \cup C_4 \subseteq (C_1 \cup C_2) \backslash y$. Hence by the induction hypothesis since $C_3 \cap C_4 \neq \emptyset$, there exists a circuit C_5 containing both x and z. C_5 is our required circuit. ∎

By letting M be the cocycle matroid of a graph we get:

COROLLARY. *If e, f, g are distinct edges of a graph G such that $\{e, f\}$ is contained in a cocycle and $\{f, g\}$ is contained in a cocycle, then $\{e, g\}$ is contained in a cocycle.*

2. CONNECTIVITY

There is no concept in a matroid which corresponds exactly to a vertex in a graph and accordingly there is no property of matroids in general corresponding to ordinary graphical connection. However Whitney [35] introduced a notion of separability in matroids which corresponds to 2-connection in graph theory. He called a matroid M on S connected or non-separable if for every proper subset A of S

$$\rho(A) + \rho(S \backslash A) > \rho(S).$$

We proceed differently by defining a matroid M on S to be *connected* or *non-separable* if for every pair of distinct elements x and y of S there is a circuit of M containing x and y. Otherwise it is *disconnected* or *separable*.

An immediate consequence of Theorem 1.1 is the important result.

THEOREM 1. *A matroid M is connected if and only if its dual M^* is connected.*

Suppose we now define a relation R between the elements of S by $x\,R\,y$ if and only if $x = y$ or there is a circuit C of the matroid M which contains both x and y. R has the properties:

(i)   (Reflexive): $x\,R\,x$,
(ii) (Symmetric): $x\,R\,y$ implies $y\,R\,x$;
      and by Theorem 1.2,
(iii) (Transitive): $x\,R\,y$ and $y\,R\,z$ imply $x\,R\,z$.

Thus R is an equivalence relation on S and we call the equivalence classes under R the *connected components* or, more briefly, *components* of the matroid M.

Because of Theorem 1.2 the relation R^* defined by $x\,R^*\,y$ if and only if $x = y$ or there is a cocircuit of M containing x and y is also an equivalence relation and

$$x\,R\,y \Leftrightarrow x\,R^*\,y, \qquad x, y \in S.$$

Hence we have proved:

THEOREM 2. *The components of a matroid M are disjoint subsets of S and coincide with the components of the dual matroid M^*.*

From the definition of component we see:

(1)   If x is a loop or coloop of M, $\{x\}$ is a component of M.

We call a subset T of S, *connected* in the matroid M on S if the matroid $M\,|\,T$ is connected.

(2)   If A is a component of the matroid M on S then A is a connected subset of S.

Proof. The circuits of $M\,|\,A$ are the circuits of M which are contained in A. Thus if C is a circuit of M containing $x, y \in A$ then A connected $\Rightarrow C \subseteq A \Rightarrow C$ a circuit of $M\,|\,A$.

It is illuminating to interpret connection for graphic matroids. We define a graph G to be a *block* if its cycle matroid is connected. Note that this differs slightly from the definition of Harary [69] inasmuch as we regard the graph consisting of a single loop to be a block.

Then as in Harary [69, p. 27] it is not difficult to prove.

THEOREM 3. *Let G be a connected graph with no loops and at least three vertices. Then the following statements are equivalent.*

(a)   *G is a block.*

(b)   *G is 2-connected.*
(c)   *Every 2 vertices of G lie on a common cycle.*
(d)   *Every vertex and edge of G lie on a common cycle.*

We close this section by showing that our definition of a connected matroid coincides with the original definition of Whitney [35] by proving:

THEOREM 4. *A matroid M on S is not connected if and only if there exists a proper subset A of S, such that*

$$\rho(A) + \rho(S \backslash A) = \rho(S).$$

Proof. Suppose M is not connected. Let A be a component of M. Then every circuit C of M is contained either in A or in $S \backslash A$.

But then by (4.2.4) and (4.3.2), the submatroids $M | A$ and $M . A$ have the same collection of circuits. Thus $M . A = M | A$. But

$$\rho(M . A) = \rho(S) - \rho(S \backslash A),$$

$$\rho(M | A) = \rho(A).$$

Hence A is a proper subset of S with

$$\rho(A) + \rho(S \backslash A) = \rho(S).$$

Conversely suppose A is a proper subset of S such that

$$\rho(A) + \rho(S \backslash A) = \rho(S).$$

If M is not disconnected then for any elements $x \in A$ and $y \in S \backslash A$ there exists a circuit C containing $\{x, y\}$. Let

$$C_1 = C \cap A, \qquad C_2 = C \cap \bar{A},$$

where $\bar{A}$ denotes $S \backslash A$.

Following Whitney let

$$\Delta(X, Y) = \rho(X \cup Y) - \rho(X),$$

and then we can write

$$\rho(S) = \Delta(A \cup C_2, \bar{A} \backslash C_2) + \Delta(C, A \backslash C_1) + \rho(C).$$

From the submodularity of ρ we have

$$\Delta(X \cup Y, Z) \leqslant \Delta(Y, Z),$$

and if $X \subseteq Y, \Delta(X, Z) \geqslant \Delta(Y, Z)$. Hence

$$\rho(S) \leqslant \Delta(C_2, \bar{A} \backslash C_2) + \Delta(C_1, A \backslash C_1) + \rho(C)$$
$$= \rho(\bar{A}) - \rho(C_2) + \rho(A) - \rho(C_1) + \rho(C).$$

But since $\rho(A) + \rho(\bar{A}) = \rho(S)$ we see that

$$\rho(C_1) + \rho(C_2) = \rho(C)$$

where $C_1 \cap C_2 = \emptyset$ and $C_1 \cup C_2 = C$.

This gives the contradiction

$$|C_1| + |C_2| = |C| - 1,$$

and therefore no such circuit C can exist. ∎

EXERCISES 2

Throughout M will denote a matroid on S.

1. Prove that a connected component of M is a flat of M.

2. Prove that if $X \subseteq S$ is a component of M then for any $T \subseteq S$, $X \cap T$ is a component of $M \mid T$ and $M \cdot T$.

3. Prove that if X, Y are connected subsets of S and $X \cap Y \neq \emptyset$ then $X \cup Y$ is a connected set.

4. Prove that a rank 3, simple matroid is connected if and only if in the diagram of its Euclidean representation it is possible to move along lines from any point to any other point.

5. Prove that a subset X of S is the union of components of M if and only if $M \mid X \simeq M \cdot X$. (Tutte [65])

6. Call a matroid M on S *critically connected* if for any element $e \in S$, $M \mid (S \backslash e)$ is disconnected, but M is connected. Prove:

   (a) if M is critically connected then $|S| \leqslant 2\rho(M) - 2$;
   (b) if M is critically connected and $|S| = 2\rho(M) - 2$ then M is the cycle matroid of the complete bipartite graph $K_{2, n-2}$. (Murty [74])

7. If M is connected and $e \in S$ then not both $M \mid (S \backslash e)$ and $M \cdot (S \backslash e)$ can be disconnected. (Crapo [67b], Murty [74] and Smith [72]).

8. Call X *dense* in A if $A \subseteq X \subseteq \sigma A$. Prove that if A is a connected subset of M and X is dense in A then X is a connected subset of M. Deduce that if A is connected σA is connected.

   (Note the analogue with connection in point set topology—see Sierpinski [52]).

3. DIRECT PRODUCT OF LATTICES, DIRECT SUM OF MATROIDS

Let M_1, M_2 be matroids on disjoint sets S_1, S_2. The *direct sum* of M_1 and M_2, written $M_1 + M_2$ is the matroid on $S_1 \cup S_2$ which has as its bases all sets of the form $B_1 \cup B_2$ where B_1 is a base of M_1, B_2 is a base of M_2. It is trivial to verify that $M_1 + M_2$ is a matroid, its rank function ρ is given in terms of the rank functions ρ_i of M_i by

$$\rho A = \rho_1(A \cap S_1) + \rho_2(A \cap S_2)$$

for all subsets $A \subseteq S_1 \cup S_2$.

The set of circuits of $M_1 + M_2$ is given by

$$\mathscr{C}(M_1 + M_2) = \mathscr{C}(M_1) \cup \mathscr{C}(M_2).$$

Its closure function σ is given for $A \subseteq S_1 \cup S_2$ by

$$\sigma A = \sigma_1(A \cap S_1) \cup \sigma_2(A \cap S_2)$$

where σ_i is the closure function of M_i $(i = 1, 2)$.

If $(\mathscr{L}_1, \overset{1}{\leqslant})$ and $(\mathscr{L}_2, \overset{2}{\leqslant})$ are lattices their *direct product* is the lattice $(\mathscr{L}_1 \times \mathscr{L}_2, \leqslant)$ where

$$\mathscr{L}_1 \times \mathscr{L}_2 = \{(a, b) \colon a \in \mathscr{L}_1, b \in \mathscr{L}_2\}$$

and the ordering is given by

$$(a, b) \leqslant (c, d) \Leftrightarrow a \overset{1}{\leqslant} c, b \overset{2}{\leqslant} d.$$

It is easy to check that this construction gives a lattice (see Birkhoff [67]).

Indeed, routine verification gives the following result.

THEOREM 1. *If* $\mathscr{L}_1$, $\mathscr{L}_2$, *are geometric lattices their direct product* $\mathscr{L}_1 \times \mathscr{L}_2$ *is a geometric lattice.*

The set of atoms of $\mathscr{L}_1 \times \mathscr{L}_2$ is the set

$$\{(a, o); a \text{ an atom of } \mathscr{L}_1\} \cup \{(o, b); b \text{ an atom of } \mathscr{L}_2\},$$

and when $\mathscr{L}_1$, $\mathscr{L}_2$ are geometric with associated simple matroids M_1, M_2 it is easy to see that $\mathscr{L}_1 \times \mathscr{L}_2$ is isomorphic to $\mathscr{L}(M_1 + M_2)$. Thus the lattice construction of direct product is essentially the same as the matroid operation of direct sum.

A lattice $\mathscr{L}$ is *irreducible* if it cannot be expressed as the direct product of two non-trivial lattices.

Now suppose the matroid M on S is disconnected with components say $M_1, \ldots, M_k$. Then clearly

$$M = M_1 + M_2 + \ldots + M_k.$$

It is now easy to see:

THEOREM 2. *A geometric lattice* $\mathscr{L}$ *is irreducible if and only if its associated simple matroid is connected.*

EXERCISES 3

1. Prove that if M_1, M_2 are matroids on disjoint sets S_1 and S_2 the hyperplanes of their direct sum are of the form $H_1 \cup S_2, S_1 \cup H_2$ where H_i is a hyperplane of M_i.

4. DESCRIPTIONS OF MATROIDS

Suppose that we wish to describe matroids to a computer (for example we might wish to test a conjecture). A very natural question arises—on average what is the most concise way of doing so?

"Concise" is a rather loose term in this context and clearly depends on the facilities of the computer in question. However, it can be generally agreed that the following statements are true:

(1) The set of bases is a more concise description of a matroid than the collection of independent sets.
(2) The set of hyperplanes is a more concise description of a matroid than the collection of flats.

However there is no obvious way of comparing the descriptions in terms of bases and say circuits. Intuitively we conjecture:

(3) The set of circuits is a more concise description of a matroid than the set of bases in the sense that on average a matroid has fewer circuits than bases. We see no way of proving such a result.

However, one very interesting result in this connection is the following theorem of Lehman [64].

It says that given the information that M is connected we do not need to know the full set of circuits of M in order to be able to specify M completely.

THEOREM 1. *Let M be a connected matroid on S and let e be a fixed element of S. The collection of circuits of M which contain e uniquely determines M.*

Proof. Let $\mathscr{C}$ be the collection of circuits of M which contain e. For each subset X of S let

$$D(X) = X \backslash \bigcap \{C \in \mathscr{C} : C \subseteq X\}$$

We shall prove that the circuits of M not containing e are precisely the minimal sets of the form $D(C_1 \cup C_2)$ where C_1, C_2 are distinct members of $\mathscr{C}$.

Let C_1, C_2 be distinct members of $\mathscr{C}$. Then there is a circuit $C \subseteq (C_1 \cup C_2) \backslash e$. For each $y \in C$ there is a circuit $C \in \mathscr{C}$ such that $C' \subseteq (C_1 \cup C_2) \backslash y$ prove! Hence $C \subseteq D(C_1 \cup C_2)$ and so $D(C_1 \cup C_2)$ is dependent in M.

Now let C be any circuit of M which does not contain e. Since M is connected there is a $C_1 \in \mathscr{C}$ such that C_1 meets C, choose C_1 such that $C \cup C_1$ is minimal. Then there exists $C_2 \in \mathscr{C}$ such that $C_2 \subseteq C \cup C_1$ and $C_2 \neq C_1$. But for any $C' \in \mathscr{C}$ such that $C' \subseteq C_1 \cup C_2$, C' meets C, and so by the minimality of $C \cup C_1$, $C \cup C'$ is not strictly contained in $C \cup C_1$. But

$$C \cup C' \subseteq C \cup C_1 \cup C_2 = C \cup C_1$$

and so $C \cup C' = C \cup C_1$. Hence

$$C \cup C_1 \cup C_2 = C \cup C_1 = C \cup (\bigcap \{C' \in \mathscr{C}: C' \subseteq C_1 \cup C_2\}).$$

Hence C contains and so equals $D(C_1 \cup C_2)$. ∎

EXERCISES 4

1. Prove that if M is a connected matroid on S and e is a fixed element of S the collection of cocircuits of M which contain e uniquely determines M.

2. A generalization of the *reconstruction problem* for graphs (see Harary [69]) is that if M_1 and M_2 are simple matroids on S and for each $x \in S$, $M_1|(S\setminus x) \simeq M_2|(S\setminus x)$ then $M_1 \simeq M_2$. Show that this is not true. (Brylawski [74a] [75]).

5. THE CIRCUIT GRAPH OF A MATROID; A LATTICE CHARACTERIZATION OF CONNECTION

An attractive characterization of matroid connection is given by Tutte [65], in terms of what we call the circuit graph of a matroid.

If M is a matroid on S, define the *circuit graph* $G(M)$ of M to have as its vertices the circuits of M and join the distinct circuits C_i, C_j by an edge if and only if

(a) $C_i \cup C_j$ is a connected subset of M,

(b) $\rho(C_i \cup C_j) = |C_i \cup C_j| - 2$.

Examples. Let M_1 be the cycle matroid of the graph G_1. Then $G(M_1)$ is as shown in Figure 1.

Now it is easy to see that if M is disconnected $G(M)$ is a disconnected graph. The converse is harder, we refer the reader to Tutte [65].

THEOREM 1. *A matroid M without coloops is connected if and only if its circuit graph $G(M)$ is a connected graph.*

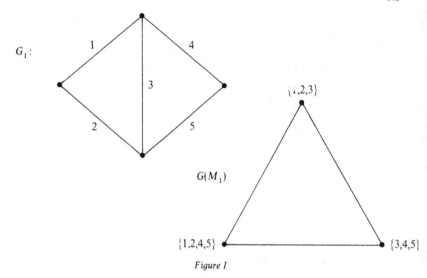

G_1:

$G(M_1)$

$\{1,2,3\}$

$\{1,2,4,5\}$ $\{3,4,5\}$

Figure 1

We now show how this theorem can be used to give a good "algorithmic" test of whether the matroid determined by a geometric lattice is a connected matroid.

Let $\mathscr{L}$ be the geometric lattice associated with a simple matroid M of rank r on S. Let $\mathscr{H}$ be the set of hyperplanes of M, let $\mathscr{F}_{r-2}$ be the set of $(r-2)$-flats of $\mathscr{L}$. Let $\mathscr{F}'_{r-2}$ be those members of $\mathscr{F}_{r-2}$ which are contained in more than two hyperplanes.

Let $\Delta(\mathscr{L})$ be the bipartite graph obtained from $\mathscr{L}$ by taking as its set of vertices $\mathscr{H} \cup \mathscr{F}'_{r-2}$ and joining a hyperplane H to a member F of $\mathscr{F}'_{r-2}$ by an edge if and only if $F \subseteq H$.

THEOREM 2. *$\mathscr{L}$ is the geometric lattice of a connected matroid if and only if $\Delta(\mathscr{L})$ is a connected bipartite graph.*

Proof. By Theorem 2.1 M is connected if and only if its dual matroid M^* is connected, and hence by Theorem 1: M is connected if and only if $G(M^*)$ is a connected graph. Now C^* is a circuit of M^* if and only if $S \backslash C^*$ is a hyperplane of M and therefore we can regard the vertices of $G(M^*)$ as the set of hyperplanes $\mathscr{H}$ of M and join H_i to H_j by an edge if and only if:

(a) $(S \backslash H_i) \cup (S \backslash H_j)$ is connected in M^*,
(b) $\rho^*((S \backslash H_i) \cup (S \backslash H_j)) = |(S \backslash H_i) \cup (S \backslash H_j)| - 2$.

But a simple application of (2.1.5.) shows that statement (b) holds for H_i, H_j

if and only if $\rho(H_i \cap H_j) = \rho S - 2$; and $(S \backslash H_i) \cup (S \backslash H_j)$ is connected in M^* if and only if $M^* | (S \backslash (H_i \cap H_j))$ is a connected matroid. This is so if and only if

$$(M^* | (S \backslash (H_i \cap H_j)))^* = M . (S \backslash (H_i \cap H_j))$$

is a connected matroid. But $H_i \cap H_j$ is a flat of M and hence we only join H_i to H_j in $G(M^*)$ if the following conditions are satisfied:

(b') $H_i \cap H_j$ is a flat F_{ij} of rank $r - 2$.
(a') the interval sublattice $[F_{ij}, S]$ is the geometric lattice of a connected matroid.

(We have used here the isomorphism between the interval sublattice $[F_{ij}, S]$ and the geometric lattice determined by $M . (S \backslash (H_i \cap H_j))$.)

But $[F_{ij}, S]$ is of height 2 and it is easy to check that it will correspond to a connected matroid if and only if it has at least 3 atoms.

Thus in $G(M^*)$ we join H_i to H_j if and only if they intersect in a member of $\mathscr{F}'_{r-2}$. It is now trivial to check that the bipartite graph $\Delta(\mathscr{L})$ is connected if and only if $G(M^*)$ is connected and hence Theorem 2 is proved. ∎

This theorem is interesting in that it is our first example of what we shall call the "scum principle". Roughly speaking this says that anything interesting which "happens" in a geometric lattice also happens somewhere near the top of the lattice. We return to this topic in Chapter 17.

6. TUTTE'S THEORY OF n-CONNECTION: WHEELS AND WHIRLS

In this section we sketch a theory of n-connection for matroids developed by Tutte [66].

A graph G has *connectivity* $\kappa(G)$ if $\kappa(G)$ is the minimum number of vertices whose removal results in a disconnected or trivial graph. Thus the connectivity of a disconnected graph is 0. Since the complete graph K_p cannot be disconnected by removing any number of points, but the trivial graph results after removing $p - 1$ points, $\kappa(K_p) = p - 1$. A graph G is therefore n-connected for any $n \leqslant \kappa(G)$. Thus G is connected if and only if it is 1-connected and if G is n-connected it is n'-connected for any $n' \leqslant n$.

Now some embarrassment arises when G has loops. Provided G has no loops we can assert that G is 2-connected if and only if the cycle matroid $M(G)$ is a connected matroid. If however G has a loop the statement is not true since $M(G)$ is not connected but the presence of loops is irrelevant to the above definition of 2-connection.

Consider therefore the following definition. A graph G is k-*separated* if there exist k vertices such that the removal of these k vertices from G "separates" k edges of G from k other edges of G. In other words if U is any subset of edges of G we write $\eta(U)$ for the number of common vertices of the subgraphs spanned by U, and $(E(G)\backslash U)$. Then for k a positive integer G is k-separated if and only if G is connected and there exists $U \subseteq E(G)$ with

(1) $|U| \geqslant k, \qquad |E(G)\backslash U| \geqslant k, \qquad \eta(U) = k.$

We say G is 0-*separated* if it is not connected. If there exists a least non-negative integer k such that G is k-separated we call it the T-*connectivity* of G and denote it by $\lambda(G)$. If there is no such integer we write $\lambda(G) = \infty$. We say G is n-T-*connected* where n is a positive integer if $n \leqslant \lambda(G)$.

We note that describing a graph as n-T-connected is not the same as calling it n-connected.

Example. Let G be the graph of Fig. 1. It can be verified that G is 3-connected. However by taking $U = \{e_1, e_1\}$, $\eta(U) = 2$, G is 2-separable and hence is not 3-T-connected.

G:

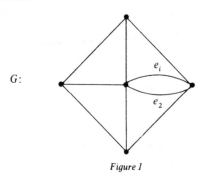

Figure 1

We can summarize the relation between $n - T$-connection and n-connection as follows:

(2) G is $1 - T$-connected $\Leftrightarrow$ G is connected.
(3) G is $2 - T$-connected $\Leftrightarrow$ $M(G)$ is connected.
(4)   If G is simple, G is $3 - T$-connected if and only if G is 3-connected.

Even for simple graphs, for $n > 3$ the concepts of n-connection and $n - T$-connection do not agree.

However by placing less emphasis on vertices it is capable of an attractive generalization to matroids.

If M is a matroid on S with rank function ρ define a function $\xi: 2^S \to \mathbb{Z}$ by:

(5) $\xi T = \rho T + \rho(S\backslash T) - \rho S + 1 \qquad (T \subseteq S).$

Call the matroid M *k-separated* for k a positive integer if there exists $T \subseteq S$, $|T| \geqslant k$, $|S\backslash T| \geqslant k$, with $\zeta T = k$. If there exists a least positive integer k such that M is k-separated we call it the *connectivity* of M and denote it by $\lambda(M)$. If there is no such integer we say $\lambda M = \infty$.

We say that M is *n-connected* for any $n \leqslant \lambda(M)$.

(6)   Clearly M is 1-separated $\Leftrightarrow M$ is a connected matroid or equivalently:

(7) $\lambda(M) = 1 \Leftrightarrow M$ is a disconnected matroid.

Also, simple calculation shows that:

(8) $\xi^* T = \rho^* T + \rho^*(S\backslash T) - \rho^* S + 1 = \xi T,$

and hence:

(9) M is k-separated $\Leftrightarrow M^*$ is k-separated.

Equivalently:

(10) $\lambda(M) = \lambda(M^*)$, for any matroid M.

Now consider the case where M is the cycle matroid $M(G)$ of the graph G. Suppose that G is connected. (For any graphic matroid M we can find a connected graph G such hat $M \simeq M(G)$.) Suppose also that G is a k-separated graph ($k \geqslant 1$). Then there exist disjoint sets of edges X, Y of G with $|X| = |Y| = k$, such that X can be separated from Y by the removal from G of a set W of vertices with $|W| = k$. Let U be the set of edges of G which can be reached from an edge $e \in X$ by a path P not containing any vertex of W except possibly as an endpoint. Then if V_U is the vertex set of the graph $G|U$, and V'_U is the vertex set of the graph $G|(E(G)\backslash U)$ we have

$$V_U \cup V'_U = V,$$
$$V_U \cap V'_U = W.$$

Hence

$$|V_U| + |V'_U| = |V| + k.$$

Now consider the cycle matroid $M(G)$,

$$\xi U \leqslant |V_U| - 1 + |V'_U| - 1 - |V| + 1 + 1 = k.$$

Also since $X \subseteq U$, $Y \subseteq E\backslash U$, $|U| \geqslant k$, $|E\backslash U| \geqslant k$ and therefore $M(G)$ is a k-separated matroid. Thus we have proved the first (easy) half of the following theorem of Tutte [66].

THEOREM 1. *A connected graph G is k-separated if and only if its cycle matroid is k-separated. That is $\lambda(G) = \lambda(M(G))$ for any connected graph G.*

It is this theorem which justifies the definition of $\lambda(M)$ as a measure of matroid connectivity.

For 3-connected matroids Tutte has developed a theory corresponding to a remarkable theory which he had earlier discovered for 3-connected graphs.

The *wheel* $\mathcal{W}_n$ of order n ($n \geqslant 3$) is obtained from the circuit C_n (called the "rim") by adjoining a new vertex called the "hub" and then joining each vertex of the rim to the hub by a single edge called a "spoke". $\mathcal{W}_5$ is shown in Fig. 1.

W_5:

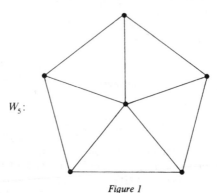

Figure 1

If G is a graph, write G'_e for the graph obtained from G by deleting the edge e from G, but not deleting its incident vertices. Write G''_e for the graph obtained from G'_e by identifying the endpoints of e in G.

Call e an *essential* edge of G if G is simple and $3 - T$-connected but neither G'_e nor G''_e is both simple and $3 - T$-connected.

Note first:

(11)   Every wheel is simple and $3 - T$-connected.
(12)   In a wheel every edge is essential.

Tutte's result [61a] was

THEOREM 2. *Let G be a simple 3-connected graph with at least 4 vertices and in which every edge is essential. Then G is a wheel.*

It is natural to ask whether this theorem has a matroid counterpart. We define an element e to be *essential* in a 3-connected matroid M on S if and only if neither $M \mid T$ nor $M \cdot T$ is 3-connected where $T = S \backslash e$, and ask what matroids have the property that every element is essential?

It is natural to ask whether the cycle matroids of wheels are the only 3-connected matroids in which every element is essential. This is not the case.

The *whirl* $\mathscr{W}^n$ of order n is a matroid got from the wheel $\mathscr{W}_n$ of order n, by letting the circuits of $\mathscr{W}^n$ be the following sets of edges:

(a) any cycle of $\mathscr{W}_n$ except the rim,
(b) any set of edges formed by adding one spoke to the set of edges of the rim.

It can be shown that this is a matroid (self-dual) and that it is not graphic.

THEOREM 3. *A 3-connected matroid has every element essential if and only if it is either the cycle matroid of a wheel or it is a whirl.*

For a proof, see Tutte [66].

EXERCISES 6

1. Prove that the dual of a whirl is a whirl, and that the cocycle matroid of the wheel is the cycle matroid of a wheel.

2. Note that the above definition of connectivity is rather strange in that there exist cases where the deletion of an edge from a graph improves the connectivity!

NOTES ON CHAPTER 5

The basic theory of matroid connection was developed by Whitney [35]. An interesting alternative approach not explicitly for matroids is given by Sierpinski [52]. This is a theory of connection for arbitrary sets endowed with a closure operator and thus it is of interest as showing how much of "connection theory" is independent of the exchange axiom and holds equally well for other closure systems—such as is found in point set topology.

A theory of semi-separation of matroids has been developed by Smith [74]. A matroid M on S is *semi separated* if a partition (X, Y) of S exists with $\rho X + \rho Y + 1 = \rho S$.

Such a semi separation defines in X and Y a pair of matroids or *patroid*. For properties of patroids and their occurrence see Smith [72], [74].

Lehman's Theorem 4.1 is very interesting; its graphical significance seems not to have been widely recognized.

R. A. Main (unpublished) has made a study of the circuit graph of a matroid and has an alternative proof of Theorem 5.1.

A theory of decomposing matroids based on Tutte's connectivity has been developed by Bixby [72].

CHAPTER 6

Matroids, Graphs and Planarity

1. UNIQUE REPRESENTATIONS OF GRAPHIC MATROIDS— THE IMPORTANCE OF 3-CONNECTION

In Chapter 1 we saw that a graphic matroid M can be the cycle matroid of several non-isomorphic graphs. For example when M is a matroid of rank n on a set of n elements, M is the cycle matroid of any forest with n edges. It would be natural to hope that provided certain restrictions are placed on a graph G, then if the cycle matroid of G is isomorphic to the cycle matroid of another graph H then G and H are isomorphic graphs.

We say a matroid M *uniquely determines* the graph G if $M \simeq M(G)$ and if for any other graph H, $M \simeq M(H)$ then G and H are isomorphic graphs.

Now if M is a disconnected graphic matroid, it will in general be the cycle matroid of several non-isomorphic graphs—for example if M has a loop the edge of the graph can be attached as a loop at any vertex of the remainder. Thus a necessary condition that M uniquely determines G is that M be connected. However this is not sufficient.

Example. The two non-isomorphic graphs of Fig. 1 have the same connected cycle matroid.

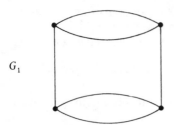
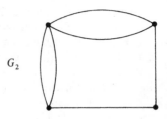

G_1 G_2

Figure 1

82

However using the connectivity theorems of Whitney [32a] we can prove the following fundamental result.

THEOREM 1. *If the matroid M has no loops and is the cycle matroid of the 3-connected graph G then M uniquely determines G, up to isolated vertices.*

Before proving this theorem we make the following observations.

We call two graphs G, H, *cycle isomorphic* if there is a bijection $\phi: E(G) \to E(H)$ such that X is the set of edges of a cycle in G if and only if ϕX is the edge set of a cycle in H. Clearly $M(G) \simeq M(H)$ if and only if G and H are cycle isomorphic.

A bijection $\phi: E(G) \to E(H)$ is an *edge isomorphism* if e_1, e_2 are incident edges in G if and only if $\phi(e_1)$, $\phi(e_2)$ are incident in H.

We have the following basic lemma of Whitney [32a].

LEMMA. *Let G, H be connected simple graphs which are edge isomorphic. Then provided G, H are not the graphs $K_3, K_{1,3}$ of Fig. 2, G, H are isomorphic.*

Proof. Harary [69, p. 72].

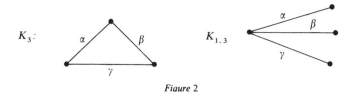

Figure 2

We can now prove our main theorem.

Proof. Suppose that there exists a loop-free, 3-connected graph G_1 and a graph G_2 not isomorphic to G_1 and having no isolated vertices such that $M(G_1) \simeq M(G_2)$.

Clearly G_2 is loop-free and since $M(G_2)$ is connected, G_2 is at least 2-connected and is therefore a block.

Our first assertion is that if $\phi: E(G_1) \to E(G_2)$ is the cycle isomorphism which is not an isomorphism then there must exist a pair of edges e_1, e_2 which are incident in G_1 but with $\phi(e_1)$, $\phi(e_2)$ not incident in G_2. Suppose not, then since $G_1 \not\simeq G_2$, there exist nonincident edges x, y of G_1 such that $\phi(x)$ and $\phi(y)$ are incident in G_2. Since G_1 is 3-connected, and a fortiori 2-connected, there exists a cycle C containing x and y. Let the edges of C, in order, be $x, c_1,$

$c_2, \ldots, c_k, y, c_{k+1}, \ldots, c_t, x$. Then by hypothesis $\phi(x)$ is incident with $\phi(c_1)$, $\phi(c_1)$ with $\phi(c_2)$ and so on. Moreover since $M(G_2) \simeq M(G_1)$, the set $\{\phi(x), \phi(c_1), \phi(c_2), \ldots, \phi(c_t), \phi(x)\}$ are an ordered set of edges of a cycle in G_2. It is thus impossible for $\phi(x), \phi(y)$ to be incident and we have proved our assertion.

Hence let $\phi: E(G_1) \to E(G_2)$ be a cycle isomorphism.

Let e_1, e_2 be incident edges of G_1 and let $\phi(e_1)$, $\phi(e_2)$ be non-incident edges of G_2. Say

$$e_1 = (a, b), \qquad e_2 = (b, c)$$

$$\phi(e_1) = (a', b'), \quad \phi(e_2) = (c', d').$$

By the connectivity assumption the graph G_1' obtained from G_1 by deleting from G, b and all the edges incident with b is 2-connected and by Theorem 5.2.3. there is a cycle in G_1' through the vertices a, c.

Let B and C be the (disjoint) paths joining a, c in G_1. Since $e_1 \cup B \cup e_2$ is a cycle in G_1 it must have a cycle as its image in G_2. This cycle must contain

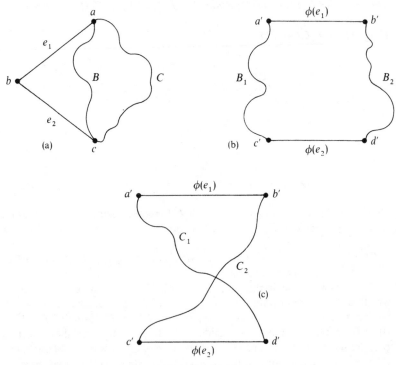

Figure 3(a), (b), (c)

$\phi(e_1)$ and $\phi(e_2)$. Since $\phi(e_1), \phi(e_2)$ are not adjacent the only possibility (up to isomorphism) is shown in Fig. 3(b) where $B_1 \cup B_2 = \phi B$.

Consider now the cycle $e_1 \cup C \cup e_2$ of G_1. In $\phi(e_1 \cup C \cup e_2)$ we must have a' connected to d' before c' as in the figure as otherwise a proper subset of $\phi(B) \cup \phi(C)$ would be a cycle in G_2 and hence give a contradiction. We show this in Fig. 3(c). But in G_2 this implies $\phi(e_1) \cup B_2 \cup C_1$ is a cycle in G_2. Thus its inverse image is a cycle of G_1. But e_1 together with a subset of $B \cup C$ is not a cycle of G_1. This is a contradiction. Hence $\phi(e_1), \phi(e_2)$ must be incident in G_2.

Thus we have shown that if ϕ is an isomorphism between $M(G_1)$ and $M(G_2)$ then it must be an edge isomorphism of G_1, G_2.

G_1, G_2 have no loops. Let $\hat{G}_1, \hat{G}_2$ be the simple graphs got from G_1, G_2 by identifying mutually parallel edges with a single edge. By the above lemma $\hat{G}_1, \hat{G}_2$ must be isomorphic and hence G_1, G_2 are isomorphic which proves the theorem. ∎

It is easy to see that the condition that G be 3-connected in order that $M(G)$ uniquely determines G is not necessary.

Example. Each of the graphs G_1, G_2 of Fig. 4 is uniquely determined by its cycle matroid. However they are not 3-connected.

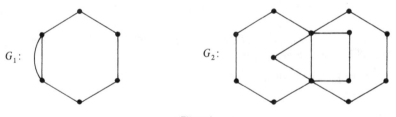

Figure 4

Suppose that G is a block graph. Let u, v be two vertices such that $G\backslash\{u, v\}$ is disconnected. Let G_1, G_2 be the components of $G\{u, v\}$. The graph G' is got from G by *twisting* at u, v if it consists of $G'_1 = G_1 \cup \{u, v\}$, and $G'_2 = G_2 \cup \{u, v\}$ but G'_1 and G'_2 are attached with the separating vertices u, v interchanged. If there is an edge (u, v) in G we assign it arbitrarily to G'_1.

We illustrate this twisting procedure in Fig. 5.

It is easy to see that if H is obtained from the block graph G by a succession of twistings then $M(G) \simeq M(H)$. The converse result is proved by Whitney [33b].

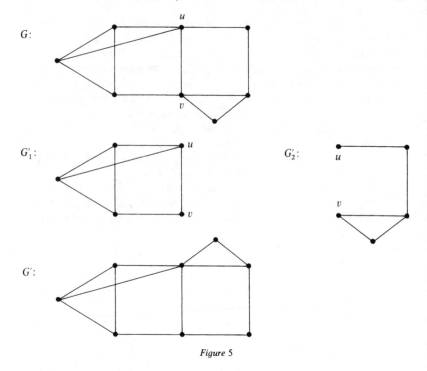

Figure 5

THEOREM 2. *If G and H are blocks and $M(H) \simeq M(G)$ then H can be obtained from G by a succession of twistings.*

The proof is not easy and will be omitted.

EXERCISES 1

1. Prove that if a simple graph is 3-connected then every set of 3 vertices lie on a cycle.
○2. Find a characterization of those matroids which are the cycle matroids of 3-connected graphs.

2. HOMEOMORPHISM OF GRAPHS AND MATROID MINORS

A graph G is *homeomorphic from* a graph H written $G \prec H$ if $H = G$ or G can be obtained from H by a finite sequence of edge subdivisions of H.

The following remarks are obvious.

(1)   If G is homeomorphic from H then G is contractible to H.
(2)   If G is a graph with every vertex of degree $\geqslant 3$ then G is not homeomorphic from any graph other than itself.
(3)   If G is homeomorphic from H and H is homeomorphic from K then G is homeomorphic from K.

We illustrate these definitions in the following example:

G_1: 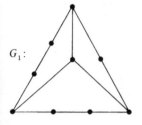 G_2: 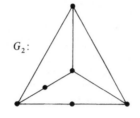 H: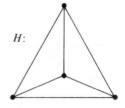

Figure 1

G_1 and G_2 are both homeomorphic from H.
    We first note the rather obvious remark.

(4)   If G contains a subgraph homeomorphic from H then $M(G)$ contains $M(H)$ as a minor.

However the converse is not true.
A trivial example of this is the pair of graphs G, H of Fig. 2.

G: 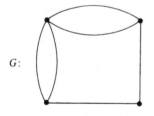 H: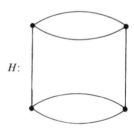

Figure 2

$M(G) \simeq M(H)$ and hence *a fortiori* has $M(H)$ as a contraction minor but G is not homeomorphic from H.
    This example again illustrates the fact that the statements "G is contractible to H" and "$M(G)$ contains $M(H)$ as a contraction minor" are not equivalent.
    We first prove the following key lemma.

LEMMA. *Let H be a graph with every vertex of degree $\leqslant 3$ and let K be a graph which is contractible to H. Then K has a subgraph homeomorphic from H.*

Proof. If $|E(K)\backslash E(H)| = 1$, let e be the edge of K which when contracted from K gives H. If e is a loop then $K\backslash e \simeq H$, otherwise let the endpoints of r be u and v. The degrees of u, v in K must be $\leqslant 3$ and at least one (say v) must have degree $\leqslant 2$ in K, for otherwise on contraction of e we will have a graph H with a vertex of degree > 3. Removing v and identifying its incident edges shows that K is homeomorphic from H.

We use induction on $p = |E(K)| - |E(H)|$. Suppose the result is true for $p < N$. Let $e_1, \ldots, e_N$ be the edges of K which are contracted to give H. Let K' be the graph got by contracting $e_1, \ldots, e_{N-1}$ from K. Then K' is contractible to H by contracting the edge e_N. Hence K' has a subgraph $U \prec H$. Also K' and hence U has no vertex of degree > 3. Since K is contractible to K', by induction K has a subgraph $T \prec K'$, hence $T \prec U \prec H$ and the lemma is proved.

THEOREM 1. *If a graph H has every vertex of degree $\leqslant 3$ and $M(H)$ uniquely determines H then for any graph G the following statements are equivalent.*

(a)   *G contains a subgraph homeomorphic from H.*
(b)   *$M(G)$ contains $M(H)$ as a minor.*

Proof. That (a) implies (b) is obvious.

Conversely suppose (b) holds. Then there exist $E_1 \subseteq E_2 \subseteq E(G)$ such that

$$(M(G)|E_2) . E_1 \simeq M(H).$$

Let $G' = G|E_2$. Then $M(G'.E_1) \simeq M(H)$ and since $M(H)$ uniquely determines H, $G'.E_1$ must be isomorphic with H. By the preceding lemma G' is homeomorphic from H and thus (a) holds. This proves the theorem. ∎

EXERCISES 2

1. Say that a graph H has the *contraction/homeomorphism* (CH) property if for all graphs G, G has a subgraph homeomorphic to H whenever G has a subgraph contractible to H. Prove that a graph has the (CH) property if and only if it has maximum vertex degree 3. (McDiarmid [74])

2. Prove that there are only two matroids on the edge set $E(G)$ of a graph G such that their circuits are connected subgraphs forming one or more equivalence classes

under homeomorphism; one is the cycle matroid $M(G)$, the other is the matroid $H(G)$ in which the circuits are the subgraphs homeomorphic from any of the three graphs shown, (see Klee [71] or Simões-Pereira [73])

3. PLANAR GRAPHS AND THEIR GEOMETRIC DUALS

A graph G is said to be *planar* if it is isomorphic to some graph G' whose vertices lie in a plane π and whose edges are Jordan curves in π such that any two distinct edges of G' do not intersect except possibly at their endpoints. Clearly a planar graph can be represented in a plane in many ways, each such representation is called a *planar representation* of G.

If G' is a planar representation of G in π then we let an equivalence relation $\sim$ be defined on the points of the plane whch are not contained in the set of vertices and edges of G' as follows. Let $x \sim y$ if and only if x can be connected to y by a Jordan curve lying in π and which does not intersect any vertex or edge of G'. The equivalence classes under $\sim$ are called the *faces* of G'.

It is well known that associated with any planar representation H of a planar graph G is a dual graph H^* constructed as follows. Associate with each face F_i of H a point $P_i \in F_i$. If two faces F_i and F_j are adjacent in the sense that the topological closures of F_i and F_j have edges $e_1, \ldots, e_k$ say of H in common then we join the points P_i and P_j by edges $e'_1, \ldots, e'_k$ with e'_i crossing e_i. With an edge e which is only in the boundary of one face F_i of H (that is e is a pendant edge of G), we associate a loop e' at the vertex P_i of H^*. The graph H^* with vertex set $\{P_i\}$ and edge set $\{e'_i\}$ we call a *geometric dual graph of G* resulting from the planar representation H of G.

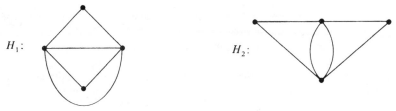

$H_1:$ $H_2:$

Figure 1

It is important to realize that a graph G may have several planar representations say $H_1, H_2, \ldots, H_k$ and these in turn can give distinct geometric dual graphs $H_1^*, H_2^*, \ldots, H_k^*$ which are *not* isomorphic. An example of this is shown by the two isomorphic planar representations H_1 and H_2 of Figure 1, which have non-isomorphic dual graphs $H_1^* = G_1$, $H_2^* = G_2$, shown in Figure 1.1.

However we do have:

THEOREM 1. *If G is a planar graph and H^* is any geometric dual of G, the cycle matroid of G and the cycle matroid of H^* are dual matroids.*

COROLLARY 1. *If G is a planar graph and H_1^*, H_2^* are any two geometric duals of G, not necessarily isomorphic, then*

$$M(H_1^*) \simeq M(H_2^*) \simeq (M(G))^*.$$

Proof of Theorem 1. When G is a tree the proof is trivial. Let G be a planar graph (not a tree) and let H be a planar representation of G. If G is disconnected we will treat the connected components separately. Hence we assume that G and therefore H is connected. Let H^* be the dual graph derived from H. Suppose that G is the set of edges of a spanning tree of H. Let T' denote the corresponding edges in H^*. We will prove that $T^* = E(H^*) \backslash T'$ is the set of edges of a spanning tree of H^*.

First notice that T^* must be incident with every vertex of H^* for otherwise a cycle of H would be contained in T and T would not be a tree.

We now assert that T^* contains no cycle, for if say C is a cycle of H^*, with $C \subseteq T^*$, then in H, the plane area "inside" C must contain at least one vertex v of H and either there is no edge of T incident with v, or T is disconnected.

Thus T^* is a set of edges of H^*, containing no cycle of H^*, and spanning in H^*. It must therefore be a spanning tree of H^*. Thus we have shown that if T is a base of $M(H)$, $E(H^*) \backslash T'$ is a base of $M(H^*)$. A similar argument shows that every base of $M(H^*)$ arises in this way and thus $M(H^*)$ is the dual matroid of $M(H)$ and the proof of Theorem 1 is complete. ∎

COROLLARY 2. *If G is a planar graph and H^* is any geometric dual of G, then if C is the set of edges of a cycle (cocycle) in G, the edges corresponding to C are the edges of a cocycle (cycle) in H^*.*

COROLLARY 3 (Euler's Formula). *If G is a connected planar graph and $F(G)$ is its set of faces*

$$|V(G)| - |E(G)| + |F(G)| = 2.$$

Proof. Follows immediately from $\rho(M(G)) + \rho(M(G)^*) = |E(G)|$. ∎

COROLLARY 4. *If T is the set of edges of a spanning tree of the planar graph G then the edges corresponding to T in any geometric dual G^* of G, form a cotree.*

EXERCISES 3

1. If G is a connected planar graph and G^* is any geometric dual of G prove that G and G^* have the same number of spanning trees.

4. PLANAR GRAPHS AND THEIR ABSTRACT DUALS

We say that a graph $\hat{G}$ is an *abstract dual* of a graph G if there is a bijection $\phi : E(G) \to E(\hat{G})$ such that X is a cycle in G if and only if ϕX is a cocycle in $\hat{G}$.

THEOREM 1. *The following statements about two graphs G and $\hat{G}$ are equivalent.*

(1) $\hat{G}$ *is an abstract dual of G.*
(2) *The cocycle matroid of $\hat{G}$ is isomorphic to the cycle matroid of G.*
(3) *The cycle matroid of $\hat{G}$ is isomorphic to the cocycle matroid of G.*
(4) *G is an abstract dual of $\hat{G}$.*

Proof. That (1) ⇒ (2) is obvious. Since $M^*(\hat{G}) \simeq M(G)$, by taking matroid duals we get $M^{**}(\hat{G}) \simeq M^*(G)$, that is $M(\hat{G}) \simeq M^*(G)$. Thus (2) ⇒ (3). (4) is merely a restatement of (3) and together they imply (1) by matroid duality. ∎

Now by Theorem 3.1 we know that any geometric dual G^* of a planar graph G has $M(G^*) \simeq M^*(G)$ and hence by the above result we have as a corollary the following statement.

(5)   If G^* is a geometric dual of the planar graph G then G^* is an abstract dual of G.

Not every abstract dual of a graph G is obtained in this way.

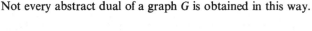

G':

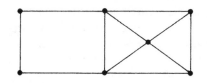

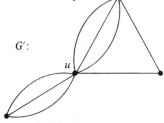

G:

u

Figure 1

Example (D. R. Woodall). Let G, G' be the graphs of Fig. 1. We assert that G' is not a geometric dual of G since there exists no embedding of G in the plane such that it has a face with 7 edges.

Hence G' is not a geometric dual of G. However G' is an abstract dual of G for the following reason.

The geometric dual G^* of G obtained from the above planar representation is shown in Fig. 2. There is an obvious map $\psi: E(G^*) \to E(G')$ which maps

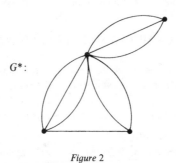

G^*:

Figure 2

cycles into cycles. Let $\phi: E(G) \to E(G^*)$ be the duality map mapping cocycles of G into cycles of G^*. Then the composite map $\phi\psi: E(G) \to E(G')$ shows that G and G' are abstractly dual.

It is not difficult to find (larger) examples which contain no parallel edges. However provided G is sufficiently well connected the next theorem shows that abstract duality coincides with the more well known geometric duality.

THEOREM 2. *If G is a planar block a graph G^* without isolated vertices is an abstract dual of G if and only if it is a geometric dual of G.*

Proof. If G^* is a geometric dual of G then by Theorem 3.1 we know that $M(G^*)$ is the dual matroid of $M(G)$ and hence G^* is an abstract dual of G.

Now let G^* be an abstract dual of G. Then $M(G^*) \simeq [M(G)]^*$. Thus since G is a block, $M(G)$ is a connected matroid and hence $(M(G))^*$ is a connected matroid. Thus $M(G^*)$ is a connected matroid and G^* must be a 2-connected graph. Hence in the terminology of Ore [67] G, G^* are abstract duals without separating vertices and by Theorem 3.4.1 of Ore [67] G can be represented in the plane such that G and G^* are geometric duals. This completes the proof. ∎

EXERCISES 4

1. Prove that a disconnected planar graph has at least two non-isomorphic abstract duals.
2. Prove that a 3-connected simple planar graph has a unique abstract dual. (Whitney [33b])

5. CONDITIONS FOR A GRAPH TO BE PLANAR

It is easy to check that the graphs K_5 and $K_{3,3}$ are not planar but that every proper subgraph of them is. One of the most famous and useful results of graph theory is the following theorem of Kuratowski [30].

THEOREM 1. *A graph G is planar if and only if it does not contain a subgraph homeomorphic from either K_5 or $K_{3,3}$.*

In one direction the proof is very easy, but showing that every non planar graph must contain a subgraph homeomorphic from either K_5 or $K_{3,3}$ is quite intricate. A proof is given in Harary [69, p. 109].

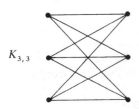
$K_{3,3}$

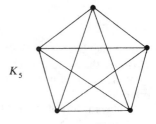
K_5

Figure 1

From the viewpoint of matroids a more useful characterization of planar graphs is the following result, which is a restatement in matroid language of a graph theorem obtained independently by Harary and Tutte [65] and Wagner [37].

THEOREM 2. *A graph G is planar if and only if its cycle matroid M(G) does not contain either of the matroids $M(K_5)$ or $M(K_{3,3})$ as a minor.*

Proof. One way is clear for if G is not planar it contains a subgraph homeomorphic from K_5 or $K_{3,3}$ and hence $M(G)$ contains $M(K_5)$ or $M(K_{3,3})$ as a minor.

We shall now show the converse that if $M(G)$ contains $M(K_5)$ or $M(K_{3,3})$ as a minor then G has a subgraph homeomorphic from K_5 or $K_{3,3}$.

Suppose not. Then by Kuratowski's theorem G must be planar. But if $M(G)$ contains $M(K_5)$ or $M(K_{3,3})$ as a minor. the graph G contains a subgraph H such that H is contractible to a graph G' and $M(G')$ is isomorphic with $M(K_5)$ or $M(K_{3,3})$. Now if $M(G') \simeq M(K_5)$ then since K_5 is 3-connected, by Theorem 1.1 $G' \simeq K_5$. Similarly if $M(G') \simeq M(K_{3,3})$, $K_{3,3}$ is 3-connected and hence $G' \simeq K_{3,3}$. Hence since G is planar, H is planar and any graph got by contracting a sequence of edges from H is planar (Prove!). Thus either K_5 or $K_{3,3}$ is planar which is a contradiction. ∎

Example. Let G be the Petersen graph of Fig. 2. The Petersen graph is

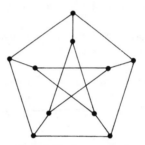

Figure 2 The Petersen graph

non-planar because its cycle matroid clearly contains $M(K_5)$ as a minor. However the reader can check that it contains no subgraph homeomorphic from K_5. It does however contain a subgraph homeomorphic from $K_{3,3}$. (Find it!)

A different sort of characterization of planar graphs is the following result of Whitney [33c].

THEOREM 3. *The following statements about a graph G are equivalent.*

(a)   *G is planar.*
(b)   *G possesses a geometric dual.*
(c)   *G possesses an abstract dual.*

Proof. That (a) ⇒ (b) is trivial. We showed that (b) ⇒ (c) in Theorem 4.2. Hence our proof will be complete if we show that if G has an abstract dual it cannot possess a subgraph homeomorphic from K_5 or $K_{3,3}$.

Let G be a graph which has an abstract dual G^* and let $\phi: E(G) \to E(G^*)$ be the map which sends cycles of G into cocycles of G^* and vice versa. We will call ϕ the *dual map* between G and G^*.

LEMMA 1. *If H is a subgraph of G then H has an abstract dual.*

Proof. Suppose $e \in E(G)$. Consider the graph $(G^*)''$ obtained by contracting $\phi(e)$ from G^*. It is easy to check that $(G^*)''$ is an abstract dual of $G \backslash e$.

Repeating this process we see that if $H = G|T$ then H has an abstract dual $G^* . \phi(T)$.

LEMMA 2. *If H is contractible to K and H has an abstract dual then K has an abstract dual.*

Proof. Let H^* be the abstract dual of H under the map ϕ. The graph K is obtained from H by successively removing either a loop or removing a vertex u of degree 2 and its incident edges (v, u), (u, w) and replacing them by a single edge (v, w). Consider one such operation on H, giving a new graph H_1. Let H_1^* be obtained from H^* by removing the edge $\phi(v, u)$. Since $\{(v, u), (u, w)\}$ is a cocycle in H, the edges $\phi(v, u)$, $\phi(u, w)$ must be in parallel in H^* and it is now easy to check that the obvious map $E(H_1) \rightarrow E(H_1^*)$ is a dual map. Removing a loop is easily dealt with and since K is obtained by a sequence of such operations it must have an abstract dual.

LEMMA 3. *Neither K_5 nor $K_{3,3}$ has an abstract dual.*

Proof. Suppose $K_{3,3}$ has an abstract dual G^*. Since $K_{3,3}$ only has cycles of lengths 4 or 6 each cocycle of G^* must be of cardinality 4 or 6. Since $K_{3,3}$ has no cocycle of size 2, G^* can have no parallel edges. Since the set of edges incident with any vertex of G^* is a cutset, and therefore contains a cocycle each vertex of G^* must be joined to at least 4 distinct vertices. Thus $|V(G^*)| \geqslant 5$, and since each vertex degree of G^* is at least 4, G^* must have at least 10 edges. But $|E(G^*)| = |E(K_{3,3})| = 9$ which is a contradiction. A similar argument shows K_5 has no abstract dual.

We can now complete the proof of Theorem 3. Suppose G is non-planar with an abstract dual G^*. By Theorem 2 $M(G)$ has a minor isomorphic to either $M(K_5)$ or $M(K_{3,3})$. By Lemmas 1, 2 either K_5 or $K_{3,3}$ has an abstract dual. This contradicts Lemma 3, and the theorem is proved. ∎

Note. In connection with Theorem 3 it is worth noting that the equivalence of (a) and (b) is a "non-theorem" in the sense that it is impossible to use the existence of a geometric dual as a test of planarity, since in order to get a geometric dual we must already have the graph embedded in the plane. Henceforth we shall use "dual of a graph" to mean "abstract dual" unless otherwise specified.

Part of Theorem 3 may be restated in matroid terminology.

THEOREM 3'. *The following statements about a graph G are equivalent.*

(a)   *G is planar.*
(b)   *Its cycle matroid M(G) is cographic.*
(c)   *Its cocycle matroid M*(G) is graphic.*

A corollary to this is:

(1)   A matroid is both graphic and cographic if and only if it is the cycle matroid of a planar graph.

Immediate consequences of the above theory are:

(2)   The smallest matroids which are graphic but not cographic are $M(K_5)$ and $M(K_{3,3})$.
(3)   The smallest matroids which are cographic but not graphic are $M^*(K_5)$ and $M^*(K_{3,3})$.

EXERCISES 5

1. Supply the counting argument which shows that K_5 cannot have an abstract dual. (see Lemma 3).

2. Prove that $M(K_5)$ is isomorphic to the matroid of the 3-dimensional Desargues' configuration.

3. Prove that a block graph is planar if and only if the vector space over GF(2) generated by the incidence vectors of the edge sets of its cycles has a basis $Z_1, \ldots, Z_m$ such that each edge belongs to at most 2 of the Z_i. (Maclane [37]).

NOTES ON CHAPTER 6

The bulk of this chapter is a matroid treatment and interpretation of a series of extremely significant papers by Whitney in the period (1931)–(1935). The existence of graphs having abstract duals which could not be geometric duals was brought to my notice by D. R. Woodall.

An alternative proof of Whitney's basic Theorem 1.1 is given by Sachs [70]. For extensions of Whitney's conditions for a graph to be planar to other polyhedral surfaces see Edmonds [65c] and Fournier [73].

What we call here the abstract dual is called a *W-dual* or *Whitney-dual* by Ore [67] and a *combinatorial dual* by Harary [69]. For a good discussion of the relation between these dual concepts see Ore [67], Chapter 3.

CHAPTER 7

Transversal Theory

1. TRANSVERSALS, PARTIAL TRANSVERSALS, THE RADO–HALL THEOREMS

Interest in matroids and their combinatorial applications accelerated rapidly when Edmonds and Fulkerson [65] proved that the set of partial transversals of a finite family of sets formed the set of independent sets of a matroid. This was also proved independently, and extended to infinite sets by Mirsky and Perfect [67]. About this time the significance of an hitherto unnoticed result of Rado [42] was realized. This chapter is mainly concerned with the applications of Rado's theorem as a tool in transversal theory. A comprehensive survey of this field is the recent book by Mirsky [71],

Throughout this chapter I denotes a finite index set and $\mathscr{A}$ denotes the family $(A_i : i \in I)$ of subsets of S. A family $(x_i : i \in I)$ of elements of S is a *system of representatives* (SR) of $\mathscr{A}$ if there exists a bijection $\pi : I \to I$ such that $x_i \in A_{\pi(i)}$. A *system of distinct representatives* (SDR) of $\mathscr{A}$ is a system of representatives $(x_i : i \in I)$ such that $x_i \neq x_j$, $i \neq j$. A set T is a *transversal* of $\mathscr{A}$ if there is a bijection $\pi : T \to I$ such that

$$x \in A_{\pi(x)}, \qquad \forall x \in T.$$

Thus the family $(x_i : i \in I)$ is an SDR of $\mathscr{A}$ if and only if the set $\{x_i : i \in I\}$ is a transversal of $\mathscr{A}$. A set X is a *partial transversal* (PT) of $\mathscr{A}$ if X is a transversal of some subfamily $(A_i : i \in J)$ of $\mathscr{A}$. In this case $|X|$ is called the *length* of the partial transversal. The *defect* of the partial transversal X is $|I| - |X|$. (The reason for the introduction of "defect" seems to be historical, though many of the theorems about the existence of partial transversals have a neater conclusion when expressed in terms of defects rather than lengths.)

The first and most basic problem in transversal theory was finding conditions for a family of sets to have a transversal. This was answered by P. Hall in 1935, and for many years this basic result appeared to be the central theorem of transversal theory. It has been proved in many different ways, by many different authors. Recently, however, the significance of the above mentioned matroid extension of Hall's theorem proved by Rado [42] has been noticed by several authors. Hall's theorem is a very special case of Rado's theorem, and while it is true that many theorems in transversal theory can be deduced by manipulating Hall's result, there seems no easy route from Hall's theorem to Rado's theorem nor from Hall's theorem to the covering and packing theorems obtained from Rado's theorem in the next chapter. At the time of writing therefore, it would appear that Rado's theorem is the most fundamental in this area of combinatorics.

All of the theorems of this chapter are of "Hall-type", that is they concern systems of representatives of finite families of subsets of a fixed finite set S.

If $(A_i: i \in I)$ is a family of sets we write for any $J \subseteq I$,

$$A(J) = \bigcup (A_i: i \in J).$$

THEOREM 1 (Hall). *The finite family of sets $(A_i: i \in I)$ has a transversal if and only if for all $J \subseteq I$,*

$$|A(J)| \geqslant |J|.$$

Several ad hoc proofs of this result have been given since Hall's original proof (see Mirsky [71]). We shall deduce it and the next three theorems from the more general Theorem 2.1 of the next section.

THEOREM 2 (Rado). *If M is a matroid on the set S with rank function ρ the finite family of subsets $(A_i: i \in I)$ of S has a transversal which is independent in the matroid M if and only if for all $J \subseteq I$,*

$$\rho(A(J)) \geqslant |J|.$$

Theorem 1 follows trivially from Theorem 2 by taking M to be the matroid in which every subset of S is independent.

THEOREM 3 (Perfect). *Let $\mathscr{A} = (A_i: i \in I)$ be a finite family of subsets of S and let M be a matroid on S with rank function ρ. If d is a non-negative integer $(d \leqslant |I|)$, $\mathscr{A}$ has a partial transversal X with defect d which is independent in M if and only if for all $J \subseteq I$,*

$$\rho(A(J)) \geqslant |J| - d.$$

The case $d = 0$ is Rado's theorem. When we take M to be the free matroid in which every set is independent we get a theorem of Ore [55], which can be regarded as the defect version of Hall's theorem.

THEOREM 4. *The finite family* $(A_i : i \in I)$ *of subsets of S has a partial transversal with defect d if and only if for all* $J \subseteq I$,

$$(1) \qquad\qquad |A(J)| \geqslant |J| - d.$$

In each of these theorems it is easy to verify that conditions such as (1) are necessary. The difficulty is in proving their sufficiency. We shall see how all four theorems are just special cases of a more general theorem about submodular functions which we prove in the next section.

EXERCISES 1

1. Prove that if $\mathscr{A} = (A_i : i \in I)$ is a finite family of subsets of S a set X is a partial transversal of $\mathscr{A}$ if and only if every subset Y of X intersects at least $|Y|$ of the A_i.

2. Hall's theorem gives $2^n - 1$ conditions for a family of sets $(A_i : 1 \leqslant i \leqslant n)$ to have a transversal. Prove that these conditions are independent of each other. (Rado [42])

3. Let $\mathscr{A} = (A_i : 1 \leqslant i \leqslant n)$ be a family of subsets of S. Let $\mathscr{A}' \subseteq \mathscr{A}$ and $T \subseteq S$. Prove that the following statements are equivalent.

   (i) There exists a set T_0 and a family $\mathscr{A}_0$ with

   $$T \subseteq T_0 \subseteq S, \qquad \mathscr{A}' \subseteq \mathscr{A}_0 \subseteq \mathscr{A}$$

   such that T_0 is a transversal of $\mathscr{A}_0$.
   (ii) $\mathscr{A}'$ possesses a transversal and T is a partial transversal of $\mathscr{A}$. (Mendelsohn and Dulmage [58], Mirsky [68]).

   (N.B. A short proof of this has also been given by Kundu and Lawler [73].)

4. Let M be a simple matroid on S and suppose that $\mathscr{A} = (A_1, \ldots, A_m)$ has a transversal independent in M. Then there exists A_i such that any element $x \in A_i$ may be chosen to represent A_i in some independent transversal of $\mathscr{A}$.

5. Prove that if $\mathscr{A} = (A_i : 1 \leqslant i \leqslant n)$ has a transversal independent in the matroid M then if A_i is any base of $M | A_i$, $\mathscr{A}^* = (A_i^* : 1 \leqslant i \leqslant n)$ has a transversal which is independent in M. (E. C. Milner).

6. Let $\mathscr{A} = (A_1, \ldots, A_n)$ have a transversal which is independent in M and let $\rho A_i \geqslant t$ for each i. Prove that the number of independent transversals of $\mathscr{A}$ in M is at least

   $$t! \qquad (t \leqslant n),$$

   $$t!/(t - n)! \qquad (t > n).$$

   (This result is an extension of a result of M. Hall [48] for ordinary transversals, see Greene, Kleitman and Magnanti [74]).

7. Call polynomials $a_1, \ldots, a_k$ with coefficients in some field F *independent* if no polynomial $f(\omega_1, \ldots, \omega_k)$ exists with coefficients in F not all zero such that

$f(a_1, \ldots, a_k) = 0$ identically. Suppose that $A_1, \ldots, A_m$ are sets of polynomials. If any k of the sets A_v contain between them at least k independent polynomials $k = 1, 2, \ldots, n$ then it is possible to select one polynomial a_i from each A_i in such a way that $a_1, \ldots, a_n$ are independent. (Rado [42]).

2. A GENERALIZATION OF THE RADO–HALL THEOREMS

In this section we prove a general theorem which has Theorems 1.1–4 as special cases.

THEOREM 1. *If* $\mathscr{A} = (A_i : i \in I)$ *is a finite family of non-empty subsets of* S *and* $\mu : 2^S \to \mathbb{Z}^+$ *satisfies for* X, Y, $\subseteq S$ *the conditions:*

(i)
$$X \subseteq Y \Rightarrow \mu X \leqslant \mu Y,$$

(ii)
$$\mu X + \mu Y \geqslant \mu(X \cup Y) + \mu(X \cap Y),$$

then $\mathscr{A}$ *has a system of representatives* $(x_i : i \in I)$ *such that*

(1)
$$\mu\{x_i : i \in J\} \geqslant |J| \qquad (J \subseteq I)$$

if and only if

(2)
$$\mu(A(J)) \geqslant |J| \qquad (J \subseteq I).$$

[Note: $\{x_i : i \in J\}$ is the set of distinct elements of the family $(x_i : i \in J)$]

Proof of Hall's Theorem 1.1. Take $\mu X = |X|$ in Theorem 1.

Proof of Rado's Theorem 1.2. Take $\mu X = \rho X$ the rank function of the matroid M in Theorem 1.

Proof of Ore's Theorem 1.4. Take $\mu X = |X| + d$ in Theorem 1.

Proof of Perfect's Theorem 1.3. Take $\mu X = \rho X + d$ in Theorem 1.

It remains only to prove Theorem 1.

Proof of Theorem 1. Suppose that for each $J \subseteq \{1, \ldots, n\} = I$ the condition (2) is satisfied.

If each A_i is a singleton set then there is clearly nothing to prove. Without loss of generality let $|A_1| \geqslant 2$. We assert that there exists $x \in A_1$ such that the family of subsets $A_1 \backslash x, A_2, \ldots, A_n$ also satisfies the conditions (2) for all

$J \subseteq \{1, \ldots, n\}$. For if not, taking x_1, x_2 to be any two distinct elements of A_1 there must exist subsets J_1, J_2 of $\{2, \ldots, n\}$ such that

$$\mu((A_1 \backslash x_1) \cup A(J_1)) < |J_1| + 1,$$
$$\mu((A_1 \backslash x_2) \cup A(J_2)) < |J_2| + 1.$$

(Notice that either J_1 or J_2 or both may be null.) Using the fact that μ satisfies the submodular inequality we get

$$|J_1| + |J_2| \geq \mu(A_1 \cup A(J_1 \cup J_2)) + \mu(A(J_1 \cap J_2)).$$

Since $\mathscr{A}$ satisfies (2) this implies that

$$|J_1| + |J_2| \geq |J_1 \cup J_2| + 1 + |J_1 \cap J_2|$$

which is a contradiction.

Hence we may successively delete elements from A_1 until we arrive at a singleton subset, and then continue to delete elements from A_2 and so on until we arrive at a family of singleton sets satisfying (2) for all $J \subseteq \{1, \ldots, n\}$. This family of singletons is the required system of representatives.

The necessity of the conditions (2) is obvious since μ is a non-decreasing set function and the proof is complete. ∎

It is easy to obtain a defect version of Theorem 1.

COROLLARY. *If $\mathscr{A} = (A_i : i \in I)$ is a finite family of non-null subsets of S and $\mu : 2^S \to \mathbb{Z}^+$ is non decreasing and submodular then for any integer* $d \leq |I|$, $\mathscr{A}$ *has a system of representatives $(x_i : i \in I)$ such that for all $J \subseteq I$,*

$$\mu\{x_i : i \in J\} \geq |J| - d,$$

if and only if for all $J \subseteq I$,

$$\mu(A(J)) \geq |J| - d.$$

Proof. Define the integer valued set function λ by

$$\lambda(X) = \mu(X) + d \qquad (X \subseteq S).$$

Then if μ is submodular so too is λ and the result follows by applying Theorem 1 to λ. ∎

Systems of representatives with repetition

We now show how to apply Theorem 1 to obtain other well known results in transversal theory. Suppose we consider the submodular function $\mu : 2^S \to \mathbb{Z}^+$ defined by

$$\mu X = k|X| \qquad (X \subseteq S)$$

where k is a positive integer. Then if we apply Theorem 1 to a collection $\mathscr{A} = (A_i : i \in I)$ of subsets of S we get the result that $\mathscr{A}$ has a system of representatives $(x_i : i \in I)$ with

$$k|\{x_i : i \in I\}| \geqslant |J|$$

for all $J \subseteq I$ if and only if for all $J \subseteq I$

$$k|A(J)| \geqslant |J|.$$

But in such a system of representatives clearly no element can occur more than k times. Hence we have proved the following result first proved by Rado (see Mirsky and Perfect [66]), by ad hoc methods.

THEOREM 2. *The finite family* $(A_i : i \in I)$ *of subsets of S has a system of representatives in which no element occurs more than k times if and only if for all $J \subseteq I$,*

$$k|A(J)| \geqslant |J|.$$

Systems of common representatives of two families

If $\mathscr{A} = (A_i : i \in I)$ and $\mathscr{B} = (B_i : i \in I)$ are two finite families of sets, the family $(x_i : i \in I)$ of elements is a *system of common representatives* (SCR) of $\mathscr{A}$ and $\mathscr{B}$ if it is a system of representatives of both $\mathscr{A}$ and $\mathscr{B}$.

The definitive solution to the problem of finding necessary and sufficient conditions for two families of sets to have an SCR was solved by P. Hall [35].

THEOREM 3. *Two finite families* $\mathscr{A} = (A_i : i \in I)$ *and* $\mathscr{B} = (B_i : i \in I)$ *of finite sets have a system of common representatives if and only if the union of any k of the A_i intersects at least k of the B_i.*

We show how Theorem 3 follows easily from our general Theorem 1.

Proof. Define $\mu : 2^S \to \mathbb{Z}^+$ by taking μX to be the number of sets B_i with which X has non-null intersection. It is clear that μ is integer valued, non-negative and non-decreasing and it is easy to check that it is submodular. Hence by Theorem 1 $(A_i : i \in I)$ has an SR $(x_i : i \in I)$ with the property that for any $J \subseteq I$, $\{x_i : i \in J\}$ intersects $\geqslant |J|$ of the B_i if and only if $\mu(A(J)) \geqslant |J|$ for all $J \subseteq I$. Now this implies that $x_i : i \in I$ is an SR of $\mathscr{B}$ and hence is our required SCR. ∎

In view of the simplicity of the above proof it is rather surprising that there is no known set of necessary and sufficient conditions for *three* families of sets to have an SCR.

EXERCISES 2

1. Show that Theorem 1 is not true when μ is not integer valued.

○2. Hautus [70] has given an algorithm for producing a transversal of a given family of sets which is independent in a matroid M. Find a "fast" algorithm for finding an $SR(x_i: i \in I)$ of $(A_i: i \in I)$ such that for each $J \subseteq I$,

$$\mu\{x_i: i \in J\} \geqslant |J|$$

where μ is a function satisfying the conditions of Theorem 1.

3. TRANSVERSAL MATROIDS

As mentioned at the start of this chapter, in addition to the influence of Rado's theorem, transversal theory made a big step forward with the discovery of the following result.

THEOREM 1. *If $\mathscr{A}$ is a family of finite subsets of a set S the collection of partial transversals of $\mathscr{A}$ is the set of independent sets of a matroid on S.*

We shall give two proofs of this theorem. The first is an easy consequence of Hall's theorem and the theory of submodular functions and is found in Section 8.2 the second is a special case of the more general theory of matchings in Section 14.6. A straightforward ad hoc proof can be found in Mirsky's book [71, p. 102].

An arbitrary matroid M on S is called a *transversal matroid* if there exists some family $\mathscr{A}$ of subsets of S such that $\mathscr{I}(M)$ is the family of partial transversals of $\mathscr{A}$. We sometimes denote this matroid by $M[\mathscr{A}]$ or $M[(A_i: i \in I)]$, and call $\mathscr{A}$ a *presentation* of the matroid M.

Notice that if the collection $\mathscr{A}$ does have a transversal the bases of $M[\mathscr{A}]$ are the transversals of $\mathscr{A}$, otherwise they are the maximal partial transversals of $\mathscr{A}$.

An immediate consequence of Theorem 1 is that all maximal PTS of a family of sets have the same cardinality.

Example 1. Let $S = \{1, 2, \ldots, 5\}$, let $\mathscr{A}$ be the family

$$A_1 = \{1, 2, 3\}, A_2 = \{1, 2\}, A_3 = \{1, 5\}.$$

Then the bases of $M[\mathscr{A}]$ are the sets

$$\{1, 2, 5\}, \{3, 1, 5\}, \{3, 2, 1\}, \{3, 2, 5\}.$$

Example 2. Let M be the matroid on $\{1, 2, \ldots, 6\}$, with bases all 2-sets except $\{1, 2\}, \{3, 4\}, \{5, 6\}$. It is quite easy to check that M cannot be transversal.

We now show that in a natural way a transversal matroid can be regarded not as a matroid on the ground set S but as a matroid on the index set I.

To do this we look at transversal theory as the study of matchings in bipartite graphs.

Let $\mathcal{A} = (A_i : 1 \leqslant i \leqslant m)$ be a family of subsets of $S = \{x_1, x_2, \ldots, x_n\}$. Construct an associated bipartite graph $\Delta = \Delta(\mathcal{A})$ as follows. The vertex set of Δ is $S \cup \{A_i : 1 \leqslant i \leqslant m\}$ and join x_i to A_j by an edge if and only if $x_i \in A_j$ The graph Δ is called the *bipartite graph of the family* $\mathcal{A}$.

Example 3. Take $S = \{x_1, x_2, x_3, x_4, x_5\}$ and $\mathcal{A}$ to be the sets

$$A_1 = \{x_1, x_2\}, \qquad A_2 = \{x_2, x_3, x_4\}, \qquad A_3 = \{x_3, x_5\}.$$

Then the associated bipartite graph is shown in Fig. 1.

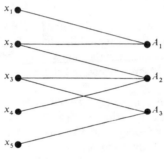

Figure 1

A *matching* of a graph G is a set X of edges such that no two members of X have a common endpoint. If U, W are disjoint subsets of the vertex set $V(G)$ of a graph we say U can be *matched into* W if there is a matching X of G with the property that every edge of the matching has one endpoint in U and the other in W and each vertex of U is the endpoint of one edge in X.

Thus if Δ is the bipartite graph determined by the family $(A_i : i \in I)$ of subsets of S, $\mathcal{A}$ has a transversal if and only if I can be matched into S in Δ.

An alternative way of looking at bipartite graphs is to regard them as incidence structures.

An *incidence structure* is a triple (p, B, R) where p, B, R are sets such that

$$p \cap B = \varnothing, \qquad R \subseteq p \times B.$$

The *dual incidence structure* (p^*, B^*, R^*) is defined by

$$p^* = B, \qquad B^* = p,$$

and an element $(x, y) \in R^*$ if and only if $(y, x) \in R$. Taking $p = S$ and B to be the collection of sets A_i there is now an obvious duality principle which says that to every theorem about points and sets and containment there is a dual theorem about sets, points and containing. This duality (which we emphasize is far removed from matroid duality), is extremely simple, and implies for example the following result which is merely the point-set dual statement of Theorem 1.

THEOREM 2. *Let $(A_i : i \in I)$ be a finite family of subsets of S and let $\mathscr{Q}$ be the collection of subsets J of I defined by $J \in \mathscr{Q}$ if and only if the subfamily $(A_i : i \in J)$ has a transversal. Then the collection $\mathscr{Q}$ is the set of independent sets of a matroid on I.*

Thus associated with any family $(A_i : i \in I)$ of subsets we have two matroids, which we will denote by $M[\mathscr{A}]$ on S and $Q[\mathscr{A}]$ on I.

The matroid $Q[\mathscr{A}]$ is a transversal matroid, its rank equals that of $M[\mathscr{A}]$. This common rank is known as the *term rank* of $\mathscr{A}$.

The defect versions of Hall's theorem enable us to obtain the rank functions ρ, ρ' of $M[\mathscr{A}]$ and $Q[\mathscr{A}]$.

A set $X \subseteq S$ has rank k in $M[\mathscr{A}]$ if and only if the family of sets $(X \cap A_i : i \in I)$ has a PT of length at least k, that is (by Theorem 1.4) if and only if

$$|A(J) \cap X| \geqslant |J| - (|I| - k) \qquad \forall J \subseteq I.$$

Hence we have proved

THEOREM 3. *The rank function ρ of the transversal matroid $M[(A_i : i \in I)]$ is given for $X \subseteq S$ by*

$$\rho X = \min_{J \subseteq I} (|A(J) \cap X| - |J| + |I|).$$

Similarly, we have

THEOREM 4. *The rank function ρ' of the transversal matroid $Q[(A_i : i \in I)]$ is given for all $J \subseteq I$ by*

$$\rho' J = \min_{K \subseteq J} (|A(K)| - |K| + |J|).$$

EXERCISES 3

1. Prove that the Fano matroid is not transversal.

2. Prove that the restriction of a transversal matroid is transversal but that in general the contraction of a transversal matroid is not transversal.

3. Show that neither the dual nor truncation of a transversal matroid need be transversal.

4. Find a transversal matroid which has distinct presentations.

5. If M is a transversal matroid and $(A_1, \ldots, A_n)$ is a presentation of M prove that each A_i is dependent in M^*.

6. Prove that $M(K_5)^*$ and $M(K_{3,3})^*$ are not transversal matroids.

7. Give an example of a transversal matroid M on a set S such that for each $x \in S$, $M \cdot (S \backslash x)$ is not transversal. (J. H. Mason)

8. Prove that on a set of cardinal n every matroid of rank $\geqslant n - 2$ is transversal. (Ingleton and Piff [73])

4. TRANSVERSALS WITH PRESCRIBED PROPERTIES— APPLICATIONS OF RADO'S THEOREM

In this section we show how to use Rado's theorem to get many results in transversal theory which were hitherto obtained by ad hoc, sometimes difficult methods. As a simple example consider the following result of Hoffman and Kuhn [56].

THEOREM 1. *If $\mathscr{A} = (A_i : i \in I)$ is a finite family of subsets of S and Y is a subset of S, $\mathscr{A}$ has a transversal containing Y if and only if* (i) $\mathscr{A}$ *has a transversal* (ii) Y *is a partial transversal of $\mathscr{A}$.*

As observed by Mirsky and Perfect [67], Theorem 1 is a trivial consequence of the fact that the PTS of $\mathscr{A}$ are the independent sets of a matroid, and hence if Y is a PT it can be augmented to a base (=maximal partial transversal) of the matroid $M[\mathscr{A}]$. A more interesting result than Theorem 1 is the next theorem.

If a set X is a transversal of both $\mathscr{A} = (A_i : i \in I)$ and $\mathscr{B} = (B_i : i \in I)$ it is called a *common transversal* (CT) of $\mathscr{A}$ and $\mathscr{B}$.

THEOREM 2. *Let $\mathscr{A} = (A_i : i \in I)$ and $\mathscr{B} = (B_i : i \in I)$ be two finite families of subsets of S. Then $\mathscr{A}$ and $\mathscr{B}$ have a common transversal which contains a prescribed set C if and only if for all subsets J, K of I,*

(1) $|(A(J) \cap B(K)) \backslash C| + |A(J) \cap C| + |B(K) \cap C| \geqslant |J| + |K| + |C| - |I|$.

As a corollary we get the following well-known result of Ford and Fulkerson [58], by taking $C = \varnothing$.

THEOREM 3. *Two finite families* $\mathscr{A} = (A_i : i \in I)$ *and* $\mathscr{B} = (B_i : i \in I)$ *of sets have a common transversal if and only if for all* $J \subseteq I$, $K \subseteq I$,

$$|A(J) \cap B(K)| \geq |J| + |K| - |I|.$$

Proof of Theorem 2. Let M be the transversal matroid of $\mathscr{B}$. Define a new matroid M' to have as its bases all bases of M which contain the prescribed set C. (It is easy to verify that this construction does give a matroid.) Then clearly $\mathscr{A}$ and $\mathscr{B}$ have a common transversal containing C if and only if $\mathscr{A}$ has a transversal which is a base of M'. Now if ρ is the rank function of M' it is not difficult to see that for any $X \subseteq S$, $\rho(X) \geq k$ if and only if $X \cup C$ contains a partial transversal of $\mathscr{B}$ of cardinal not less than

$$k + |C| - |C \cap X|.$$

This means that $\{(X \cup C) \cap B_i : i \in I\}$ must have a partial transversal with defect

$$|I| - k - |C| + |C \cap X|$$

By Theorem 1.4, necessary and sufficient conditions for this are that for any subset K of I,

$$|(X \cup C) \cap B(K)| \geq |K| - |I| + k + |C| - |C \cap X|.$$

Hence since a transversal with desired properties exists if and only if for all $J \subseteq I$,

$$\rho(A(J)) \geq |J|$$

we see that necessary and sufficient conditions for such a transversal to exist are that for any subsets J, K of I,

$$|(A(J) \cup C) \cap B(K)| \geq |J| + |K| + |C| - |I| - |C \cap A(J)|$$

and it is trivial to put this in the symmetric form (1). ∎

The following stronger version of the basic Ford–Fulkerson theorem was obtained by Perfect [68] using Menger's graph theorem.

A set X is a *common partial transversal* of *length* p of $\mathscr{A}$ and $\mathscr{B}$ if $|X| = p$, and X is a partial transversal of both $\mathscr{A}$ and $\mathscr{B}$.

THEOREM 4. *The two finite families* $(A_i : i \in I)$ *and* $(B_i : i \in I')$ *have a common partial transversal of length* p *if and only if the inequality*

$$|A(J) \cap B(J')| \geq |J| + |J'| - |I| - |I'| + p$$

holds for all $J \subseteq I$ *and* $J' \subseteq I'$.

It is not difficult to prove this from Perfect's Theorem 1.4 using a technique similar to the proof of Theorem 2. We leave the details to the reader.

It is natural to ask whether the techniques used above can be extended to give corresponding results for more than two collections of sets. The natural analogue of Theorem 3 is:— the three collections of sets $(A_i : i \in I)$, $(B_i : i \in I)$ and $(C_i : i \in I)$ have a common transversal if and only if

(2) $|A(J) \cap B(K) \cap C(L)| \geqslant |J| + |K| + |L| - 2|I|$

holds for any three subsets J, K, L of I. The conditions (2) are clearly necessary but the following example given by Mirsky and Perfect [67] shows that they are not sufficient.

Example. Let $S = \{1, 2, 3\}$ and define $\mathscr{A}$, $\mathscr{B}$, $\mathscr{C}$ as follows.

$$\mathscr{A} = \{1\}, \{2, 3\},$$
$$\mathscr{B} = \{2\}, \{1, 3\},$$
$$\mathscr{C} = \{3\}, \{1, 2\}.$$

Then $\mathscr{A}$, $\mathscr{B}$, $\mathscr{C}$ satisfy (2), but they do not have a common transversal. The reader will see that this problem is just a particular case of the still unsolved and apparently very difficult problem of deciding when three matroids have a common independent set of prescribed cardinality, and that the basic reason that none of the above techniques "work" for this 3-family problem is that the common partial transversals of two families of sets do not in general form the independent sets of a matroid.

EXERCISES 4

1. Let $\mathscr{A} = (A_1, \ldots, A_n)$ be a family of subsets of S and let $X \subseteq S$. Prove that $\mathscr{A}$ has a transversal which contains X if and only if:

   (a) for each $T \subseteq S$, T contains at most $|T|$ of the A_i,

   (b) each $Y \subseteq X$ intersects at least $|Y|$ of the A_i.

2. Let M be a matroid on S and let $U \subseteq S$. Prove that the finite family $\mathscr{A} = (A_i : 1 \leqslant i \leqslant n)$ of subsets of S has a transversal T which is independent in M and which has non-empty intersection with U if and only if, for every family $(I_k : 1 \leqslant k \leqslant n)$ of subsets of $\{1, .., , n\}$

$$\sum_{k=1}^{n} \rho[(\bigcup_{j \in I_k} A_j) \cap \delta_{k, I_k}(U)] \geqslant \sum_{k=1}^{n} |I_k| - n + 1,$$

where $\delta_{k,I}$ is defined for $I \subseteq \{1, \ldots, n\}$, $k \in \{1, \ldots, n\}$ by

$$\delta_{k,I} X = \begin{cases} X & \text{if } k \in I, \\ S & \text{if } k \notin I. \end{cases} \qquad (X \subseteq S).$$

(Perfect [72])

5. GENERALIZED TRANSVERSALS

If $\mathscr{A} = (A_i : 1 \leqslant i \leqslant n)$ is a family of subsets of S and $p = (p_1, \ldots, p_n)$ is an n-tuple with non-negative integer coordinates we say that a subset X is a *p-transversal* of $\mathscr{A}$ if and only if, writing $I = (1, \ldots, n)$

$$X = X_1 \cup \ldots \cup X_n$$

where

$$X_i \cap X_j = \varnothing, \qquad (i \neq j);$$
$$|X_i| = p_i, \qquad (i \in I);$$
$$X_i \subseteq A_i.$$

Thus a $(1, 1, \ldots, 1)$-transversal is a transversal in the usual sense.

THEOREM 1. *For fixed p, the collection of p-transversals of $\mathscr{A}$ form the bases of a matroid on S.*

Proof. Let $\mathscr{A}^*$ be the collection of subsets of S consisting of p_1 copies of A_1 p_2 copies of $A_2, \ldots, p_n$ copies of A_n. It is easy to see that X is a p-transversal of $\mathscr{A}$ if and only if X is a transversal of $\mathscr{A}^*$. Hence by Theorem 3.1. the p-transversals of $\mathscr{A}$ are the bases of a matroid on S. ∎

Similarly we may extend Rado's theorem to

THEOREM 2. *If M is a matroid and $\mathscr{A}$ is a finite family of subsets of S then $\mathscr{A}$ has a p-transversal which is independent in M if and only if for all $J \subseteq I$,*

$$\rho(A(J)) \geqslant \sum_{i \in J} p_i$$

where ρ is the rank function of M.

Proof. Let $\mathscr{A}^*$ be the collection of subsets of S defined in the proof of Theorem 1. Then $\mathscr{A}$ has an independent p-transversal if and only if $\mathscr{A}^*$ has an independent transversal, and the result follows from Rado's theorem. ∎

If $Z = (z_1, z_2, \ldots, z_n)$ is a family of not necessarily distinct elements of S and $e \in S$ we denote the number of times that e occurs in Z by $f(e; Z)$.

If $\mathscr{A}$ is any collection of subsets of S and k is any non-negative integer, then we say that the subset $X = \{x_1, \ldots, x_m\}$ of distinct elements of S is a *k-transversal* of $\mathscr{A}$, if there exists a set $Y = \{y_i; i \in I\}$ of not necessarily distinct elements such that

$$y_i \in A_i, \qquad (i \in I),$$
$$1 \leqslant f(x_i; Y) \leqslant k, \quad (1 \leqslant i \leqslant m).$$

Thus a 1-transversal is a transversal in the usual sense. k-transversals are more usually described as *systems of representatives with repetition*, see Ford and Fulkerson [62] or Mirsky [71].

In general the collection of k-transversals of a family $\mathscr{A}$ of subsets of S do not form the bases of a matroid. However we generalize Rado's theorem to get:

THEOREM 3. *If M is a matroid with rank function ρ then $\mathscr{A}$ has a k-transversal with rank not less than t if and only if*:

(1) $$k|A(J)| \geqslant |J|,$$

(2) $$\rho(A(J)) \geqslant |J| + t - |I|,$$

for all subsets $J \subseteq I$.

Before proving this we need a preliminary lemma. For notational reasons let

$$S = \{e(1), e(2), \ldots, e(m)\}.$$

Define S^k to be the set of km elements,

$$S^k = \{e(1, 1), \ldots, e(1, k), e(2, 1), \ldots, e(2, k), \ldots, e(m, 1), \ldots, e(m, k)\}$$

and let $g: S^k \to S$ be the natural map

$$g(e(i, j)) = e(i).$$

If now M is a matroid on S, let $\mathscr{G}$ be the collection of those subsets X of S^k which satisfy

(a) $|X| = \rho(S)$

(b) $g(X)$ is a base of M.

LEMMA. *$\mathscr{G}$ is the set of bases of a matroid on S^k which we denote by $M(k)$.*

Proof. From (a) no member of $\mathscr{G}$ properly contains another. Let X, Y be two members of $\mathscr{G}$, say:

$$X = e(i_1, j_1), \ldots, e(i_r, j_r)$$
$$Y = e(k_1, l_1), \ldots, e(k_r, l_r).$$

Let $B_1 = g(X)$, $B_2 = g(Y)$, so that B_1, B_2 are bases of M. Consider

$$X' = X \backslash e(i_t, j_t), \qquad (1 \leqslant t \leqslant r).$$

Then

$$g(X') = B_1 \backslash e(i_t).$$

But since B_1, B_2 are bases of M, there exists u such that

$$B_3 = B_1 \backslash e(i_t) \cup e(k_u)$$

is also a base of M. Hence

$$X \backslash e(i_t, j_t) \cup e(k_u, l_u)$$

is a member of $\mathscr{G}$, and therefore $\mathscr{G}$ satisfies the base axioms, which proves the lemma.

Proof of Theorem 3. Define the collection $\mathscr{A}^k = \{A_i^k . i \in I\}$ of subsets of S^k by

$$e(i, j) \in A_u^k \Leftrightarrow e(i) \in A_u, \qquad (u \in I).$$

Suppose that $\mathscr{A}$ has a k-transversal X such that $\rho(X) \geqslant t$.
Then $\mathscr{A}^k$ has a transversal Y such $g(Y) = X$ and $\rho_k(Y) \geqslant t$.
But from the extension of Rado's theorem this implies that for any $J \subseteq I$

$$\rho_k(A^k(J)) \geqslant |J| + t - |I|.$$

Hence

$$\rho(A(J)) \geqslant |J| + t - |I|.$$

The necessity of the condition (1) is obvious.
    Conversely if (2) holds, then for any $J \subseteq I$

$$\rho_k(A^k(J)) \geqslant |J| + t - |I|.$$

Hence $\mathscr{A}^k$ has a partial transversal Y of rank $\geqslant t$ in $M(k)$. But by (1), $\mathscr{A}^k$ has a transversal and since any partial transversal may be augmented to a transversal (if it exists), there exists $Z \subseteq S^k$ such that

(i) Z is a transversal of $\mathscr{A}^k$
(ii) $Y \subseteq Z$,

and therefore $\rho_k(Z) \geqslant t$.

Hence $g(Z)$ is a k-transversal of $\mathscr{A}$ and $\rho(g(Z)) \geqslant t$, which proves the theorem. ■

By choosing M suitably in the above theorems we can get analogues for p-transversals and k-transversals of results previously obtained for ordinary transversals. For example:

(3) $\mathscr{A} = (A_i : i \in I)$ has a p-transversal if and only if for all $J \subseteq I$,

$$|A(J)| \geqslant \sum_{i \in J} p_i$$

Proof. Take M to be the free matroid in which every set is independent.

(4) $\mathscr{A} = (A_i : i \in I)$ has a p-transversal which contains a prescribed set U if and only if for all $J \subseteq I$,

$$|A(J) \cap U| \geqslant \sum_{i \in I \setminus J} p_i + |U|$$

Proof. Let M be the matroid which has a single base U. Then we are looking for a transversal whose rank in M is at least $|U|$.

Finally we obtain the analogue of the Ford–Fulkerson theorem about common transversals.

THEOREM 4. *Let* p, q *be non-negative integer n-vectors such that*

$$\sum_{i=1}^{n} p_i = \sum_{i=1}^{n} q_i.$$

Then if $\mathscr{A}$ and $\mathscr{B}$ are two families of subsets of S, $\mathscr{A}$ has a p-transversal which is a q-transversal of $\mathscr{B}$ if and only if for all subsets J, K of $\{1, 2, \ldots, n\}$

$$|A(J) \cap B(K)| \geqslant \sum_{i \in J} p_i + \sum_{i \in K} q_i - \sum_{i=1}^{n} p_i.$$

Proof. Let $M[p, \mathscr{A}]$ be the matroid whose bases are the p-transversals of $\mathscr{A}$. Then a set of the desired type exists if and only if $\mathscr{B}$ has a q-transversal which is independent in the matroid $M[p, \mathscr{A}]$. But if $X \subseteq S$, it is easy to see that X has rank at least t in $M[p, \mathscr{A}]$ if and only if

$$|A(J) \cap X| \geqslant \sum_{i \in J} p_i - \sum_{i \in I} p_i + t.$$

The result follows by taking $X = B(K)$. ■

The reader will note that Theorem 4 is essentially the *supply–demand theorem* familiar in operations research and transportation theory.

Consider n sources of supply $U_1, \ldots, U_n$ and n sources of demand $V_1, \ldots, V_n$. Join U_i to V_j by a directed edge if there is a route from U_i to V_j. Suppose there is an integer amount s_i of material at source U_i and the V_j demands an integer amount d_j of material. Then let A_i be the set of edges coming out of U_i, let B_j be the set of edges coming into V_j. Then the set of supplies will satisfy the set of demands by sending at most 1 unit of material along any edge if and only if $\mathscr{A}$ has a $(s_1, \ldots, s_n)$-transversal which is also a $(d_1, \ldots, d_n)$-transversal of $\mathscr{B}$.

(*Note*: this assumes that the supplies and demands are integers. However provided they are rational, they can be reduced to the integer case and if they are irrational the problem can be solved by approximation.)

EXERCISES 5

1. Prove that the family $(A_i : 1 \leqslant i \leqslant n)$ of subsets of S has d disjoint transversals if and only if for each subset $J \subseteq \{1, \ldots, n\}$,

$$|A(J)| \geqslant d|J|.$$

(Ore [55])

2. Let $\mathscr{A} = (A_i : 1 \leqslant i \leqslant n)$ be a family of subsets of S and let d be a positive integer. Then it is possible to partition $\mathscr{A}$ into d subfamilies each of which possesses a transversal if and only if for all $J \subseteq \{1, \ldots, n\}$,

$$d|A(J)| \geqslant |J|.$$

(R. Rado)

6. A CONVERSE TO RADO'S THEOREM

P. Hall's basic theorem is not the best possible in the following sense: the class of set functions f such that a finite family $\mathscr{A} = (A_i : i \in I)$ has a transversal X with $f(X) \geqslant |I|$ if and only if

$$f(A(J)) \geqslant |J| \qquad (J \subseteq I),$$

contains functions other than the cardinality function $fX = |X|$.

The following theorem shows that Rado's theorem is essentially the best possible.

THEOREM 1. *Let $\mathscr{G}$ be a finite non-empty family of subsets of S such that for any finite family $\mathscr{A} = (A_i : i \in I)$ of subsets of S the following statements are equivalent.*

(i) $\mathscr{A}$ has *a transversal which belongs to* $\mathscr{G}$.

(ii) *For* $k \leqslant |I|$, *the union of any* k *members of* $\mathscr{A}$ *contains a member* Y *of* $\mathscr{G}$ *with* $|Y| \geqslant k$.

Then $\mathscr{G}$ *is the collection of independent sets of a matroid on* S.

Similarly there is a converse to Theorem 2.1 which shows that that too is best possible.

Let $\mathscr{F}$ be the collection of functions $f: 2^S \to \mathbb{R}^+$. Call $f \in \mathscr{F}$ a *representative function* on S if for any finite family $(A_i: i \in I)$ of non empty subsets of S the following statements are equivalent:

(1) $(A_i: i \in I)$ has an $SR(x_i: i \in I)$ such that for any $J \subseteq I$,

$$f\{x_i: i \in J\} \geqslant |J|.$$

(2)   For any $J \subseteq I$, $f(A(J)) \geqslant |J|$.

THEOREM 2. *A function* $f: 2^S \to \mathbb{R}^+$ *is a representative function if and only if* $[f]$ *is non-decreasing and submodular.*

($[f]$ denotes the function: $2^S \to \mathbb{R}^+$ defined by $[f](x) = [f(x)]$ where as usual $[a]$ is the greatest integer $\leqslant a$.)

We shall only prove Theorem 1, a proof of Theorem 2 can be found in Welsh [71a].

Proof of Theorem 1. Suppose $X \in \mathscr{G}$ and $Y \subseteq X$. Suppose

$$X = \{x_1, x_2, \dots, x_m\}$$
$$Y = \{x_1, x_2, \dots, x_p\}.$$

Let $\mathscr{A} = (A_i: 1 \leqslant i \leqslant m)$ be the family of subsets defined by

$$A_i = \{x_i\} \qquad (1 \leqslant i \leqslant m).$$

Then $\mathscr{A}$ satisfies (i), and hence satisfies (ii). Thus taking $J = \{1, \dots, p\}$, $Y = A(J)$ contains a subset of cardinality $|J|$ which is a member of $\mathscr{G}$. Thus $Y \in \mathscr{G}$.

Now let $X = \{x_1, \dots, x_m\} \in \mathscr{G}$, and let $Y = \{y_1, \dots, y_{m+1}\} \in \mathscr{G}$. We wish to show there exists an element y_i of $Y \backslash X$ such $X \cup y_i \in \mathscr{G}$.

Take $\mathscr{A} = (A_i: 1 \leqslant i \leqslant m + 1)$ to be the family of sets

$$A_i = \{x_i\} \qquad (1 \leqslant i \leqslant m),$$
$$A_{m+1} = Y.$$

Then it is easy to see that $\mathscr{A}$ satisfies (ii) Hence $\mathscr{A}$ satisfies (i).

Thus $\mathscr{A}$ has a transversal $\{x_1, \ldots, x_m, y_i\}$ which belongs to $\mathscr{G}$, and therefore $X \cup y_i \in \mathscr{G}$ and $\mathscr{G}$ is the set of independent sets of a matroid. ■

NOTES ON CHAPTER 7

For a much fuller treatment of the material of this chapter see the book of Mirsky [71]. The basic Theorem 2.1 was proved by Welsh [71] but as pointed out there it leans heavily on the very elegant proof by Rado [67] of Hall's theorem. The first to pinpoint the significance of Rado's theorem were Mirsky and Perfect [67]. Previously many of the non-matroid theorems of this chapter had been proved by methods of linear programming or network flows—see for example Hoffman and Kuhn [63] and chapter 2 of the book by Ford and Fulkerson [62]. An interesting simple proof of Hall's theorem using the methods of linear algebra is given by Edmonds [67], though the first paper to use linear algebra techniques in transversal theory seems to be that of Perfect [66] which gives a simple linear algebra proof of a hitherto complicated theorem of Mendolsohn and Dulmage [58]. Many applications of Rado's theorem are given in Mirsky [71]. Section 5 is based on Welsh [69b], again the "non-matroid" results can be obtained by direct methods—see for example Mirsky [71]. Theorem 6.1, the converse of Rado's theorem is due to Rado [42], see also Ulltang [72] for related results. For an extremely interesting account of matching theory interpreted in its widest sense we refer to the paper of Harper and Rota [71]. For other extensions of these ideas in particular to infinite structures we refer to the papers of Brualdi [69b], [71f], [71g], Mirsky [71] and Damerell and Milner [75].

In Chapter 13 we characterize the dual of transversal matroids. Chapter 14 contains conditions for a matroid to be transversal and further properties of transversal matroids and their presentations.

CHAPTER 8

Covering and Packing

1. SUBMODULAR FUNCTIONS

In this chapter we will show how matroid theory has led to the discovery of many new results of a packing or covering type, and has also made very simple the proof of many earlier apparently unrelated results in combinatorics.

Recall that a function $\mu: 2^S \to \mathbb{R}$ is *submodular* if for any subsets A, B of S

$$\mu A + \mu B \geqslant \mu(A \cup B) + \mu(A \cap B).$$

The theory of matroids is closely connected to the theory of submodular set functions—the rank function of a matroid is submodular. In this chapter we aim to show a reverse relationship, namely how to generate matroids from arbitrary submodular set functions.

We start with a straightforward result, which seems to be implicit in the work of Dilworth [44] who first noticed this connection with submodular functions.

THEOREM 1. *Let* $\mu: 2^S \to \mathbb{Z}^+$ *be submodular, and nondecreasing (that is* $A \subseteq B \Rightarrow \mu A \leqslant \mu B$*), with the further property that* $\mu X \leqslant |X| \, \forall X \in 2^S$*. Then* $\mathscr{I}(\mu) = \{A : A \subseteq S, \ \mu A = |A|\}$ *is the collection of independent sets of a matroid on S. Moreover this matroid has rank function* μ*.*

This theorem is essentially the characterization of a matroid by its submodular rank function (see the set of axioms ((R1′)–(R3′)). The proof is straightforward and is left to the reader.

Let $\mathscr{C}$ be a *lattice of subsets* of S which is ordered by inclusion, is closed under intersection, and contains $\varnothing$, S. Thus if $A, B \in \mathscr{C}$ then $A \wedge B = A \cap B$, and, $A \vee B \supseteq A \cup B$.

We say that a function $\mu: \mathscr{C} \to \mathbb{R}$ is *submodular* if for $A, B \in \mathscr{C}$

$$\mu A + \mu B \geqslant \mu(A \vee B) + \mu(A \wedge B).$$

116

Define

$$\mathscr{I}(\mathscr{C};\mu) = \{X : X \subseteq S, \mu C \geqslant |X \cap C| \, \forall C \in \mathscr{C}\}.$$

THEOREM 2. *If μ is a non-negative integer-valued submodular function with $\mu(\varnothing) = 0$, $\mathscr{I}(\mathscr{C}, \mu)$ is the collection of independent sets of a matroid on S with rank function ρ given for $X \subseteq S$ by*

$$\rho X = \inf_{C \in \mathscr{C}} (\mu C + |X \backslash C|).$$

We denote this matroid by $M(\mathscr{C}, \mu)$ and call it the matroid *induced* by $\mathscr{C}$ and μ. When $\mathscr{C} = 2^S$ we abbreviate $M(\mathscr{C}, \mu)$ to $M(\mu)$.

Proof. Define a function υ on 2^S by, for $X \subseteq S$,

$$\upsilon X = \inf_{C \in \mathscr{C}} (\mu C + |X \backslash C|).$$

It is straightforward to check that υ is nondecreasing. We assert that υ is is submodular. For if X, Y, A, B be subsets of S then it can be checked that

$$|X \backslash A| + |Y \backslash B| \geqslant |(X \cup Y) \backslash (A \cup B)| + |(X \cap Y) \backslash (A \cap B)|,$$

and hence

$$\begin{aligned}
\upsilon X + \upsilon Y &= \inf_{A, B \in \mathscr{C}} \{\mu A + |X \backslash A| + \mu B + |Y \backslash B|\} \\
&\geqslant \inf_{A, B \in \mathscr{C}} \{\mu(A \vee B) + \mu(A \wedge B) + |(X \cup Y) \backslash (A \cup B)| \\
&\quad + |(X \cap Y) \backslash (A \cap B)|\} \\
&\geqslant \inf_{A, B \in \mathscr{C}} \{\mu(A \vee B) + \mu(A \wedge B) + |(X \cup Y) \backslash (A \vee B)| \\
&\quad + |(X \cap Y) \backslash (A \wedge B)|\}.
\end{aligned}$$

The right-hand side above is not less than $\upsilon(X \cup Y) + \upsilon(X \cap Y)$, and hence υ is a submodular set function.

Now clearly $\upsilon X \leqslant \mu(\varnothing) + |X| = |X|$ and since

$$\begin{aligned}
X \in \mathscr{I}(M(\mathscr{C}; \mu)) &\Leftrightarrow \mu C \geqslant |X \cap C| \qquad \forall C \in \mathscr{C} \\
&\Leftrightarrow \mu C + |X \backslash C| \geqslant |X| \qquad \forall C \in \mathscr{C} \\
&\Leftrightarrow \upsilon X \geqslant |X|,
\end{aligned}$$

we have by Theorem 1 that $M(\mathscr{C}, \mu)$ is a matroid with rank function υ. ∎

As corollaries of this main theorem it is easy to prove

COROLLARY 1. *If $\mu : 2^S \to \mathbb{Z}^+$ is non-decreasing, submodular and $\mu(\varnothing) = 0$*

then $\{A: \mu C \geqslant |C| \, \forall C \subseteq A\}$ is the collection of independent sets of a matroid $M(\mu)$ on S. The rank function of $M(\mu)$ is given for $X \subseteq S$ by

$$\rho X + \inf_{C \subseteq X} (\mu C + |X \backslash C|).$$

This last result was a theorem announced by Edmonds and Rota [66].

COROLLARY 2. If $\mu: 2^S \to \mathbb{Z}^+$ is non-decreasing and submodular then $M(\mu)$ is still a matroid on S but with rank function given for $X \subseteq S$ by

$$\rho X = \min(|X|, \inf_{C \subseteq X} (\mu C + |X \backslash C|)).$$

Negative submodular functions

Suppose now that we try to generate the cycle matroid $M(G)$ of a graph G by some simple submodular function μ. The μ which springs to mind is the following:

$$\mu A = |V(A)| - 1 \qquad A \subseteq E(G)$$

where $V(A)$ is the set of vertices in the subgraph of G generated by the set of edges A. Now μ is a perfectly respectable increasing submodular function and

$$\mathcal{I}(M'(\mu)) = \{X: \mu A \geqslant |A| \, \forall A \subseteq X, A \neq \varnothing\}$$

is in fact $\mathcal{I}(M(G))$. Thus even though μ takes a negative value on the empty set we get a matroid from it. This is a special case of the following theorem.

THEOREM 3. *Let $\mathscr{C}$ be a collection of subsets of S closed under unions and intersections and let $\mu: \mathscr{C} \to \mathbb{Z}$ be submodular and non-negative on $\mathscr{C} \backslash \{\varnothing\}$. Then*

$$\mathcal{I}''(\mathscr{C}; \mu) = \{X: \mu Y \geqslant |X \cap Y|, \qquad Y \in \mathscr{C} \backslash \{\varnothing\}\}$$

is the set of independent sets of a matroid on S.

Moreover its rank function ρ_μ is given for $A \subseteq S$ by

$$\rho_\mu A = \min\left(|A|, \sum_{i=1}^{k} \mu(X_i) + |A \cap (S \backslash \bigcup X_i)|\right),$$

where the minimum is taken over all collections $\{X_1, \ldots, X_k\}$ of disjoint sets in $\mathscr{C} \backslash \{\varnothing\}$.

When $\mathscr{C} = 2^S$ and μ is increasing the above theorem becomes a theorem announced by Edmonds and Rota [66]. For a proof see Edmonds [70].

EXERCISES 1

1. If $\mathscr{C}$ is a collection of subsets of S closed under unions and intersections is it a distributive lattice under the inclusion ordering?

2. Let μ_1, μ_2 be non-negative increasing functions: $2^S \to \mathbb{R}$ which are submodular. Prove that

   (a) $\lambda: 2^S \to R$ defined by
   $$\lambda A = \mu_1(A) + \mu_2(S\backslash A)$$

   is submodular.

   (b) $\gamma: 2^S \to R$ defined by
   $$\gamma A = \min_{X \subseteq A} (\mu_1 X + \mu_2(A\backslash X))$$

   is not in general submodular.

   (c) $\delta: 2^S \to R$ defined by
   $$\delta A = \max_{X \subseteq A} (\mu_1(X) + \mu_2(A\backslash X))$$

   is submodular when the μ_i are the rank functions of matroids.

$^{\circ}$3. Is δ submodular when the μ_i are not the rank functions of matroids (see Pym and Perfect [70])?

2. FUNCTIONS OF MATROIDS; INDUCING MATROIDS BY BIPARTITE GRAPHS

We now apply the theory of submodular functions developed in the last section to obtain two very useful constructions of matroid theory.

Let $\mathscr{A} = (A_i : i \in I)$ be a finite family of subsets of S and let M be a matroid on S with rank function ρ. Then define $\mu: 2^I \to \mathbb{Z}^+$ by

$$\mu J = \rho(A(J)) \qquad (J \subseteq I).$$

It is easy to check that μ satisfies the conditions of Corollary 1.1 and hence

$$\mathscr{I} = \{J : (A(K)) \geqslant |K| \forall K \subseteq J\}$$

is the collection of independent sets of a matroid on I. But by Rado's theorem $\mathscr{I}$ is exactly the collection of sets $J \subseteq I$, such that $(A_i : i \in J)$ has a transversal which is independent in M. Thus we have proved the following extension of Theorem 7.3.2.

THEOREM 1. *Let $\mathscr{A} = (A_i : i \in I)$ be a finite family of subsets of S and let M be a matroid on S. Then the collection $Q(M, \mathscr{A})$ of subsets J of I with the property that $(A_i : i \in J)$ has a transversal which is independent in M is the set of independent sets of a matroid on I.*

When M is the free matroid in which everything is independent $Q(M, \mathscr{A})$ is the matroid $Q(\mathscr{A})$ of Theorem 7.3.2 and hence using point set duality a corollary of Theorem 1 is the Edmonds–Fulkerson result Theorem 7.3.1.

COROLLARY 1. *The collection of partial transversals of a finite family of sets is the set of independent sets of a matroid.*

Now let Δ denote a directed bipartite graph with vertex set $S \cup T$ and edges e_{ij} joining $s_i \in S$ to $t_j \in T$. If $X \subseteq S$ and $Y \subseteq T$ we say $X = \{x_1, \ldots, x_k\}$ can be *matched into* Y in Δ if there exists a set of edges joining each x_i to a distinct member of Y. In other words X is matched into Y in Δ if the subgraph determined by $X \cup Y$ has a matching which covers every vertex of X.

Then we can restate Theorem 1 in the following form.

If M is a matroid on T with $\mathscr{I}$ as its collection of independent sets let

$$\Delta(\mathscr{I}) = \{X : X \subseteq S, X \text{ can be matched in } \Delta \text{ into a member of } \mathscr{I}\}.$$

THEOREM 2. *The collection $\Delta(\mathscr{I})$ is the set of independent sets of a matroid on S.*

We call this matroid the matroid *induced* from M by Δ, and denote it by $\Delta(M)$. Its rank function $\tilde{\rho}$ is related to the rank function ρ of M by

$$(2) \qquad \tilde{\rho}A = \min_{X \subseteq A} \left((\rho(\delta X)) + |A \backslash X| \right) \quad (A \subseteq S)$$

where for any set U of vertices of Δ, δU is the set of vertices which are joined to some member of U by an edge.

If S, T are sets and $f : S \to T$, for any collection $\mathscr{U}$ of subsets of S we define $f(\mathscr{U})$ to be the collection of sets $\{f(X); X \in \mathscr{U}\}$.

COROLLARY 2. *If M is a matroid on S and $f : S \to T$ then if $\mathscr{I}$ is the collection of independent sets of M, $f(\mathscr{I})$ is the collection of independent sets of a matroid $f(M)$ on T. When f is a surjection the rank function ρ_f of $f(M)$ is given by*

$$\rho_f A = \min_{Y \subseteq A} \left(\rho f^{-1}(Y) + |A \backslash Y| \right) \qquad (A \subseteq T),$$

where ρ is the rank function of M.

Proof. Choose Δ in the obvious way to be the bipartite graph $(S, T; E)$ where $e \in E$ joins s to t if and only if $f(s) = t$. Applying Theorem 2 gives the required result. ∎

Note. For any directed bipartite graph Δ and any matroid M, $\Delta(M)$ has rank at most equal to $\rho(M)$. There seems to be no known "combinatorial type" operation ψ which is "dual" to inducement in the sense that

$$M_1 = \psi(M_2) \Leftrightarrow M_2 = \Delta(M_1)$$

for some bipartite graph Δ, however see Perfect [73].

EXERCISES 2

1. Show by example that if M is the cycle (cocycle) matroid of a graph $\Delta(M)$ need not be the cycle (cocycle) matroid of a graph.

2. Find a matroid M such that for any uniform matroid U there does not exist a directed bipartite graph Δ such that $M \simeq \Delta(U)$.

3. If M is a matroid on S and $f: S \to T$ then for $A \subseteq S$ which of the following statements are true?

   (a) $f(M \mid A) \simeq f(M) \mid f(A)$.

   (b) $f(M \cdot A) \simeq f(M) \cdot f(A)$.

   (c) $f(M^*) \simeq (f(M))^*$.

4. If M_1, M_2 are two matroids of rank r there exists a bipartite graph Δ such that $M_1 \simeq \Delta(M_2)$ and $M_2 \simeq \bar{\Delta}(M_1)$ if and only if between them M_1 and M_2 have at least r coloops. ($\bar{\Delta}$ is the directed bipartite graph obtained from Δ by reversing the orientation of the edges.) (Perfect [73])

3. THE UNION OF MATROIDS

Because of its simplicity, the operation of taking the direct sum of two matroids has very few combinatorial applications. A more interesting construction is the following:

THEOREM 1. *Let* $M_1, M_2, \ldots, M_m$ *be matroids on S. Let*

(1) $\mathscr{I} = \{X : X = X_1 \cup X_2 \cup \ldots \cup X_m; X_i \in \mathscr{I}(M_i)(1 \leqslant i \leqslant m)\}.$

Then $\mathscr{I}$ *is the collection of independent sets of a matroid on S whose rank function* ρ *is given by*

(2) $\rho A = \min_{X \subseteq A}\left(\sum_{i=1}^{m} \rho_i X + |A \backslash X| \right) \qquad (A \subseteq S),$

where ρ_i *is the rank function of* M_i $(1 \leqslant i \leqslant m)$.

We call this matroid the *union* of $M_1, \ldots, M_m$ and denote it by $M_1 \vee \ldots \vee M_m$ or $\bigvee_{i=1}^{m} M_i$.

Proof of Theorem 1. Let $I = \{1, \ldots, m\}$ and let $\{S_i : i \in I\}$ be disjoint sets each of the same cardinal as S. More explicitly let

$$S = \{1, 2, \ldots, n\}$$
$$S_i = \{1^i, 2^i, \ldots, n^i\} \qquad (i \in I).$$

Let M^i on S^i be a matroid isomorphic to S under the map $x \to x^i$. Then the direct sum $M^1 + \ldots + M^m$ on $\bigcup_{i \in I} S_i$ has rank function $\rho^1 + \rho^2 + \ldots + \rho^m$ in an obvious notation. Now consider the function $f : \bigcup_{i \in I} S_i \to S$ defined by $f(x^i) = x \, (1 \leqslant x \leqslant n)$. The image of $M^1 + \ldots M^m$ under f is the matroid $\bigvee_{i \in I} M_i$ and hence by Corollary 2.2 it is a matroid with rank function given by (2). ∎

Since the rank functions ρ_1, ρ_2 of M_1, M_2 respectively are non-decreasing and submodular, so is $\rho_1 + \rho_2$. Hence by Theorem 1.2 the matroid $M(\rho_1 + \rho_2)$ has rank function ρ_{12} given by

$$\rho_{12} A = \min_{X \subseteq A} (\rho_1 X + \rho_2 X + |A \backslash X|).$$

But this is exactly the rank function of $M_1 \vee M_2$. Hence since $M_i = M(\rho_i)$ we have

$$M(\rho_1) \vee M(\rho_2) = M(\rho_1 + \rho_2).$$

In fact this is just a special case of the more general result of Pym and Perfect [70] which we state without proof.

THEOREM 2. *If μ_1, μ_2 are non-decreasing, integer valued, non-negative submodular functions on 2^S, then*

$$M(\mu_1 + \mu_2) = M(\mu_1) \vee M(\mu_2).$$

However it is interesting that if μ_1, μ_2 are not non-decreasing this relation need not hold.

Example (C. J. H. McDiarmid). Take $S = \{a, b\}$ and let μ_1, μ_2 be defined as in the following table.

	μ_1	μ_2	$\mu_1 + \mu_2$
$\{a, b\}$	0	1	1
$\{a\}$	0	1	1
$\{b\}$	1	0	1
$\varnothing$	0	0	0

Then μ_1 and μ_2 are submodular, and

$$\mathscr{I}(M(\mu_1)) = \{\varnothing\},$$

$$\mathscr{I}(M(\mu_2)) = \{\varnothing, \{a\}\}.$$

Hence

$$\mathscr{I}(M(\mu_1) \vee M(\mu_2)) = \{\varnothing\{a\}\},$$

whereas

$$\mathscr{I}(M(\mu_1 + \mu_2)) = \{\varnothing, \{a\}, \{b\}\}.$$

The intersection of matroids

If $\mathscr{B}_1$ and $\mathscr{B}_2$ denote respectively the sets of bases of M_1 and M_2 on S then the bases of $M_1 \vee M_2$ are the maximal members of the family $(B_1 \cup B_2 : B_1 \in \mathscr{B}_1, B_2 \in \mathscr{B}_2)$. Since a base is a minimal spanning set of a matroid we can define the intersection of two matroids as follows.

Let $\mathscr{S}_1$ and $\mathscr{S}_2$ denote the collection of spanning sets of M_1 and M_2 respectively. Let

$$\mathscr{S}_1 \wedge \mathscr{S}_2 = \{X_1 \cap X_2 : X_1 \in \mathscr{S}_1, X_2 \in \mathscr{S}_2\}.$$

Then define the *intersection* $M_1 \wedge M_2$ of the matroids M_1, M_2 to be the matroid on S whose spanning sets are the collection $\mathscr{S}_1 \wedge \mathscr{S}_2$. It is a routine exercise in duality to check that $M_1 \wedge M_2$ is a matroid. The easiest proof is to show directly the stronger result.

THEOREM 3. *The intersection $M_1 \wedge M_2$ is a matroid on S and*

$$M_1 \wedge M_2 = (M_1^* \vee M_2^*)^*.$$

In fact the operations $\wedge, \vee$ commute in a nice algebraic way with the contraction and deletion operations.

It is easy to check that if M_1, M_2 are matroids on S and $T \subseteq S$ then

(5) $(M_1 \vee M_2) | T = (M_1 | T) \vee (M_2 | T).$

Then using Theorem 3 and duality we get

$$(M_1 . T) \wedge (M_2 . T) = ((M_1 . T)^* \vee (M_2 . T)^*)^*$$
$$= ((M_1^* | T) \vee (M_2^* | T))^*$$
$$= ((M_1^* \vee M_2^*) | T)^*$$
$$= (M_1^* \vee M_2^*)^* . T$$
$$= (M_1 \wedge M_2) . T.$$

EXERCISES 3

1. Prove that a transversal matroid is the union of matroids of rank 1.

2. Prove that if $M_1 \vee M_2$ has no coloops then a base of $M_1 \vee M_2$ is the union of disjoint bases of M_1 and M_2.

3. Prove that if M_1, M_2 are transversal matroids then so is $M_1 \vee M_2$, but that $M_1 \wedge M_2$ is not in general.

4. Prove that the rank function ρ of $M_1 \vee M_2$ is given in terms of ρ_1, ρ_2 the rank functions of M_1, M_2 respectively by, for all $X \subseteq S$,

$$\rho X = \max_{Y \subseteq X} (\rho_1(Y) + \rho_2(X \backslash Y))$$

(Pym and Perfect [70])

5. If M, M_1, M_2 are matroids on S and $f : S \to T$ prove
   (a) $f(M_1 \vee M_2) = f(M_1) \vee f(M_2)$
   (b) M transversal $\Rightarrow f(M)$ transversal.

(A. P. Heron)

6. Let X, Y be independent sets of a matroid M and let $X_1 \cup \ldots \cup X_k$ be a partition of X. Then there exists a partition $Y_1 \cup \ldots \cup Y_k$ of Y such that for $1 \leqslant i \leqslant k$,

$$X_i \cap Y_i = \varnothing, \qquad X_i \cup Y_i \in \mathscr{I}(M).$$

(Woodall [74], McDiarmid [75a])

7. Call a matroid *irreducible* if it cannot be expressed as the union of non-trivial matroids. Prove that $M(K_4)$ and $M(\text{Fano})$ are irreducible matroids.

8. Prove that the cycle matroid of a graph is irreducible if and only if the removal of any edge leaves the graph 2-connected.

(Lovász and Recski [73])

○9. Characterize irreducible matroids.

4. COVERING AND PACKING THEOREMS

Let M be a matroid on S with rank function ρ. Then M has k pairwise disjoint bases if and only if the union of M with itself k times, (denoted by $M^{(k)}$) has rank at least $k\rho(S)$. But by Theorem 3.1 the rank of $M^{(k)}$ is $k\rho(S)$ if and only if for all $A \subseteq S$,

$$(1) \qquad\qquad k\rho A + |S \backslash A| \geqslant k\rho S.$$

Thus we have proved the first "packing" theorem of Edmonds [65a].

THEOREM 1. *A matroid M on S has k disjoint bases if and only if* (1) *holds for each subset A of S.*

Also, M will have k bases whose union covers S if and only if the rank of $M^{(k)}$ is $|S|$.

In this way we get the second "covering" theorem of Edmonds [65a].

THEOREM 2. *A matroid on S with rank function ρ has k independent sets whose union is S if and only if*

$$(2) \qquad\qquad k\rho A \geqslant |A| \qquad (A \subseteq S).$$

Note that as with many theorems of this type (e.g. the Hall–Rado theorems) the necessity of the conditions (1) and (2) is obvious. However, the apparent simplicity with which they are now deduced is in some ways misleading. The original proofs by Edmonds in 1965 were intricate. From Theorems 1 and 2 we can deduce other hitherto very difficult results in combinatorial theory.

An algebraic theorem of Horn.

When S is a subset of a vector space V and M is the matroid induced on S by linear independence in V we see that Theorem 2 gives the algebraic theorem of Horn [55] who found that conditions (2) were necessary and sufficient for a vector space to have k bases with the desired covering property. It is interesting that his proof seems incapable of generalization to arbitrary matroids, and is in fact more difficult than the existing proofs of the more general theorem.

Two graphical theorems of Tutte and Nash-Williams.

Another application of Theorems 1 and 2 is to deduce the following theorems about graphs first proved by ad-hoc, and rather intricate methods by Tutte [61] and Nash-Williams [61], [64].

The problem considered was to find necessary and sufficient conditions for a graph to have k edge-disjoint spanning trees. The solution is as follows.

Let $V(G)$, $E(G)$ denote respectively the vertex and edge sets of G. For a partition $\mathbb{P}$ of the vertex set $V(G)$ let $E(\mathbb{P}; G)$ denote the number of edges of G which join two distinct classes of the partition $\mathbb{P}$.

THEOREM 3. *A simple connected graph G has k edge disjoint spanning trees if and only if for any partition $\mathbb{P}$ of $V(G)$*

(3) $$E(\mathbb{P}; G) \geqslant k(|\mathbb{P}| - 1),$$

where $|\mathbb{P}|$ denotes the number of distinct classes of the partition $\mathbb{P}$.

Proof. Suppose there exist edge disjoint spanning trees $T_1, \ldots, T_k$. Then if $E(\mathbb{P}; T_i)$ denotes the number of edges of T_i which join distinct classes of $\mathbb{P}$

$$E(\mathbb{P}; T_i) \geqslant |\mathbb{P}| - 1.$$

Hence

$$E(\mathbb{P}; G) \geqslant \sum_{i=1}^{k} E(\mathbb{P}; T_i) \geqslant k(|\mathbb{P}| - 1).$$

Conversely suppose that (3) holds, then if it is clearly sufficient to show that the cycle matroid $M(G)$ has k pairwise disjoint bases. By (1) this is so if and only if for all $A \subseteq E = E(G)$.

(4) $$k\rho A + |E \backslash A| \geqslant k\rho E$$

where ρ is the rank function of $M(G)$.

Suppose $\exists A \subseteq E(G)$ with

$$|E \backslash A| < k(\rho E - \rho A).$$

Let $k(A)$ be the number of connected components in the subgraph $G | A$. Let $D_1, D_2, \ldots, D_{k(A)}$ denote the vertex sets of these connected components.

If there are t vertices $v_1, v_2, \ldots, v_t$ of G which are not incident with any edge of A take $\mathbb{D}$ to be the partition

$$\mathbb{D} = (D_1, D_2, \ldots, D_{k(A)}, \{v_1\}, \ldots, \{v_t\})$$

so that

$$|\mathbb{D}| = k(A) + t$$

$$E(\mathbb{D}; G) = |E \backslash A|.$$

Then since

$$\rho E = |V(G)| - 1, \qquad \rho A = |V(G)| - t - k(A),$$

we have

$$E(\mathbb{D}; G) < k(|V(G)| - 1 - |V(G)| + t + k(A))$$
$$= k(|\mathbb{D}| - 1)$$

which contradicts (4) and hence we have proved the theorem. ∎

THEOREM 4. *If G is a simple graph its edges can be coloured with k colours in such a way that no cycle has all its edges the same colour if and only if for each set A of vertices the set E(A) of edges with both endpoints in A satisfies*

$$(5) \qquad\qquad |E(A)| \leqslant k(|A| - 1).$$

Proof. Suppose that the edges are so coloured with k colours $1, 2, \ldots, k$. Let $E_i(A)$ be the set of edges of colour i which have both endpoints in A. Since these edges must not contain a cycle

$$|E_i(A)| \leqslant |A| - 1$$

so that

$$|E(A)| \leqslant \sum_{i=1}^{k} E_i(A) \leqslant k(|A| - 1).$$

Conversely if (5) holds for all $A \subseteq V(G)$ it is easy to see that it implies that for any set X of edges of G,

$$k\rho X \geqslant |X|$$

where ρ is the rank function of the cycle matroid of G. Thus by (1) it is possible to partition $E(G)$ into k disjoint forests, which in turn says that a k-colouring exists in which no cycle has all its edges the same colour. ∎

Existence of zero-one matrices and digraphs with prescribed vertex degrees.

Gale [57] and Ryser [57] first considered the following existence theorem for zero-one matrices. Given two sequences of non-negative integers $(r_1, \ldots, r_m)$ and $(c_1, \ldots, c_n)$ does there exist an $m \times n$ matrix, all of whose entries are either zero or one, such that the ith row sum is r_i and the jth column sum is c_j?

We will use Theorem 3.1 to prove

THEOREM 5. *If $(r_1, r_2, \ldots, r_m)$ and $(c_1, \ldots, c_n)$ are two sequences of non-negative integers there exists an $m \times n$, zero-one matrix having r_i as its ith row sum and c_j as its jth column sum if and only if*

(a) $\displaystyle\sum_{k=1}^{m} \min(|J|, r_k) \geqslant \sum_{i \in J} c_i, \qquad (J \subseteq \{1, \ldots, n\}),$

(b) $\displaystyle\sum_{i=1}^{m} r_i = \sum_{i=1}^{n} c_i.$

Proof. Let $A_i (1 \leqslant i \leqslant n)$ be disjoint sets each containing m elements as follows,

$$A_i = \{x_{i1}, x_{i2}, \ldots, x_{im}\} \qquad (1 \leqslant i \leqslant n).$$

Let $S = \bigcup(A_i : 1 \leqslant i \leqslant n)$. Let B_j $(1 \leqslant j \leqslant m)$, be the disjoint subsets of S defined by

$$B_j = \{x_{1j}, x_{2j}, \ldots, x_{nj}\}.$$

Let M_j be the matroid defined on S which has as its independent sets all subsets of B_j which have cardinality $\leqslant r_j$. Then a zero-one matrix with the desired row and column sums exists, if and only if $\mathscr{A}$ has a $(c_1, c_2, \ldots, c_n)$-transversal which is independent in the matroid $M_1 \vee M_2 \vee \ldots \vee M_m$. By Theorem 7.5.2 and Theorem 3.1 this is true if and only if, letting ρ_i be the rank function of M_i

(6) $$\rho_1(Y) + \ldots + \rho_m(Y) + |A(J) \backslash Y| \geqslant \sum_{i \in J} c_i,$$

for all subsets J of $\{1, 2, \ldots, n\}$, and all subsets Y of $A(J)$. It is not difficult to see that the left hand side of (6) takes its minimum values for a fixed J when $Y = A(J)$ and that then

$$\rho_k(A(J)) = \min(|J|, r_k) \qquad (1 \leqslant k \leqslant m). \qquad \blacksquare$$

A term rank identity.

If $\mathscr{A} = (A_i : i \in I)$ is a collection of subsets of S, the *term rank function* of $\mathscr{A}$ is the function v on I in which $v(J)$ is the maximum number of sets in the collection $(A_i : i \in J)$ which have a transversal. Using Theorem 2 we get the following result of Edmonds and Fulkerson [65].

THEOREM 6. *If $(A_i : i \in I)$ is a finite family of subsets of S, then I can be partitioned into as few as k subsets $I_1, \ldots, I_k$, such that for each j $(1 \leqslant j \leqslant k)$ the collection $(A_i : i \in I_j)$ has a transversal, if and only if for all $J \subseteq I$*

$$kv(J) \geqslant |J|.$$

Proof. By Theorem 7.3.2, v is just the rank function of the matroid $Q(\mathscr{A})$ on I in which J is independent if $(A_i : i \in J)$ has a transversal, and then applying Theorem 2 to $Q(\mathscr{A})$ we get the result. $\qquad \blacksquare$

THEOREM 7. *For any finite family of subsets $\mathscr{A} = (A_i : i \in I)$, and any non-negative integer k the following statements are equivalent.*

(i) $kv(J) \geqslant |J|$ *for all $J \subseteq I$.*
(ii) $k|A(J)| \geqslant |J|$ *for all $J \subseteq I$.*

(iii) $\mathscr{A}$ *has a system of representatives in which no element occurs more than*
       k times†.
(iv)   *The index set I can be partitioned into as few as k subsets* $I_1, \ldots, I_k$ *such*
       that for each j($1 \leqslant j \leqslant k$) the collection of sets $(A_i : i \in I_j)$ *has a transversal.*

Note†. The number of occurrences of an element x in the family $(x_i : i \in I)$
is defined as the cardinal of the set $\{i \in I : x_i = x\}$.

Theorem 7, when $k = 1$ is just Hall's theorem. The statement for arbitrary k
is due to Edmonds and Fulkerson [65].

Proof. It is trivial to see that (iii) and (iv) are equivalent. By Theorem 7.5.3
(iii) is equivalent to (ii) and by Theorem 6 (i) is equivalent to (iv). ∎

EXERCISES 4

1. Prove that a necessary and sufficient condition for a graph G to be the union of k
   forests is that for every non-empty subset X of $V(G)$, at most $k(|X| - 1)$ edges of
   G have both endpoints in X. (This was the original version of Nash-Williams'
   theorem [61].)
2. The *arboricity* of a simple graph G, denoted by $\gamma(G)$, is the minimum number of
   forests whose union is G. Thus Nash-Williams' theorem can be restated as follows.
   Let $q_n(G)$ be the maximum number of edges in any subgraph of G with n vertices.
   Then

$$\gamma(G) = \max_n \{q_n(G)/(n - 1)\},$$

   where $\{x\}$ denotes the least integer greater than or equal to x.
3. Prove that

$$\gamma(K_n) = \left\{\frac{n}{2}\right\}, \qquad \gamma(K_{m,n}) = \left\{\frac{mn}{m + n - 1}\right\}.$$

4. Prove that if $J_k(1 \leqslant k \leqslant m)$ are pairwise disjoint independent subsets of M there
   exist pairwise disjoint bases $B_k(1 \leqslant k \leqslant m)$ of M such that $J_k \subseteq B_k$ if and only
   if for all $A \subseteq S'$, $S' = S \setminus \bigcup(J_k : 1 \leqslant k \leqslant m)$,

$$|A| \geqslant \sum_k [\rho(S) - \rho((S' \setminus A) \cup J_k)].$$

5. Let M_1, M_2 be matroids on S with closure functions σ_1, σ_2 respectively. Prove that
   if X, Y are subsets of S which are independent in both M_1 and M_2 there exists
   $Z \subseteq X \cup Y$ such that Z is independent in both M_1 and M_2 and

$$\sigma_1 X \subseteq \sigma_1 Z, \qquad \sigma_2 Y \subseteq \sigma_2 Z.$$

(Kundu and Lawler [73])

5. EDMONDS' INTERSECTION THEOREM

From the rank function of the union of two matroids, it is easy to deduce the following theorem of Edmonds [70].

THEOREM 1. *Let M_1, M_2 be matroids on S with rank functions ρ_1, ρ_2 respectively. Then M_1, M_2 have a common independent set of cardinal k if and only if for all $A \subseteq S$*

$$\rho_1 A + \rho_2(S \backslash A) \geqslant k.$$

Proof. Suppose X is independent in both M_1 and M_2. Then $S \backslash X$ must contain a base of M_2^* and thus the rank of $M_1 \vee M_2^*$ cannot be less than $|X| + \rho_2^*(S)$. Conversely if $\rho(M_1 \vee M_2^*) \geqslant k + \rho_2^*(S)$, then since any base B^* of M_2^* is independent in $M_1 \vee M_2^*$, there must exist a subset X of $S \backslash B^*$ which is independent in M_1 and has cardinality not less than k. Thus M_1 and M_2 have a common independent set of cardinality k if and only if

$$\rho(M_1 \vee M_2^*) \geqslant k + \rho_2^*(S)$$

which is equivalent to the statement that for any $A \subseteq S$,

$$\rho_1(A) + \rho_2^*(A) + |S \backslash A| \geqslant k + \rho_2^*(S),$$

and by (2.1.5) this reduces to

$$\rho_1(A) + \rho_2(S \backslash A) \geqslant k,$$

and completes the proof. ∎

In the case of three or more matroids the corresponding problem is unsolved. Consider matroids M_i ($i = 1, 2, 3$) with rank functions ρ_i. It is natural to conjecture that the clearly necessary condition

(1) $$\rho_1(A) + \rho_2(B) + \rho_3(C) \geqslant k$$

for all partitions of S into disjoint sets A, B, C is also sufficient. This is not so.

Example. Take $S = \{1, 2, 3\}$ and let M_i be defined to have the following sets as bases

$$M_1 : \{1, 2\}, \{1, 3\}$$
$$M_2 : \{2, 3\}, \{2, 1\}$$
$$M_3 : \{3, 1\}, \{3, 2\}.$$

Then (1) is satisfied with $k = 2$, but M_1, M_2, M_3 have no common independent set of cardinal 2.

It is clear by taking M_i to be transversal matroids, that the problem here is a generalization of the unsolved problem discussed in Section 7.4 of finding necessary and sufficient conditions for three families of sets to have a common transversal. At the moment there is not even a conjectured set of possible necessary and sufficient conditions for this problem.

It is likely that a solution of this problem would be a major breakthrough in this area of combinatorics. The reason for this is that from it one can easily obtain a solution to the apparently intractable graph-problem, of deciding whether or not a digraph G has a path which passes through each vertex of G exactly once. Such a path is called a *Hamiltonian path* of the digraph.

The "Hamiltonian path problem" as a 3-matroid intersection problem

Let G be a digraph with vertex set $v_0, \ldots, v_{n+1}$. Let A_i be the set of edges having v_i as an endpoint, let B_i be the set of edges having v_i as an initial point. Then consider any set X of edges which satisfies

(a) X is a transversal of $(B_0, \ldots, B_n)$

(b) X is a transversal of $(A_1, \ldots, A_{n+1})$

(c) X is independent in the cycle matroid of the undirected graph $\hat{G}$ determined by G.

Then it is clear that X is a path from v_0 to v_{n+1} in G and X passes through each vertex of G exactly once; in other words it is a Hamiltonian path of G which starts at v_0 and ends at v_{n+1}. It is not difficult to see that the general problem of testing whether a graph has a Hamiltonian can be reduced to the special case of finding whether there is a Hamiltonian path with prescribed endpoints in a digraph. By the above argument this is now a problem of deciding whether the two transversal matroids $M[\mathscr{A}]$, $M[\mathscr{B}]$ have a common base which is also a base of the cycle matroid $M(\hat{G})$.

We now show how the intersection theorem leads to a proof of a theorem of Brualdi [70] which Mirsky [71] calls "a symmetrized version of Rado's theorem."

THEOREM 2. *Let Δ be a bipartite graph with edges joining the disjoint vertex sets S, T. Let M_1, M_2 be respectively matroids on S, T and let them have rank functions ρ_1 and ρ_2. Then there exists $X \subseteq S$ with $|X| = k$, such that*

(i) *X is independent in M_1,*

(ii) *X can be matched into an independent set Y of M_2,*
   if and only if for all $A \subseteq S$

(2) $\rho_1(S\backslash A) + \rho_2(\partial A) \geqslant k.$

(Note: Recall ∂A is the set of vertices of T which are endpoints of some edge whose other endpoint is in A.)

We use the terminology X is an *independent matching* of Δ to mean that X is the set of edges of a matching in Δ and the endpoints of X which belong to S form an independent set in M_1, and the endpoints of X which belong to T form an independent set of M_2.

Proof of Theorem 2. Let E be the edge set of G and let M_1', M_2' be matroids on E defined as follows: let X be independent in M_1' (M_2') if it is the set of edges of a matching in G and its endpoints in S (respectively T) form an independent set of M_1 (respectively M_2). It is easy to check that M_1', M_2' are matroids on E and that M_1, M_2 have an independent matching of cardinal k if and only if M_1' and M_2' have a common independent set of cardinal k. But by the intersection Theorem 1, this is so if and only if for all $X \subseteq E$

$$(3) \qquad \rho_1' X + \rho_2'(E \backslash X) \geqslant k$$

where ρ_i' is the rank function of M_i'. Now suppose that condition (2) holds but that (3) fails for some $X_0 \subseteq E$. Letting $V_S(X_0)$ be the set of endpoints of X_0 belonging to S, and $V_T(E \backslash X_0)$ those endpoints of $E \backslash X_0$ belonging to T we have

$$\rho_1(V_S(X_0)) + \rho_2(V_T(E \backslash X_0)) = \rho_1'(X_0) + \rho_2'(E \backslash X_0) < k.$$

But then

$$\rho_1(V_S(X_0)) + \rho_2(\partial(S \backslash V_S(X_0))) \leqslant \rho_1(V_S(X_0)) + \rho_2(V_T(E \backslash X_0)) < k$$

which contradicts (2) with $A = S \backslash V_S(X_0)$. The conditions (2) are clearly necessary and the proof is complete. ∎

We now show how the famous theorem of König [31] is a special case of Theorem 2. If A is a finite rectangular matrix, all of whose entries are either zero or one we say that two non-zero entries x, y are *unrelated* if they do not lie in the same row or column. The *scatter number* $s(A)$ of a matrix A is the maximum number of pairwise unrelated non-zero entries. The *cover number* $c(A)$ of a matrix A is the minimum number of *lines* (line = row or column) which contain all the non-zero elements of A.

THEOREM 3 (König). *For any zero-one matrix A the scatter number and the cover number are equal.*

For bipartite graphs König's theorem has the following equivalent interpretation:

THEOREM 3'. *If Δ is a bipartite graph the minimum number of vertices needed* to "*cover*" all the edges of Δ equals the maximum number of edges no two of which have a vertex in common.

Example. The circled entries of the matrix form a scattered set of elements of maximum cardinal, and correspond to the circled lines of the associated bipartite graph Δ. There exist several sets of four lines which cover A.

We close this section by showing how König's theorem follows easily from Brualdi's theorem.

Deduction of König's Theorem

Let Δ be a bipartite graph $Δ = (S, T; E)$. Let the maximum number of edges of Δ which have no vertex in common be k. Then clearly we need at least k vertices to cover all the edges. Suppose that we need $k + 1$ vertices to cover Δ. Now since the maximum number of disjoint edges is k, taking M_1 and M_2 on S, T respectively to be the free matroid in which every set is independent, we know by (2) that there exists $X_0 \subseteq S$ such that

$$|\partial X_0| + |S \backslash X_0| < k + 1.$$

But $(\partial X_0) \cup (S \backslash X_0)$ covers all the edges of Δ and we have found a cover of cardinal $< k + 1$. The contradiction proves the theorem. ■

EXERCISES 5

1. If M_1, M_2 are matroids of common rank on S and $M_i^{(d)}$ denotes the d-fold union of M_i with itself prove that a necessary condition that M_1 and M_2 have d pairwise disjoint common bases is that $M_1^{(d)}$ and $M_2^{(d)}$ have a common base of cardinality $d\rho(M_i)$. Give an example to show that this is not a sufficient condition.

○2. If M_1, M_2 are matroids on a set S find necessary and sufficient conditions that M_1, M_2 have d pairwise disjoint bases in common.

3. Prove that if $\mathscr{A} = (A_i : i \in I)$ and $\mathscr{B} = (B_i : i \in I)$ are two families of subsets of S and d is a positive integer then $\mathscr{A}$ and $\mathscr{B}$ have d pairwise disjoint common transversals if and only if for all $J, K \subseteq I$,

$$|A(J) \cap B(K)| \geqslant d(|J| + |K| - |I|)$$

(This has been proved by distinct methods by Fulkerson [71], Davies [71]).

○4. Given two matroids M_1, M_2 on S and positive integers k_1 and k_2 find necessary and sufficient conditions for there to exist a set X whose rank in M_1 is k_1 and whose rank in M_2 is k_2.

5. Prove that an algorithm for testing whether or not a digraph has a Hamiltonian path starting and ending at prescribed vertices can be modified to test whether or not a digraph has a Hamiltonian path.

NOTES ON CHAPTER 8

The first explicit derivation of matroids from non-negative, non-decreasing submodular, integer valued set functions that I know of is the abstract of Edmonds and Rota [66]. Theorem 1.2, the main theorem of this section was proved by McDiarmid [73]. For a survey of the relationship between submodular functions and matroids see Edmonds [70] and Pym and Perfect [70]. This last paper is particularly interesting as it also extends the ideas of this chapter to infinite sets.

The idea of a function of, and union of matroids, originates in the paper of Nash-Williams [66] where there is a particularly interesting direct proof of Corollary 2.2. The rank formula (3.2) for the union of matroids is implicit in the paper of Edmonds and Fulkerson [65], though as pointed out in Welsh [70] it can easily be deduced from the theorem of Rado [42].

Perfect [69] first noticed that matroids could be induced through bipartite graphs, though her original proof of Theorem 2.2 was different and longer than the one given here. She also extends these inducing ideas to infinite structures.

The proof of Brualdi's theorem from Edmonds' intersection theorem and a simple deduction of both via duality from Rado's theorem is taken from Welsh [70]. Another proof of Brualdi's theorem is given by Aigner and Dowling [71].

Mirsky [73] reviews various applications of the union operator, Bruno and Weinberg [71] use it to derive extensions of various graph theory results, Woodall [74] and McDiarmid [75a] use it to give simple proofs of an exchange theorem of Greene [73] (see exercises 3.6 and 1.5.3). For other more "applied" applications see Maurer [75], Bruno and Weinberg [72]. We return to these topics in Section 19.3.

The only progress on the irreducibility problem for matroids (see problem 3.8) are the papers by Lovász and Recski [73], Recski [75], in which they settle the problem for the cycle matroids of graphs (see exercise 3.9).

As mentioned earlier Dilworth's lattice embedding Theorem 3.4.2. was proved as long ago as 1941/42. The proof of this (see Crawley and Dilworth [73] Chapter 14) contains many of the ideas of submodular function theory developed in Section 1.

CHAPTER 9

The Vector Representation of Matroids

1. THE REPRESENTABILITY PROBLEM

In 1935 Whitney posed the following problem which is still unanswered. Under what conditions on a matroid M does there exist a field F and a vector space V over F so that M is isomorphic to some matroid induced on a subset of V by linear independence over F? There is some confusion in the literature surrounding this problem about the exact interpretation of representability so we shall give some explicit definitions.

Let F be a field. We say that the matroid M on S is *vectorial over* F if there exists a vector space V over F, a subset T of V, and a bijection $\phi: S \to T$ such that under ϕ, M is isomorphic to the matroid M' induced on T by linear independence in V.

In any isomorphism between two matroids M_1 and M_2 a loop of M_1 must be mapped into a loop of M_2 and conversely. Since the matroid M' on a subset T of a vector space V has at most one loop, namely the zero vector of V, we see that any matroid M with more than one loop cannot be vectorial.

If the field F is finite, similar anomalies can arise with respect to parallel elements, and the property of being vectorial is rather unnatural. For example it is not preserved under duality.

A more mathematically appealing concept is that of representability, defined as follows. We call a matroid M on S *representable over the field* F if there exists a vector space V over F and a map $\phi: S \to V$, which preserves rank.

Such a map ϕ is called a *representation* of M, and we describe a matroid M as *representable (vectorial)* if it is representable (vectorial) over some field.

The reader will observe that M on S is representable over F if and only if there exists a matrix A with entries belonging to F, such that if $C = \{c_1, \dots, c_n\}$ denotes the set of column vectors of A and $S = \{x_1, \dots, x_n\}$ the map $\phi: S \to C$ defined by $\phi(x_i) = c_i$ is a representation of M over F. Fulkerson [68] uses the term *matric matroid* for what we call a representable matroid.

136

It is clear from basic linear algebra that if M is representable over the field F then given any base $B = \{b_1, \ldots, b_r\}$ of M there exists a matrix $A = [a_{ij}]$ of the form

$$A = [I_r, D]$$

where I_r is the $r \times r$ unit matrix and D is an $r \times (|S| - r)$ matrix such that if $S = \{b_1, \ldots, b_r, b_{r+1}, \ldots, b_n\}$ then the map ϕ of S into the vector space V over F generated by the columns of A and such that $\phi(b_i)$ is the ith column of A, is a representation of M.

In other words M is isomorphic to the matroid induced on the columns of A by linear independence. We call a matrix A of the above form a *standard matrix representation* of M with respect to the base B.

The relationship between vectorial and representable matroids is summarized in the following remark.

Every vectorial matroid is representable. A representable matroid is vectorial if and only if M has a standard matrix representation A in which no two columns are identical.

Note. We should emphasize that in this chapter we have followed Whitney and confined our attention in the main to representations over (commutative) fields. The obvious extension to arbitrary division rings does not make much difference though there is an important class of "non-commutative" matroids which are representable over some division rings but not over any field.

EXERCISES 1

1. Prove that the uniform matroid $U_{2,4}$ is representable over every field except $GF(2)$.

2. Find the smallest uniform matroid which is not representable over $GF(3)$.

3. Prove that if M is a matroid on S of rank $\leqslant 2$ then M is representable over some field.

4. Prove that it is impossible to coordinatize the vector space $V(n, q)$ over the field $GF(p)$ unless q is a power of p.

2. NON-REPRESENTABLE MATROIDS

We stress that the Euclidean representation discussed in Chapter 1 has nothing to do with linear representability. However it does provide us with an easy way of presenting some matroids of rank 3 which are not linearly representable over any field. The first such matroid was discovered by MacLane [36] and independently by Ingleton [59].

Example 1 (The "Non-Pappus" matroid). Let M_0 be the matroid of rank 3 on 9 elements whose Euclidean representation is given by the diagram of Fig. 1. The configuration is familiar in that it is just the well-known "Pappus configuration" with one line (7, 8, 9) missing. Now if M_0 was representable over a field a basic result of classical projective geometry would demand that the set of points $\{7, 8, 9\}$ be linearly dependent. Thus M_0 is *not* linearly representable over any field.

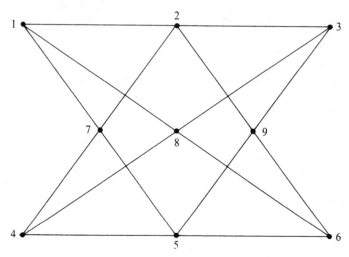

Figure 1. The Non-Pappus matroid

We note however that the above argument depends upon the commutativity of the field and that M_0 is representable over a division ring D.

Example 2. Let $S = \{1, 2, \ldots, 10\}$ and let M_1 be the rank 3 matroid on S with Euclidean representation given by Figure 2.

Figure 2 is the Desargues' configuration with one line $\{8, 9, 10\}$ missing. Since a projective geometry can be coordinatised over a division ring if and only if Desargues' Theorem holds, we would only be able to represent M_1 over a division ring if the set $\{8, 9, 10\}$ was dependent in M_1. Hence M_1 is not representable over any division ring.

Ingleton [59] made the intuitively appealing conjectures that these matroids were the smallest matroids not representable over a field or division ring respectively.

These conjectures were proved false by Vamos [68].

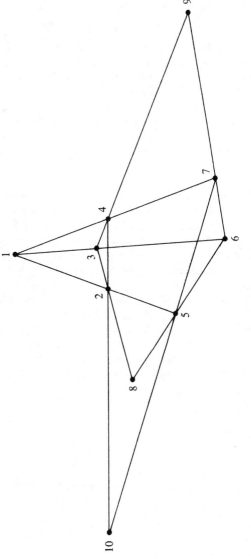

Figure 2. The "Non-Desargues" matroid

Example 3. Let $S = \{a_1, a_2, b_1, b_2, c_1, c_2, d_1, d_2\}$ and let every set of 4 elements be a base of M except for the following sets

$$\{a_1, a_2, b_1, b_2\}, \qquad \{a_1, a_2, c_1, c_2\}, \qquad \{b_1, b_2, c_1, c_2\},$$
$$\{b_1, b_2, d_1, d_2\}, \qquad \{c_1, c_2, d_1, d_2\}.$$

It can be checked that this is a matroid. To see that it is not representable over any field Vamos used the simple argument that if it were representable it would essentially represent a configuration of 4 "lines" (a_1, a_2), (b_1, b_2), (c_1, c_2), (d_1, d_2) no three of which are "coplanar", but which had the property that (a_1, a_2), (b_1, b_2), (c_1, c_2) were concurrent, and (b_1, b_2), (c_1, c_2), (d_1, d_2) were concurrent. If it were representable over a field it is not difficult to show that these conditions demand that (a_1, a_2) meets (d_1, d_2); in other words $\{a_1, a_2, d_1, d_2\}$ is a circuit. This is not the case and hence M is not representable.

Lazarson [58] constructed an infinite class of non-representable matroids. The basic idea in his method was to take a matroid M on S which was only representable over fields of characteristic p, and to take another matroid M' on S' with S' disjoint from S, which was only representable over fields of characteristic $p' \neq p$. By taking the union of M and M' he therefore obtained a matroid on $S \cup S'$ which was not representable over any field. He showed how to construct, for any prime p, a matroid which is representable only over fields of characteristic p, thus showing the existence of this infinite class of non-representable matroids. We illustrate this when $p = 2$.

Let M_1 be the Fano matroid on $S_1 = \{1, \dots, 7\}$. Let M_2 be the matroid on $S_2 = \{1', 2', \dots, 7'\}$ which has the Euclidean representation of Fig. 3.

The Fano matroid M_1 is only representable over fields of characteristic 2. This is a standard piece of projective geometry, for a formal proof see Rado [57].

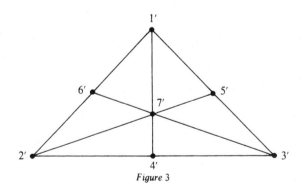

Figure 3

The matroid M_2 (the Fano matroid with one line missing) is easily shown to be representable over any field which is not of characteristic 2. This is because if the underlying field had characteristic 2, it would force $\{4', 5', 6'\}$ to be collinear.

Hence the direct sum $M_1 + M_2$ on $S_1 \cup S_2$ is not representable over any field.

EXERCISES 2

1. Let $n = pr + 1$, p being a prime, r a positive integer. Let $e_1, \ldots, e_n$ be a linearly independent sequence of vectors in a vector space V over a field of characteristic p. Let M be the vectorial matroid on $\{e_1, \ldots, e_n, u - e_1, \ldots, u - e_n, u\}$ where $u = e_1 + \ldots + e_n$.
   Prove that if M is representable over a field F the characteristic of F is a factor of pr.
   (Lazarson [57])

2. Find a matroid on a set of eight elements which is representable over the complex field but is not representable over the rationals. (MacLane [36])

3. THE REPRESENTABILITY OF MINORS, DUALS AND TRUNCATIONS

If M is a matroid on S which is representable over a field F it is obvious that for $T \subseteq S$, $M \mid T$ is also representable over F.

Now consider the contraction $M . T$. Since

$$M . T = (M^* \mid T)^*$$

if we know that M^* is representable over F whenever M is, we have the representability of $M . T$ and hence of any minor.

THEOREM 1. *If M is representable over a field F then any minor of M is representable over F.*

This follows from the following theorem which is the main theorem of this section and was first proved by Whitney [35] when the underlying field was the rationals.

THEOREM 2. *If M is representable over the field F, its dual M^* is also representable over F.*

Proof. Suppose M is of rank r on the set S, $|S| = n$, and is representable over a field F.

Let the $r \times n$ matrix $A = (a_{ij})$, $a_{ij} \in F$ be a matrix representation of M.

Now A defines a linear transformation (which we denote by ψ_A) from the vector space $V(n, F)$ to the field $V(r, F)$. Clearly

$$\text{Ker } \psi_A = \{x \in V(n, F): Ax' = 0\},$$

$$\dim \psi_A = n - r.$$

Now take any matrix B with entries from F, of size $n \times (n - r)$ such that the columns of B are linearly independent and span Ker ψ_A. That is

(1) $Ax' = 0 \Leftrightarrow x' = By'$ for some $y = (y_1, \ldots, y_{n-r}) \in V(n - r, F)$,

We assert that B' is a matrix representation of the dual matroid M^*.

To prove this we need to show that r columns of A are linearly independent over F if and only if the complementary set of $n - r$ columns of B' are linearly independent over F. Also because of the symmetrical relation between A and B, and by reordering the columns of A it is sufficient to show that the first r columns of A are linearly independent if and only if the last $n - r$ columns of B' are linearly independent.

Again it is sufficient to show that the first r columns of A are linearly dependent if and only if the remaining columns of B' are linearly dependent. Now by (1)

$$\exists y = (y_1, y_2, \ldots, y_r, 0, 0 \ldots 0) \in F^{(n)}$$

with $y \neq 0$ and such that $Ay' = 0$ if and only if

$$\exists z = (z_1, \ldots, z_{n-r}) \in F^{(n-r)}, \qquad z \neq 0$$

such that

$$y' = Bz'.$$

Now the $n \times (n - r)$ matrix B can be written in the form

$$B = (B_1, B_2)'$$

where B_1 is $r \times (n - r)$ and B_2 is $(n - r) \times (n - r)$. Hence $B_2 z' = 0$ and since $z \neq 0$, B_2 is singular. Hence its rows are linearly dependent and the theorem is proved. ∎

A useful corollary of Theorem 2 is the following result which enables us to write down from a standard representation of M with respect to a base B, the standard representation of the dual M^* with respect to the complementary cobase $S \backslash B$.

COROLLARY 1. *If M of rank r on $S = \{e_1, e_2, \ldots, e_n\}$ has the standard representation*

$$\begin{array}{c} \underline{e_1, \ldots e_r \ , \ e_{r+1} \cdots e_n} \\ \left| \begin{array}{cc} I_r, & A \end{array} \right. \end{array}$$

then M^* has the standard representation

$$\begin{array}{c} \underline{e_1, \ldots e_r \ , \ e_{r+1} \cdots e_n} \\ \left| \begin{array}{cc} -A', & I_{n-r} \end{array} \right. \end{array}$$

where I_k is the $k \times k$ identity matrix.

Proof. Show that the columns of $[-A', I_{n-r}]'$ span the space of solutions of $[I_r, A]x' = 0$, ■

Example. Let G be the graph shown in Fig. 1.

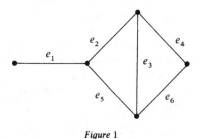

Figure 1

A matrix representation of $M(G)$ over $GF(2)$ is

	e_1	e_2	e_3	e_4	e_5	e_6
	1	0	0	0	0	0
	0	1	0	0	1	0
	0	0	1	0	1	1
	0	0	0	1	0	1.

Its dual $[M(G)]^*$ therefore has the representation

e_1	e_2	e_3	e_4	e_5	e_6
0	1	1	0	1	0
0	0	1	1	0	1

This is easily checked to be a correct representation of $[M(G)]^*$ which we know to be isomorphic to $M(G^*)$,

G^*:

Figure 2

THEOREM 3. *Let M be a matroid on S which is representable over the finite field F. Then there exists an integer N such that for every k the k-truncation of M is representable over any extension K of F with $|K| > N$.*

Proof. Let $r = \rho(M)$; it is sufficient to prove the result for the $(r - 1)$-truncation M' of M. Let $\phi: S \to V$ be a representation of M over some finite extension K of F in a vector space V of rank r. Let $X_1, X_2, \ldots, X_q$ be the independent $(r - 1)$-sets of M. Consider the hyperplanes of V spanned by $\phi(X_1), \ldots, \phi(X_q)$. If $|K| = t$, each such flat contains t^{r-1} elements of V. Thus together they contain at most qt^{r-1} elements. Thus provided $t > q$, their union is not equal to V. Choose $v \in V$ which does not lie in the union of these hyperplanes. Put $T = S \cup s$ where $s \notin S$ and let M'' be the matroid on T induced by the representation $\phi': T > V$ defined by

$$\phi'(x) = \phi(x) \qquad x \in S,$$

$$\phi'(s) = v.$$

Then the $(r - 1)$ truncation M' is just $M'' . S$, since any independent $(r - 1)$ set of M is independent in $M'' . S$ but the rank is reduced to $r - 1$ on contraction. Since the contraction of a representable matroid is representable, M' is representable. ∎

Note. Since no vector space over an infinite field is the union of a finite number of hyperplanes, we can prove similarly:

THEOREM 4. *If M is a matroid on S which is representable over the infinite field F then the k-truncation of M is representable over F.*

EXERCISES 3

1. Show that if M is representable over a field F, the truncation of M at level k need not be representable over F.

2. Prove that every proper minor of the Fano matroid is representable over every field.

4. CHAIN GROUPS

The algebra of chain groups and their applications to matroid theory was developed by Tutte [65]. As will be seen, this approach to the representability problem is dual to the more conventional and traditional approach initiated by Whitney [35].

Throughout this section R is a commutative integral domain, though in all applications R will either be the ring of integers or the field of integers mod 2.

If S is a finite set a *chain* on S over R is a map $f: S \rightarrow R$. If $x \in S$, we refer to $f(x)$ as the *coefficient* of x in f and write $\|f\|$ to denote the *support* of f, that is

$$\|f\| = \{x \in S; f(x) \neq 0\}.$$

The *zero chain* (denoted by 0) has null support. The *sum* $f + g$ of two chains f, g, is the chain on S satisfying

$$(f + g)(x) = f(x) + g(x) \qquad x \in S.$$

For $\lambda \in R$ and f a chain, the *product* λf is the chain on S satisfying

$$(\lambda f)(x) = \lambda f(x) \qquad x \in S.$$

A *chain group on S over R* is a collection of chains on S over R which is closed under the operations of sum and product. We let $\mathscr{C}(S, R)$ denote the collection of chains on S over R.

Now if N is any chain group on S over R a member f of N is an *elementary chain* of N if it is non-zero and there is no chain $g \in N$ with $\|g\|$ a proper subset of $\|f\|$.

THEOREM 1. *Let N be a chain group on S over R and let $\mathscr{D}(N)$ be the set of supports of the chains of N. Then $\mathscr{D}(N)$ is the collection of dependent sets of a matroid on S.*

Proof. It is sufficient to show that the supports of the elementary chains of N are the circuits of a matroid.

Suppose A, B are distinct sets which are supports of the elementary chains f, g respectively. Let $e \in A \cap B$. Consider the chain h defined by

$$h = g(e)f - f(e)g.$$

Clearly $\|h\| \subseteq (A \cup B)\backslash e$ and $\|h\| \neq \varnothing$. Thus the circuit axioms are satisfied. ∎

The matroid obtained in this way we call the *matroid of the chain group N* and denote it by $M(N)$.

Example. Let $S = \{1, 2, 3, 4, 5\}$. Let $R = GF(2)$ and let N be the chain group generated by f_1, f_2 where

	1	2	3	4	5
f_1	1	0	1	1	0
f_2	1	0	1	0	0
f_3	0	0	0	1	0
$0 \doteq f_4$	0	0	0	0	0

Then the elementary chains of N are f_2 and f_3 and the corresponding circuits of $M(N)$ are $\{1, 3\}$ and $\{4\}$.

We now show that when R is a field F the class of chain group matroids on S over R is exactly the class of matroids which are linearly representable over F.

If $\alpha : \mathscr{C}(S, F) \to V$ is a linear map where V is a vector space over F, there is an associated matroid M_α on S defined by

$$\{s_1, s_2, \ldots, s_k\} \in \mathscr{I}(M_\alpha) \Leftrightarrow \{\alpha_{\chi_{s_1}}, \ldots, \alpha_{\chi_{s_k}}\}$$

are linearly independent in V.

[χ_s is the indicator chain of s, that is

$$\chi_s(t) = 0 \qquad s \neq t$$
$$= 1 \qquad s = t.]$$

Conversely for any matroid M on S which is representable in V by the map ϕ we can define a linear map $\alpha : \mathscr{C}(S, F) \to V$ by

$$\alpha(f) = \sum_{e \in S} f(e)\phi(e) \qquad f \in \mathscr{C}(S, F)$$

and it is now clear that $M = M_\alpha$.

Now given a linear map $\alpha: \mathscr{C}(S, F) \to V$ let

$$N = \operatorname{Ker} \alpha = \{f \in \mathscr{C}(S, F); \quad \alpha(f) = 0\}.$$

Then N is a chain group on S over F and $\{s_1, s_2, \ldots, s_k\} \notin \mathscr{I}(M_\alpha)$ if and only if $\exists y_1, y_2, \ldots, y_k$ not all zero such that

$$y_1\, \alpha(\chi_{s_1}) + \ldots + y_k\, \alpha(\chi_{s_k}) = 0,$$

that is if and only if

$$y_1\, \chi_{s_1} + \ldots + y_k\, \chi_{s_k} \in N.$$

Hence a subset X of S is dependent in M_α if and only if there is a non-zero chain $y \in N$ with $\|y\| \subseteq X$. Hence M_α is the chain group matroid $M(N)$.

Conversely given any chain group $N \subseteq \mathscr{C}(S, F)$ we can obtain an associated chain group matroid $M(N)$ by taking the matroid M_α corresponding to the canonical map

$$\alpha: \mathscr{C}(S, F) \to \mathscr{C}(S, F)/N.$$

Thus we have proved the following basic result.

THEOREM 2. *A matroid M on S is isomorphic to the matroid $M(N)$ of a chain group N over a field F if and only if M is representable over F. Moreover if $M \simeq M(N)$ then M is isomorphic to the matroid induced on the quotient space $\mathscr{C}(S, F)/N$ by linear independence.*

It is now easy to recognize the dual of a chain group matroid $M(N)$ as the chain group matroid obtained from the orthogonal complement of N with respect to the obvious bilinear map.

We leave the reader to prove:

THEOREM 3. *Let $M = M(N)$ be the matroid of the chain group $N \subseteq \mathscr{C}(S, F)$. Then $M^* = M(N^\perp)$ where $N^\perp$ is the orthogonal complement of N with respect to the bilinear map $b: \mathscr{C} \times \mathscr{C} \to F$ defined by*

$$b(f, g) = \sum_{e \in S} f(e)\, g(e).$$

EXERCISES 4

1. Let N be a chaingroup on S over R. Tutte [65] defines a *dendroid* to be a minimal set X meeting the domain of every non-zero chain of N. Prove that

   (a) all dendroids have the same cardinality,

(b) $M(N)$ is determined by its dendroids.

How are the dendroids related to the bases of $M(N)$?

2. If N is a chaingroup on S and $T \subseteq S$ show how to construct chaingroups N_1, N_2 from N such that

$$M(N_1) \simeq M(N)|T, \qquad M(N_2) \simeq M(N).T.$$

5. REPRESENTABILITY OF GRAPHIC MATROIDS

In this section we show that a graphic matroid is representable over any field, a result first stated explicitly by Rado [57], though it is implicit in the work of Tutte [65].

THEOREM 1. *Let G be a graph, then its cycle matroid M(G) and cocycle matroid $M^*(G)$ are representable over any field.*

Proof. Since $M^*(G) = (M(G))^*$, by Theorem 3.2 it is sufficient to prove that $M(G)$ is representable over any field F.

Corresponding to each vertex v_i of G choose a vector u_i over F in such a way that the $|V(G)|$ vectors u_i are linearly independent. Order $V(G)$ by a relation $<$ and for every edge e joining v_i, v_j with $v_j > v_i$ let

$$(1) \qquad\qquad \phi(e) = u_j - u_i.$$

We assert that ϕ is a representation of $M(G)$. Suppose X is dependent in $M(G)$. Then there is a cycle $C = \{c_1, \ldots, c_k\} \subseteq X$ and it is clear that there exist numbers $\delta_i \in \{-1, 1\}$ such that

$$\sum_1^k \delta_i \phi(e_i) = 0,$$

and hence $\phi(X)$ is a linearly dependent set of vectors.

Conversely if Y is a set of edges $\{e_1, \ldots, e'_k\}$ such that $\{\phi(e_i): 1 \leqslant i \leqslant k\}$ is linearly dependent, then we may assume that for some $m \leqslant k$, there exist $a_i \neq 0$, $\forall i$, such that

$$(2) \qquad\qquad \sum_1^m a_i \, \phi(e_i) = 0.$$

Let G_0 be the subgraph of G whose edge set is $\{e_i: 1 \leqslant i \leqslant m\}$ and vertex set all vertices of G incident with $\{e_i: 1 \leqslant i \leqslant m\}$. If $v \in G_0$, is incident with a single edge of G_0, then substituting for $\phi(e_i)$ from (1) in (2) and remembering that the vectors u_i are linearly independent we get a contradiction. Hence G_0 has no vertex of degree 1 and Y must contain a cycle of G.

Hence Y is dependent in $M(G)$ which proves the theorem. ∎

Thus a further necessary condition for a matroid to be graphic or cographic is that it be representable over any field.

Notice also that in proving Theorem 1 we have in fact proved the stronger result.

THEOREM 2. *If G is a graph and F is a field then M(G) has a matrix representation A in which every entry in A is 0 or a unit of F, and each column of A has only two non-zero entries.*

Example. Take G to be the graph of Fig. 1, and order the vertices as shown. Then a matrix representation of $M(G)$ over any field F is as shown.

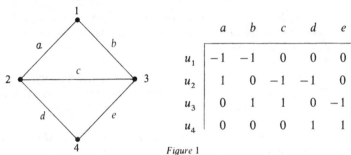

	a	b	c	d	e
u_1	-1	-1	0	0	0
u_2	1	0	-1	-1	0
u_3	0	1	1	0	-1
u_4	0	0	0	1	1

Figure 1

In the special case when F is $GF(2)$, the ring of integers mod 2, we see that the matrix representation of $M(G)$ obtained above is the vertex-edge incidence matrix of G. In the terminology of Harary [69] we have proved.

THEOREM 3. *If G is a graph on p vertices with q edges, a matrix representation of the cycle matroid of G over GF(2) is the (p × q) incidence matrix of G.*

EXERCISES 5

1. Prove that the cycle matroid of the graph K_5 is isomorphic to the vectorial matroid of the 3-dimensional Desargues' configuration.

2. Prove that the matroid of the vector space $V(n, q)$ cannot be graphic or cographic for any prime power q.

3. Let Δ be a bipartite graph, and F any field. Prove that the cycle matroid $M(\Delta)$ has a matrix representation A over F in which each non-zero entry in A is the identity element of F. (Rado [57])

6. THE REPRESENTABILITY OF INDUCED MATROIDS, UNIONS OF MATROIDS AND TRANSVERSAL MATROIDS

In this section we point out that the property of being representable over a field is preserved under various operations discussed in Chapter 8. In particular we show that if M is representable and f is any function then fM is representable. Similarly if Δ is a directed bipartite graph, the induced matroid $\Delta(M)$ will be representable whenever M is. However, usually $f(M)$ and $\Delta(M)$ will not be representable over exactly the same field as M, but over some larger extension field. This will lead to the proof that transversal matroids are always representable over any field of sufficiently large but finite cardinality.

THEOREM 1. *If M is a matroid on S there exists a finite integer N such that if M is representable over the field F and $|F| \geqslant N$, then $f(M)$ is also representable over F, for all surjections $f: S \to T$.*

Proof. We call a surjection $f: S \to T$ *simple* if $|S| = |T| + 1$. Clearly, any surjection is the composition of a finite member of simple surjections. Hence it is only necessary to prove the theorem for simple surjections.

Let M be representable over F and let ϕ be a representation of M in a vector space V over F. Let $S = \{s_1, s_2, \ldots, s_n\}$ and $T = \{t_2, t_3, \ldots, t_n\}$. Let

$$f(s_1) = f(s_2) = t_2,$$

$$f(s_i) = t_i \qquad (3 \leqslant i \leqslant n).$$

We now define an injection $\phi': T \backslash t_2 \to V$ by

$$\phi'(t_i) = \phi(s_i) \qquad (3 \leqslant i \leqslant n)$$

and we will show that there exist $\lambda, \mu \in F$ such that if $\phi'(t_2) = \lambda\phi(s_1) + \mu\phi(s_2)$ then the extended map $\phi': T \to V$ is a representation of $f(M)$ in V.

Clearly, if $Y \subseteq T \backslash t_2$ and Y is independent in $f(M)$, then $\phi'(Y)$ is a linearly independent set of vectors. The converse holds since ϕ is a representation of M. Now let Y be independent in $f(M)$ and $Y = X \cup t_2$.

We first look at the particular case where s_1 is parallel to s_2. It is obvious that if we let $\phi'(t_2) = \phi(s_1)$ then ϕ' is a representation of $f(M)$.

If s_1 is not parallel to s_2, consider the subspace $U(Y)$ of vectors of the form $\lambda\phi(s_1) + \mu\phi(s_2)$ which belong to the subspace spanned by $\phi'(X)$. Since $Y \in f(M)$, at least one of the sets $s_1 \cup f^{-1}(X)$ and $s_2 \cup f^{-1}(X)$ is independent in M, and hence, since ϕ is a representation of M, $U(Y)$ does not contain both

$\phi(s_1)$ and $\phi(s_2)$ and so $U(Y)$ is a proper subspace of the subspace spanned by $\phi(s_1)$ and $\phi(s_2)$.

Consider now all independent sets Y_i of $f(M)$ of the form $t_2 \cup X_i$ and their corresponding subspaces $U(Y_i)$. Now provided $\bigcup_i U(Y_i)$ is properly contained in the subspace generated by $\{\phi(s_1), \phi(s_2)\}$, $\exists \lambda, \mu \in F$ such that $w = \lambda\phi(s_1) + \mu\phi(s_2)$ is not linearly dependent on the vectors of $\phi'(X_i)$ for all i. We then define $\phi'(t_2) = w$ and the injection $\phi': T \to V$ is such that if $Y \in f(M)$, $\phi'(Y)$ is a linearly independent set of vectors.

But for a given M and a given simple function f it is clear that since $|S| < \infty$, there exists a finite integer $N(M, f)$ such that $\bigcup U(Y_i)$ is properly contained in the subspace generated by $\{\phi(s_1), \phi(s_2)\}$. Since the number of non-isomorphic matroids M, and simple surjections f, is finite we may take $N = \max N(M, f)$ as our lower bound for the cardinality of F.

We now prove that if Y is dependent in $f(M)$, then $\phi_1(Y)$ is a dependent set of vectors. Since every dependent set contains a circuit, we need only consider the case when Y is a circuit of $f(M)$. Clearly, if $Y \subseteq T\backslash t_2$ then $\phi'(Y) = \phi(f^{-1}(Y))$ is dependent. Now if $Y = t_2 \cup X$, the both $s_1 \cup f^{-1}(X)$ and $s_2 \cup f^{-1}(X)$ are dependent in M. Hence both $\phi(s_1)$ and $\phi(s_2)$ belong to the subspace spanned by $\phi'(X) = \phi(f^{-1}(X))$, hence $\{\lambda\phi(s_1) + \mu\phi(s_2)\} \cup \phi'(X)$ is a dependent set of vectors which completes the proof of the theorem. ∎

We can extend this result in the following way.

THEOREM 2. *Let $\Delta = (T, S; E)$ be a directed bipartite graph and let M be a matroid on S. Then there exists an integer N such that if M is representable over a field F which has cardinality at least N then the induced matroid $\Delta(M)$ on T is also representable over F.*

Proof. Call the directed bipartite graph $\Delta = (T, S; E)$ a *simple relation* if $|S| = |T| - 1$ and in the obvious way it defines a simple surjection: $T \to S$.

Now for any Δ if $M_0 = \Delta(M)$, it is easy to check that there is a sequence of matroids $M_1, M_2, \ldots$ such that

$$M_0 = \Delta_1(M_1) = \Delta_2(M_2) = \ldots = \Delta_k(M)$$

where for some i, $1 \leqslant i \leqslant k$, $\Delta_i, \Delta_{i+1}, \ldots, \Delta_k$ are simple relations and $\Delta_1, \Delta_2, \ldots, \Delta_{i-1}$ are simple surjections. Hence by Theorem 1, Theorem 2 will be proved if we can prove it for a simple relation Δ.

So let $S = \{s_1, \ldots, s_n\}$, $T = \{t_0, t_1, \ldots, t_n\}$ and let $E = \{(t_0 s_1), (t_1 s_1), (t_2 s_2), \ldots, (t_n s_n)\}$.

Let $\phi: S \to V$ be a representation of M in the vector space V over the field F.

Define $\phi': T \to V$ by

$$\phi'(t_i) = \phi(s_i) \qquad (1 \leqslant i \leqslant n)$$

$$\phi'(t_0) = a\phi(s_1),$$

where a is a non-zero element of F chosen so that $a\phi(s_1) \neq \phi(s_i)$, $(1 \leqslant i \leqslant n)$. By taking $|F| \geqslant |T| + 1$ such an a exists. It is clear that ϕ' is a representation of $\Delta(M)$, and this completes the proof. ∎

Now let M on S be the transversal matroid determined by the family of sets $\mathscr{A} = (A_i : i \in I)$. Then M is induced from the free matroid M_0 on I by the bipartite graph of $\mathscr{A}$ against S. But M_0 on I is representable over every field. Hence we have proved

THEOREM 3. *Let M be a transversal matroid and F any field. Then M is representable over every sufficiently large but finite extension of F.*

COROLLARY. *If M is transversal it is representable over fields of every characteristic.*

Theorems 1, 2, 3, were proved by Piff and Welsh [70]. An interesting proof of a weaker form of Theorem 3 was given by Mirsky and Perfect [67].

The outline of their argument was to take the collection of sets $(A_i : i \in I)$ of S and consider the incidence matrix B of I against S in which $B_{ij} = 1$ or 0 depending on whether or not A_i contained the element x_j of S. Let $\{z_{ij} : i \in I, j \in S\}$ be a collection of independent indeterminates over the field F of rationals. Consider the matrix B' got from B be replacing each non-zero entry b_{ij} of B by the indeterminate z_{ij}. Then it is not difficult to check that B' is a representation of the transversal matroid $M[(A_i : i \in I)]$ over the field $F(z_{11}, \ldots, z_{nn})$. This proves that a transversal matroid is representable, but the representation given here is, in a certain sense, extravagant.

However the basic idea of "indeterminate" representation of combinatorial structures is very appealing, and often very useful. It seems to have originated in a paper by Perfect [66].

Example. Let $S = \{1, 2, 3, 4\}$, let $A_1 = \{1, 2\}$, $A_2 = \{2, 3, 4\}$, $A_3 = \{1, 4\}$. The matrix B' representing $M[A_1, A_2, A_3]$ is given by

$$
B' = \begin{array}{c} \\ A_1 \\ A_2 \\ A_3 \end{array}
\begin{array}{cccc}
1 & 2 & 3 & 4 \\
\hline
z_{11} & z_{12} & 0 & 0 \\
0 & z_{22} & z_{23} & z_{24} \\
z_{13} & 0 & 0 & z_{34}
\end{array}
$$

Another corollary of Theorem 1 is:

THEOREM 4. *Let M_i $(1 \leqslant i \leqslant m)$ be matroids on S which are each representable over a field F. Then their union $M_1 \vee \ldots \vee M_m$ is representable over any sufficiently large (but finite) extension of F.*

Proof. Let S_i $(1 \leqslant i \leqslant m)$ be disjoint copies of S. Take M_i' on S_i to be isomorphic to M_i $(1 \leqslant i \leqslant m)$. Then the direct sum $M_1' \vee \ldots \vee M_m'$ on $S_1 \cup \ldots \cup S_m$ is representable over F. But as in the proof of Theorem 8.3.1 the union $M_1 \vee \ldots \vee M_m$ can be induced from $M_1' \vee \ldots \vee M_m'$ by an obvious function. Hence by Theorem 1 $M_1 \vee \ldots \vee M_m$ is representable over some extension of F. ∎

EXERCISES 6

1. Prove that $M(Fano)^*$ is not transversal.

2. Prove that it is impossible to find a set S and a directed bipartite graph $\Delta = (S, T; E)$ such that if M is a uniform matroid on S, $\Delta(M)$ is isomorphic to the Fano matroid.

3. Prove that a transversal matroid of rank r on a set of n elements is representable over any field F such that

$$|F| > n + \binom{n}{r-1}.$$
(Atkin [72])

4. Give an example to show that the restriction on the cardinality of F in Theorem 1 cannot be lifted. (Piff and Welsh [70])

7. THE CHARACTERISTIC SET OF A MATROID

An interesting idea introduced by Ingleton [71] is the concept of the *characteristic set* $C(M)$ of a matroid M, which is defined to be the set of primes p such that M is representable over some field of characteristic p. We interpret 0 for this purpose as a prime number and let P denote the set of all primes. (Actually Ingleton defined the characteristic set in terms of division rings but since almost all of known representability theory refers to fields we will confine our attention to fields.)

Thus if we recap our known theorems we have the following properties of characteristic sets.

(1) Non representable matroids (e.g. the non-Pappus) have $C(M) = \varnothing$.

(2)   If M_i $(i \in I)$ are matroids on S then Theorem 6.4 implies

$$C(\vee M_i) \supseteq \bigcap_{i \in I} C(M_i).$$

(3)   If the matroid M is transversal, graphic or cographic

$$C(M) = P.$$

The question asked by Ingleton was, what subsets of P are possible characteristic sets of matroids?

Rado [57] proved:

(4)   If $0 \in C(M)$ then all sufficiently large primes belong to $C(M)$.

More recently Vamos [71] has proved:

(5)   If $0 \notin C(M)$ then $|C(M)|$ is finite.

A theorem of Lazarson [58] essentially says:

(6)   Any singleton subset of P is a characteristic set.

To obtain a matroid whose characteristic set is $P \backslash \{2\}$ take the Fano matroid and omit one line, to get the matroid M_0 whose Euclidean representation is given by Fig. 1.

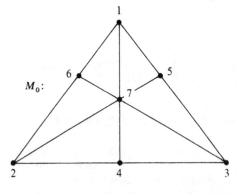

Figure 1

Consider the matroid M_1 with Euclidean representation as shown in Fig. 2.

A representation is possible over a field F (with for instance coordinates as shown) if and only if F contains a root of

$$\xi^2 + \xi + 1 = 0.$$

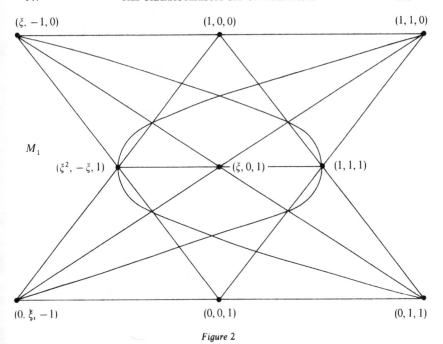

$(\xi, -1, 0)$ $(1, 0, 0)$ $(1, 1, 0)$

M_1

$(\xi^2, -\xi, 1)$ $(\xi, 0, 1)$ $(1, 1, 1)$

$(0, \xi, -1)$ $(0, 0, 1)$ $(0, 1, 1)$

Figure 2

Now $(0, 1, 0)$ is collinear with $(1, 1, 1)$, $(\xi^2, -\xi, 1)$ and $(\xi, 0, 1)$ if and only if $\xi = 1$, that is if and only if the characteristic of F is 3.

Thus by taking M_1 together with the point $(0, 1, 0)$ *not* collinear with $(\xi^2, -\xi, 1)$, $(\xi, 0, 1)$ and $(1, 1, 1)$ we get the matroid M_2 whose Euclidean representation is given in Fig. 3.

By the previous argument

$$C(M_2) = P\backslash\{3\}.$$

However it seems difficult to construct matroids with "small" prescribed characteristic set, and Ingleton [71] poses:

Problem. For what subsets Q of P does there exist a matroid with characteristic set Q; in particular is $\{p_1, p_2\}$ a possible characteristic set (p_1, p_2 distinct primes)?

R. Reid (private communication) has proved:

(7)   If $Q \subseteq P$ with $P\backslash Q$ finite, then Q is the characteristic set of some matroid.

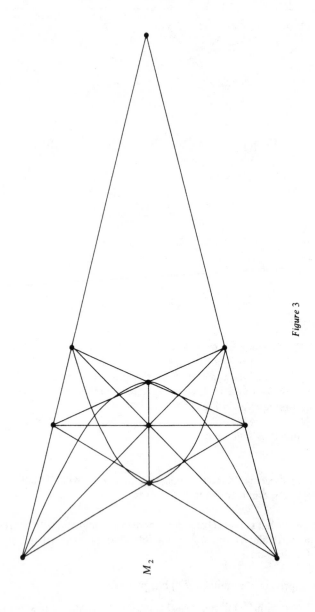

M_2

Figure 3

Even "within the characteristic set" there are interesting problems.

If M is representable over F then M is representable over any extension of F. In particular since every field of characteristic zero has prime field isomorphic to the rationals we have:

(8)   If M is representable over the rationals then M is representable over every field of characteristic zero.

The converse is not true. Maclane [36] first gave an example of a matroid on 8 elements which is representable over the complex field but not over the rationals, or indeed the reals, but the reader will realize that since $\xi^2 + \xi + 1 = 0$ has complex but no real roots the matroid M_1 with Euclidean representation given by Fig. 2 also has this property.

EXERCISES 7

1. Prove that for any integer n there exists a matroid with characteristic set

$$\{0\} \cup \{p : p \geqslant n, p \text{ prime}\}.$$

(Piff [72])

○2. Does there exist for each prime p a matroid with characteristic set $P \backslash \{p\}$?

8. CONDITIONS FOR REPRESENTABILITY

Although the "representability problem" for matroids is now 40 years old, no one has even suggested a reasonable set of conditions which might turn out to be necessary and sufficient for a given matroid to be representable over some field. Any progress in this direction must, it seems be made by first finding properties of representable matroids which are not true for matroids in general. At one time, for example, it appeared that a necessary condition for a matroid M to be representable was that the conditions

$$kr(A) \geqslant |A| \qquad \forall A \subseteq S$$

were necessary and sufficient for S to be the union of k bases of M (see for example the footnote to Edmonds and Fulkerson [65]). At the time Horn [55] had proved Theorem 8.4.2 for subsets of vector spaces and Edmonds had not proved it for matroids.

However it is not easy to find properties of representable matroids which do not hold for matroids in general. One result of this type is the basis of the following theorem of Ingleton [71b] which we state without proof.

THEOREM 1. *A necessary condition for a matroid with rank function ρ on S to be representable is that for any subsets A, B, C, D, of S,*

(1) $\rho(A) + \rho(B) + \rho(A \cup B \cup C) + \rho(A \cup B \cup D) + \rho(C \cup D)$

$$\leqslant \rho(A \cup B) + \rho(A \cup C) + \rho(A \cup D) + \rho(B \cup C) + \rho(B \cup D).$$

An example of a matroid for which (1) does not hold is the non-representable matroid constructed by Vamos (Example 2.3).

One set of necessary and sufficient conditions is found in what Crapo and Rota [70] call "the coordinatization theorem" of Tutte [65].

THEOREM 2. *A necessary and sufficient condition that a simple matroid M on S be representable over a field F is that for every hyperplane H of M there exists a function $f_H : S \to F$ such that:*

(i) Ker $f_H = H$.
(ii) *For any three hyperplanes H_1, H_2, H_3 of M which interesect in a coline (a flat of rank $= \rho M - 2$), there exist constants $a_1, a_2, a_3 \in F$, all non-zero, so that*

$$a_1 f_{H_1} + a_2 f_{H_2} + a_3 f_{H_3} = 0.$$

The necessity of the conditions is obvious for if M is representable by the map ϕ let $e_H(x) = 0$ be the equation of the hyperplane $\phi(H)$ in the vector space. Then the maps $\{ f_H \}$ defined by

$$f_H(y) = e_H(y) \qquad y \in S$$

is the required set of functions. For a direct proof of the sufficiency, see White [70].

As a practical (algorithmic) test of whether or not a given matroid is representable, Theorem 2 seems to be even more difficult than finding a direct representation, and the necessary conditions (1) of Ingleton are much more useful. Note also that the above theorems are necessary for representability over division rings. This is not true for the only other known conditions for representability of Vamos [71], which again are very unwieldy.

We briefly discuss Vamos' conditions.

Let $B \subseteq S$, be a base of M and for each $x \in S$ let

$$B_x = \begin{cases} x & x \in B \\ C(x, B) \backslash x & x \notin B, \end{cases}$$

where $C(x, B)$ is the fundamental circuit of x with respect to the base B. For each $X \subseteq S$ define

$$B_X = \bigcup_{x \in X} B_x.$$

Call X *critical* if $|B_X| = |X|$.

Now consider the incidence matrix A of $S = \{1, 2, \ldots, n\}$ and the family $(B_x : x \in S)$. Without loss we may take $B = \{1, 2, \ldots, r\}$.

$$
\begin{array}{c}
\\
\\
B_1 \\
\vdots \\
Br \\
\\
\\
B_{r+1} \\
\vdots \\
B_n
\end{array}
\begin{array}{|cc}
1, 2, \ldots, r, \quad r + 1, \ldots, n \\
\hline
\\
I_r \qquad\qquad 0 \\
\qquad\qquad\qquad\qquad = A \\
\\
A' \qquad\qquad 0 \\
\end{array}
$$

Replace each non-zero entry a_{ij} of A by an indeterminate q_{ij} and let the q_{ij} be independent indeterminates over the ring of integers $\mathbb{Z}$.

For each critical $X \subseteq \{1, 2, \ldots, n\}$ define a polynomial

$$P_X = \det A(X)$$

where $A(X)$ is the matrix with row indices $(B_x : x \in X)$ and column indices $(x : x \in X)$.

Let I be the ideal generated by those polynomials P_X such that X is critical and dependent in M.

Let J be the multiplicatively closed subset of $\mathbb{Z}(Q)$, generated by $(P_X : X$ critical and X independent in $M)$, $(Q = \{q_{ij}\})$. Then we have:

THEOREM 3. *The matroid M is representable over some field if and only if $I \cap J = \varnothing$.*

We return to the problem of representability in Chapter 10 where we solve completely the problem of representability over $GF(2)$ and also characterize those matroids which are representable over every field. Apart from these two cases little was known until very recently when Bixby [75a] and Seymour [75d] independently proved the following remarkable theorem.

THEOREM 4. *A necessary and sufficient condition that a matroid M is representable over the field $GF(3)$ is that it does not contain a minor isomorphic to one of*

$$U_{2,5}, \quad U_{3,5}, \quad M(Fano), \quad M(Fano)^*.$$

The conditions are easily seen to be necessary, but as we could expect the proof of sufficiency is quite difficult.

NOTES ON CHAPTER 9

In his paper [35] Whitney seems to be only concerned with representability over the reals. For example, he gives the Fano matroid as an example of a "non-matric matroid". Examples of matroids with interesting representation properties can be found in the review paper of Ingleton [71]. The representability theorems for minors and duals of representable matroids are due to Tutte [65], though he worked solely in the language of chaingroups. Our proofs are different and are based on unpublished work of A. W. Ingleton. Section 4 on chaingroups is based on Tutte [65] though the treatment presented here (and in particular the proof of Theorem 4.1) is due to A. W. Ingleton. Section 6 is based on the paper by Piff and Welsh [70]. P. Vamos (unpublished) had previously proved that the union of representable matroids was representable. Section 7 is based on Rado [57] and Ingleton [71]. Theorem 8.1 is interesting, in that it can be checked that dualizing it does not give a new set of conditions for representability. It would be nice to have more theorems of this type—indeed is it true that there exists an infinite "independent" set of rank conditions for a matroid to be representable?

In Chapter 10 we will characterize matroids which are representable over every field. In Chapter 11 we describe a new class of matroids discovered by Dowling [73] which have interesting characteristic sets. In Chapter 12 we describe some non-representable matroids with highly transitive groups which are not representable over any field—see Kantor [74a].

Kantor [74a], [75], has produced some interesting sufficient conditions for representability, and has studied the problem of embedding geometric lattices in modular lattices—a natural extension of the representability problem.

CHAPTER 10

Binary Matroids

1. EQUIVALENT CONDITIONS FOR REPRESENTABILITY OVER GF(2)

H. Whitney was the first to consider the problem of finding necessary and sufficient conditions for a matroid M to be graphic. An obvious starting point is to find properties of graphic matroids which do not hold for matroids in general.

For example it is well known that if C_1, C_2 are the sets of edges of two distinct cycles of a graph G, the symmetric difference $C_1 \triangle C_2 = (C_1 \cup C_2) \backslash (C_1 \cap C_2)$ must be the union of disjoint cycles of G. We illustrate this in Fig. 1, where $C_1 = \{a, b, c, d, e, f\}$ and $C_2 = \{c, h, f, g\}$, then $C_1 \triangle C_2$ is the union of $\{a, b, g\}$ and $\{d, e, h\}$.

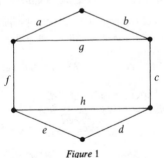

Figure 1

Thus a necessary condition for a matroid M to be graphic is that for any two circuits C_1, C_2 of M, the symmetric difference $C_1 \triangle C_2$ is the union of disjoint circuits. In an arbitrary matroid, the symmetric difference of two circuits may even be independent.

Example 1. Let $S = \{1, 2, 3, 4\}$, and let M have as its circuits all subsets of S of cardinality 3. Then $\{1, 2, 3\} \triangle \{1, 2, 4\}$ is the independent set $\{3, 4\}$.

161

Another well-known property of graphs is that if C and C^* are respectively a cycle and cocycle of a graph G then $|C \cap C^*|$ must be even. Again this is false for arbitrary matroids (see Example 1).

A definitive solution to the problem of 'when is a matroid graphic' was given by Tutte [58], who introduced the concept of a binary matroid as a matroid determined by a chain group over the field $GF(2)$ of integers modulo 2.

It turns out that there are many equivalent ways of defining a binary matroid. In view of the adjective 'binary' probably the most appropriate is to define M to be *binary* if it is representable over the field $GF(2)$.

Because of Theorems 9.3.1 and 9.3.2, we have

THEOREM 1. *A matroid M is binary if and only if its dual M^* is binary.*

THEOREM 2. *Any minor of a binary matroid is binary.*

The smallest non-binary matroid is $U_{2,4}$, the uniform matroid of rank 2 on 4 elements.

By Theorem 9.5.1 a necessary condition that a matroid M be graphic or cographic is that M is binary. However it is not sufficient. The smallest binary matroid which is neither graphic nor cographic is the Fano matroid.

We now give some alternative definitions of binary matroids. Some of the proofs are surprisingly complicated.

THEOREM 3. *The following statements about a matroid M are equivalent.*

(a) *For any circuit C and cocircuit C^*, $|C \cap C^*|$ is even.*
(b) *The symmetric difference of any collection of distinct circuits of M is the union of disjoint circuits of M.*
(c) *If C_1, C_2 are distinct circuits of M, the symmetric difference $C_1 \triangle C_2$ contains a circuit C.*
(d) *For any base B and circuit C of M, if $C \backslash B = \{e_1, \dots, e_q\}$ and if $C(e_i)$ is the fundamental circuit of e_i in the base B, then*
$$C = C(e_1) \triangle \dots \triangle C(e_q).$$
(e) *M is binary.*

Note. In view of (b) property (c) tells us nothing about a binary matroid but is a simpler way of testing whether a given matroid is binary.

Proof that (a) $\Rightarrow$ (b)

Let M satisfy (a) and let $C_1, \dots, C_k$ be distinct circuits of M. Write $A = C_1 \triangle \dots \triangle C_k$. If $A = \varnothing$ it is the vacuous union of circuits. If A contains loops $x_1, \dots, x_t$ let
$$A' = \{x_1\} \triangle \dots \triangle \{x_t\} \triangle A$$

so that A' is without loops, and is the symmetric difference of distinct circuits. Hence we can assume that A is without loops and non-null.

Let $x \in A$ and let C^* be a cocircuit of M containing x. (C^* must exist or x would be a loop of M). Since $|C_i \cap C^*|$ is even for $1 \leqslant i \leqslant k$, it is an easy exercise to check that $|A \cap C^*|$ is even, so that $C^* \cap (A \backslash x) \neq \varnothing$. Suppose that A is independent in M. Then since $C^* \backslash x$ is independent in M^*, by the dual augmentation theorem there exists a base B of M such that

$$A \subseteq B, \qquad C^* \backslash x \subseteq S \backslash B.$$

Hence $C^* \cap A = \{x\}$, contradicting $|C^* \cap A|$ even. Thus A is dependent in M, and contains a circuit, say C.

If $A = C$ we are finished, if $A \neq C$ let $A_1 = C \triangle C_1 \triangle \ldots \triangle C_k$ and apply the above argument with $A = A_1$. Since A is finite, and $A_1 = A \backslash C$ this process eventually terminates yielding a finite collection of disjoint circuits whose union is A. Thus (a) $\Rightarrow$ (b).

That (b) $\Rightarrow$ (c) is obvious.

Proof that (c) $\Rightarrow$ (a)

Suppose that M satisfies (c) but not (a). Let C, C^* be respectively a circuit and cocircuit of M such that $|C \cap C^*|$ is the smallest integer which is not even. Since in any matroid

$$|C \cap C^*| \neq 1, |C \cap C^*| \geqslant 3.$$

Let a, b, c be distinct elements of $C \cap C^*$.

By Theorem 5.1.1* there exists a circuit C_1 such that

$$C_1 \cap C^* = \{a, c\}.$$

Since $a \in C \cap C_1$ and $b \in C \backslash C_1$, there exists a circuit C_2 such that

$$b \in C_2 \subseteq (C \cup C_1) \backslash a.$$

Choose C_2 so that $C \cup C_2$ is minimal. Let C_3 be a circuit such that

$$a \in C_3 \subseteq (C \cup C_2) \backslash b,$$

and let C_4 be a circuit such that

$$b \in C_4 \subseteq (C \cup C_3) \backslash a.$$

It is clear that

$$C \cup C_4 \subseteq C \cup C_3 \subseteq C \cup C_2.$$

But $b \in C_4$ and it is easy to check that

$$C_4 \subseteq (C \cup C_1) \backslash a.$$

Hence since $C \cup C_2$ is minimal

$$C \cup C_2 \subseteq C \cup C_4 \subseteq C \cup C_3 \subseteq C \cup C_2.$$

Thus $C_3 \backslash C = C_2 \backslash C$ and therefore

$$C_2 \triangle C_3 = (C_2 \backslash C_3) \cup (C_3 \backslash C_2) \subseteq C.$$

Since $C_2 \neq C_3$ it follows from (c) that $C_2 \triangle C_3$ contains a circuit. Consequently

$$C_2 \triangle C_3 = C.$$

But by construction $|C_2 \cap C^*|$ and $|C_3 \cap C^*|$ are positive and strictly less than $|C \cap C^*|$. But if $C_2 \triangle C_3 = C$, $C \triangle C_2 = C_3$ and therefore if both $|C_2 \cap C^*|$ and $|C_3 \cap C^*|$ are even, then $|C \cap C^*|$ must be even (draw a Venn diagram). Hence one or other of $|C_2 \cap C^*|$ and $|C_3 \cap C^*|$ is odd and thus we contradict the minimum cardinal property of $C \cap C^*$. This completes the proof of the equivalence between (a), (b) and (c).

Proof that each of (a), (b), (c) *are equivalent to* (d)

Let M satisfy (a), and hence also (b) and (c). Let C be a circuit and B be a base with $C \backslash B = \{e_1, \ldots, e_t\}$. The set $Z = C(e_1) \triangle \ldots \triangle C(e_t) \supseteq \{e_1, \ldots, e_t\}$, and is otherwise contained in B. Thus $C \triangle Z \subseteq B$. But by (b) Z is the union of disjoint circuits so that the only possibility is that $C \triangle Z = \varnothing$ which is the desired result.

Conversely let M have property (d). It is sufficient to show that if D_1, D_2 are distinct circuits of M, $D_1 \triangle D_2$ is dependent. Suppose not, and let $D_1 \cap D_2 = \{x_1, \ldots, x_k\}$. Then $D_1 \triangle D_2 = (D_1 \cup D_2) \backslash \{x_1, \ldots, x_k\}$ is independent and $\exists$ a base $B \supseteq D_1 \triangle D_2$. But $D_1 \backslash B = \{x_1, \ldots, x_k\} = D_2 \backslash B$ so that by (d), if $C(x)$ is the fundamental circuit of x in B,

$$D_1 = C(x_1) \triangle \ldots \triangle C(x_k) = D_2$$

which contradicts $D_1 \neq D_2$ and completes the proof.

Proof of equivalence of (a)–(d) *and* (e)

Let M be binary and let $B = \{b_1, \ldots, b_r\}$ be a base of M. Then there exists a standard representation of M over $GF(2)$ of the form.

(1)

$$
\begin{array}{c}
b_1 \ldots b_r \quad e_1 \ldots e_q \\
\hline
\\
I_r \qquad\qquad A \\
\\
\end{array}
$$

Here I_r is the $r \times r$ identity matrix, $\{e_1, \ldots, e_q\} = S \backslash B$, and A is a (0–1)-matrix.

Let C be a circuit of M. Say $C = \{b_1, \ldots, b_t, e_1, \ldots, e_p\}$. Then $C(e_j)$, the fundamental circuit of e_j in B satisfies

$$C(e_j) \backslash e_j = \{b_i : (A)_{ij} \neq 0\}.$$

But if ϕ is the representation of M over $GF(2)$ given by the columns of the above matrix, then since C is a circuit

$$\sum_{i=1}^{t} \phi(b_i) + \sum_{j=1}^{p} \phi(e_j) \equiv 0 \qquad (\text{mod } 2).$$

But this is equivalent to the statement that

$$C = C(e_1) \triangle \ldots \triangle C(e_p).$$

Conversely suppose that (d) holds for any base B and circuit C. Let $B = \{b_1, \ldots, b_r\}$, $S \backslash B = \{e_1, \ldots, e_q\}$. Define a matrix A by:

$$A_{ij} = \begin{cases} 1 & b_i \in C(e_j) \\ 0 & b_i \notin C(e_j). \end{cases} \quad (1 \leqslant i \leqslant r, 1 \leqslant j \leqslant q.)$$

We assert that the set of columns of the matrix (1) is a representation of M over $GF(2)$.

Suppose C is a circuit of M, then by (d), if $C \backslash B = \{e_{i_1}, \ldots, e_{i_k}\}$,

$$C = C(e_{i_1}) \triangle \ldots \triangle C(e_{i_k})$$

and thus the columns of (1) corresponding to the members of C are linearly dependent over $GF(2)$. Similarly if $U = \{b_1, \ldots, b_t, e_1, \ldots, e_p\}$ is a circuit of the matroid M' induced on S by linear dependence of the corresponding column vectors of (1), then

$$U = C(e_1) \triangle \ldots \triangle C(e_p).$$

But since (d) is equivalent to (c), U contains a circuit C of M. It clearly cannot contain C properly without contradicting the fact that U is a circuit of M' and hence the matrix A is a representation of M. This completes the proof of Theorem 3. ∎

In proving Theorem 3 we showed that if M on S is binary there is an easy way of getting a standard representation of M.

Let $B = \{b_1, \ldots, b_r\}$ be any base, $S \backslash B = \{e_1, \ldots, e_q\}$ and let C_j be the fundamental circuit of e_j in B. A standard representation of M over $GF(2)$ is given by the columns of the matrix

$$\begin{array}{cc} b_1 \ldots b_r & e_1 \ldots e_q \\ \hline & \\ I_r, & A \end{array}$$

(2)

where $A_{ij} = 1$ if and only if b_i belongs to C_j, otherwise $A_{ij} = 0$.

Since the cycle matroid $M(G)$ of a graph G is binary this gives another way of representing the cycle matroid of a graphic matroid over $GF(2)$. (Compare with Section 9.5).

Example. Consider the graph K_4

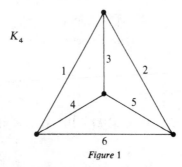

Figure 1

Take as base $\{1, 2, 3\}$. The fundamental circuits of this base are $\{4, 1, 3\}$, $\{5, 2, 3\}$ and $\{6, 1, 2\}$ so that a representation of $M(K_4)$ over $GF(2)$ is:

$$\begin{array}{cccccc} 1 & 2 & 3 & 4 & 5 & 6 \\ \hline 1 & 0 & 0 & 1 & 0 & 1 \\ 0 & 1 & 0 & 0 & 1 & 1 \\ 0 & 0 & 1 & 1 & 1 & 0 \end{array}$$

Now by Theorem 9.3.2 if (2) is a standard representation of M, a standard representation of M^* is given by:

$$\begin{array}{cc} e_1 \ldots e_q & b_1 \ldots b_r \\ \hline & \\ I_q, & A' \end{array}$$

(3)

But since a matroid has a unique standard representation with respect to a given base we have shown:

COROLLARY 1. *In the matrix representation* (2) *of M the* ith *row vector is the incidence vector of the fundamental cocircuit of* b_i *in the cobase* $\{e_1, \ldots, e_q\}$.

EXERCISES 1

1. Show that the truncation of a binary matroid is not in general binary.

2. For what values of k and n is the uniform matroid $U_{k,n}$ a binary matroid?

3. Prove that a matroid is binary if and only if for any circuit C and cocircuit C^*, $|C \cap C^*| \neq 3$. (Seymour [75])

4. Call a matroid M on S *Eulerian* if S can be expressed as the union of disjoint circuits of M. Call M *bipartite* if every circuit has even cardinality. Prove that a binary matroid M is Eulerian if and only if its dual is bipartite.

    (Note that this is a natural generalisation of Euler's theorem for graphs; it is not true in general for nonbinary matroids.) (Welsh [69a], Wilde [75])

2. AN EXCLUDED MINOR CRITERION FOR A MATROID TO BE BINARY

We now prove a theorem of Tutte [65] giving necessary and sufficient conditions for a matroid to be binary, in terms of an excluded minor.

THEOREM 1. *A matroid M is binary if and only if it has no minor isomorphic to* $U_{2,4}$, *the uniform matroid of rank* 2 *on* 4 *elements.*

Proof. Since any minor of a binary matroid is also binary, if M is binary it has no minor isomorphic to $U_{2,4}$. To prove the converse we first need the following easy lemma.

LEMMA 1. *Let M be binary on S, let* $z \in S$ *and let C be a circuit of M with* $z \in C$. *Then* $C\backslash z$ *is a circuit of* $M . (S\backslash z)$. *If* $z \notin C$ *then either C is a circuit of* $M . (S\backslash z)$ *or is the disjoint union of two circuits of* $M . (S\backslash z)$.

We leave the proof as an exercise.

So let M be a matroid on S satisfying the excluded minor condition. We use induction on $|S|$ and suppose the theorem is true for all matroids on sets of cardinality less than $|S|$. We shall show that the symmetric difference of any two circuits of M is a disjoint union of circuits.

Thus we may assume $S = C_1 \cup C_2$ where C_1, C_2 are circuits of M and $C_1 \cap C_2 \neq \emptyset$. Let $X = C_1 \cap C_2$, $Y_1 = C_1 \backslash C_2$, $Y_2 = C_2 \backslash C_1$. We show that $Y = Y_1 \cup Y_2$ is a disjoint union of circuits.

Case (a). If $|Y_1| = |Y_2| = 1$ and either $|X| = 1$ or $Y_1 \cup Y_2$ is a circuit the result is trivial.

Case (b). $Y_1 = \{y_1\}$, $Y_2 = \{y_2\}$ and Y is independent. Then for some $x_1, x_2 \in X$, $Y \cup (X \backslash \{x_1, x_2\})$ is a base of M. But then $M . \{x_1, x_2, y_1, y_2\}$ is $U_{2,4}$, a contradiction.

Case (c). $Y_1 = \{y, z, \ldots\}$; $C_1 \backslash y$ is a circuit of $M . (S \backslash y)$ and by induction C_2 is a disjoint union of at most 2 circuits of $M . (S \backslash y)$. Thus the symmetric difference of $C_1 \backslash y$ and C_2 is a disjoint union of circuits of $M . (S \backslash y)$. We may therefore write

$$Y = S_1 \cup \ldots \cup S_r \cup \ldots \cup S_t$$

where $S_i (i = 1, \ldots, t)$ is a circuit of M and if $i \neq j$,

$$S_i \cap S_j = \begin{cases} \{y\} & i, j \leqslant r \\ \emptyset & \text{otherwise.} \end{cases}$$

Suppose r is even. Then by the induction hypothesis

$$Y = T_1 \cup \ldots \cup T_h \cup \{y\},$$

where $T_1, \ldots, T_h$ are disjoint circuits of M, since symmetric differences of $S_1, \ldots, S_r$ may be taken in pairs. But by the induction hypothesis, $M | (S \backslash y)$ is binary and so for each $i = 1, \ldots, h$, $T_i \backslash y$ is a disjoint union of at most two circuits of

$$(M | (S \backslash y)) . (S \backslash \{y, z\})$$

and hence of $M . (S \backslash z)$. Thus

$$Y \backslash z = R_1 \cup \ldots \cup R_k \cup \{y\}$$

where $R_1, \ldots, R_k$ are circuits of $M . (S \backslash z)$.

But $Y \backslash z$ is the symmetric difference of $C_1 \backslash z$ and C_2, and C_2 is the disjoint union of at most two circuits of $M . (S \backslash z)$, which is binary by the induction hypothesis. Thus $C_1 \backslash z$ is the symmetric difference of at most 3 circuits of $M . (S \backslash z)$, and finally $\{y\}$ is a symmetric difference of circuits of $M . (S \backslash z)$, a contradiction.

Thus r is odd, and we may pair $S_1, \ldots, S_{r-1}$, take symmetric differences in pairs, and use the induction hypothesis. ∎

Because of the correspondence between minors of M and intervals of the associated geometric lattice $\mathcal{L}(M)$, Theorem 1 can be restated in lattice terminology:

THEOREM 1′. *A geometric lattice $\mathcal{L}$ is the lattice of a binary matroid if and only if every interval of height 2 in $\mathcal{L}$ contains at most 5 elements.*

This is because $\mathcal{L}(U_{2,4})$ is the lattice of Fig. 1.

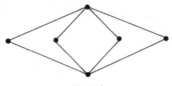

Figure 1

As we shall prove in Chapter 17, Theorem 1′ in turn is equivalent to the apparently much stronger result.

THEOREM 2. *A geometric lattice $\mathcal{L}$ of height r is the lattice of a binary matroid if and only if every element of $\mathcal{L}$ of height $r - 2$ is contained in at most 3 copoints.*

In other words, $\mathcal{L}$ is binary if and only if $\mathcal{L}(U_{2,4})$ does not appear as an interval at the 'top' of the lattice. This is an example of the 'scum principle' see Chapter 17.

Crapo and Rota [70] prove Theorem 2 by appealing to Tutte's coordinatization theory. We will prove it in Chapter 17, the reader can try to deduce it by duality principles from Theorem 1.

EXERCISES 2

1. Prove Lemma 1.

○2. Is there any excluded minor condition for a matroid to be representable over the prime field $GF(p)$, where p is >3? (For $p = 3$ see Theorem 9.8.4.).

3. THE CIRCUIT SPACE AND COCIRCUIT SPACE; ORIENTABLE MATROIDS

Let M be a binary matroid on S, $|S| = n$ and let V be the vector space of rank n over $GF(2)$. The *circuit space* of M is the subspace of V generated by the incidence vectors of the circuits of M. Similarly the *cocircuit space* of M is the subspace of V generated by the cocircuits of M.

THEOREM 1. *The rank of the circuit space of the binary matroid M is $\rho(M^*)$ and the incidence vectors of the set of fundamental circuits of any base of M form a base of the circuit space.*

Proof. It is obvious that these fundamental circuit vectors are linearly independent (over any field), since each has a 1 in a position where all the others have 0's: thus no one of these vectors is a linear combination of the others. By Theorem 1.3(d) the incidence vector of any circuit C can be written as the linear combination of the fundamental circuit vectors and thus the theorem is proved. ■

The dual statement is:

THEOREM 1*. *The rank of the cocircuit space of the binary matroid M is $\rho(M)$ and the incidence vectors of the set of fundamental cocircuits of any cobase of M form a base of the cocircuit space.*

We define the *circuit matrix* $D(M)$ of M to be the incidence matrix of circuits against elements. The *cocircuit matrix* $D^*(M)$ is the incidence matrix of cocircuits against elements. Clearly

$$D^*(M) = D(M^*)$$

Call a binary matroid M *orientable* if it is possible to assign negative signs to some of the non-zero entries in $D = D(M)$, $D^* = D^*(M)$ in such a way that the scalar product of any row of D with any row of D^* is 0 over the ring of integers.

Example. Let G_0 be the graph of Fig. 1. The circuit and cocircuit matrices of its cycle matroid are shown below.

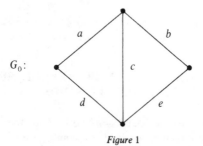

Figure 1

	a	b	c	d	e
C_1	1	1	0	1	1
$D = C_2$	1	0	1	1	0
C_3	0	1	1	0	1

	a	b	c	d	e
C_1^*	1	1	1	0	0
C_2^*	1	0	0	1	0
$D^* = C_3^*$	1	0	1	0	1
C_4^*	0	1	1	1	0
C_5^*	0	1	0	0	1
C_6^*	0	0	1	1	1

Now suppose that we assign directions to the edges of G_0 arbitrarily. Suppose $\omega(G_0)$ is the resulting digraph

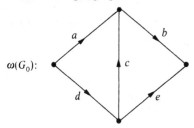

$\omega(G_0)$:

Then consider C_1, we will assign negative signs to the entries corresponding to the intersections of e and d with C_1 since these have directions contrary to the conventional positive orientation of C_1 (clockwise). Similarly consider the cocycle $C_4^* = \{b, c, d\}$, in an obvious way we orient the corresponding incidence vector $(0, 1, -1, 1, 0)$. The reader can verify that if by continuing in this way we multiply some of the elements of D and D^* by -1, we get matrices which are orthogonal row by row.

The assignation of signs to D, D^* corresponding to the above orientation ω of G_0 is shown in the matrices $\omega(D)$, $\omega(D^*)$.

	a	b	c	d	e
C_1	1	1	0	-1	-1
$\omega(D) = C_2$	1	0	-1	-1	0
C_3	0	1	1	0	-1

	a	b	c	d	e
C_1^*	1	-1	1	0	0
C_2^*	1	0	0	1	0
$\omega(D^*) = C_3^*$	1	0	1	0	1
C_4^*	0	1	-1	1	0
C_5^*	0	1	0	0	1
C_6^*	0	0	1	-1	1

The reader will recognize that all we are doing is using the following well known and easily proved property of a graph G.

If C is a cycle of a connected graph G and C^* is a cocycle which separates G into connected parts U and V, then if f^* is a chain on $E(G)$ over $\mathbb{Z}$ with

$$f^*(e) = \begin{cases} +1 & e \in C^*, \quad e \text{ oriented from } U \text{ to } V \\ -1 & e \in C^*, \quad e \text{ oriented from } V \text{ to } U \\ 0 & \text{otherwise,} \end{cases}$$

and f is a chain on $E(G)$ over $\mathbb{Z}$ with

$$f(e) = \begin{cases} +1 & e \in C, \quad e \text{ oriented clockwise} \\ -1 & e \in C, \quad e \text{ oriented anticlockwise} \\ 0 & \text{otherwise,} \end{cases}$$

then

$$\sum_{e \in S} f(e) \, f^*(e) = 0.$$

A more succinct algebraic formulation of the above is

THEOREM 2. *Any graphic matroid is orientable.*

It is obvious from the definition that:

(1) M is orientable if and only if M^* is orientable

We also know that there exist orientable, non-graphic matroids namely the cocycle matroids of $K_{3,3}$ and K_5.

Also not every binary matroid is orientable. the Fano matroid is such an example. In fact it can be shown that:

(2)   The smallest binary matroids which are not orientable are the Fano matroid and its dual.

Note also that if M is orientable on S and $T \subseteq S$, then the same orientation, restricted in the obvious way gives an orientation of $M \,|\, T$. Thus using duality we have:

(3)   If M is orientable any minor of M is orientable.

In the next sections we relate the property of being orientable with other matroid concepts.

EXERCISES 3

1. Prove assertion (2) above.

4. REGULAR MATROIDS

In this section we develop Tutte's theory of regular chain groups and matroids.

Let N be a chain group on S over the ring of integers. Recall that a chain $f \in N$ is elementary if there is no non-zero chain $g \in N$ with $\|g\|$ a proper subset of $\|f\|$. If also the coefficients of f are ± 1 or 0, f is called a *primitive* chain. The chain group N is *regular* if to every elementary chain of N there corresponds a primitive chain with the same domain. A matroid is *regular* if it is isomorphic to the matroid of a regular chain group.

We first prove the following lemmas.

LEMMA 1. *If N is a regular chain group on S and $f \neq 0$, $f \in N$, then f can be represented as a sum of primitive chains of N, each with the same support as f.*

Proof. If $Z(k)$, where k is a chain of N denotes the sum of the absolute values of the coefficients in k, choose f so that the lemma fails and $Z(f)$ has the least value consistent with this. Let g be the primitive chain with the same support as $\|f\|$ and write $h = f - g$. If $h = 0$ then $f = g$, if $h \neq 0$ it is clear that $Z(h) < Z(f)$ and $\|h\| = \|f\|$. Hence h is a sum of primitive chains of N, and each such chain has support $= \|h\| = \|f\|$. In either case the definition of f is contradicted.

If f, g are integral chains on S and $p > 2$ is an integer, we call f a *p-representative* of f if for every $x \in S$

(i) $g(x) \equiv f(x) \bmod p$

(ii) $|g(x)| < p$.

LEMMA 2. *Let N be a regular chain group on S. Then for each integer $q \geqslant 2$ and for each $f \in N$ there exists a chain of N which is a q-representative of f.*

Proof. There is at least one chain $g \in N$ which satisfies condition (i), namely f. For each such g define $Y(g)$ to be the number of elements $a \in S$ with $|g(a)| \geqslant q$. Choose a particular g satisfying (i) so that $Y(g)$ has the least possible value.

If $Y(g) > 0$ choose $b \in S$ so that $|g(b)| \geqslant q$. By Lemma 1, there exists a primitive chain $h \in N$, $\|h\| = \|g\|$ and $h(b) = \pm 1$. Write $g_1 = g - qh$. Clearly

g_1 satisfies (i) and moreover we have

$$|g_1(b)| < |g(b)|$$
$$|g(a)| < q \Rightarrow |g_1(a)| < q.$$

Hence $Y(g_1) \leqslant Y(g)$ with equality if and only if $g_1(b) \geqslant q$. If $g_1(b) \geqslant q$ we repeat the process with g_1 replacing g and with the same choice of b. Proceeding in this way we eventually obtain a chain g' of N such that g' satisfies (i) and $Y(g') < Y(g)$. This is contrary to the choice of g. Thus we deduce $Y(g) = 0$ and the lemma is proved.

LEMMA 3. *Let M be a regular matroid on S. Then for any prime p, there exists a chain group N on S over $GF(p)$ such that $M = M(N)$.*

Proof. Let M be regular. Then there exists a regular chain group N such that $M = M(N)$. For each $f \in N$, let f' be the chain over $GF(p)$ defined by

$$f'(x) \equiv f(x) \bmod p \qquad (x \in S).$$

Clearly the set of chains f' form a chain group N' on S over $GF(p)$. We will show that $M(N) \simeq M(N')$.

Let C be a circuit of M. Then since N is regular there exists a primitive chain $g \in N$ such that

$$\|g\| = C.$$

Consider $\|g'\|$, clearly $\|g'\| = C$. Hence C is dependent in $M(N')$. That is there exists a circuit C' of $M(N')$ such that

(α) $C' \subseteq C$.

Now let C' be a circuit of $M(N')$. There is a chain $f' \in N'$ such that

$$\|f'\| = C'.$$

But this means there exists $f \in N$ such that

$$C' \subseteq \|f\|$$

and if $x \in \|f\| - C'$, $f(x) \equiv 0 \bmod p$. But by Lemma 2, there exists a chain g which is a p-representative of f and clearly

$$\|g\| = \|f'\| = C'.$$

Hence there is a circuit C of M such that $C \subset C'$. Combining this with (α) we see that $M(N) = M$, and the proof is complete.

We now can prove:

(1)   If M is regular then M is representable over every field.

Proof. Since M is the matroid of a chain group over the integers it is a fortiori the matroid of a chain group over the rationals. By Theorem 9.4.2 it is therefore representable over the rationals, and hence over any field of zero characteristic. Similarly Lemma 3 and Theorem 9.4.2 show that M is representable over $GF(p)$ and hence over any field of characteristic p.

As a corollary we have the following result.

(2)   If M is regular then M is binary.

Now we know from Section 9.2, than the Fano matroid and its dual are representable only over fields of characteristic 2. They are therefore examples of binary matroids which are not regular. The main result of Tutte's work [58, 59] is the following very difficult theorem.

THEOREM 1. *A binary matroid is regular if and only if it does not contain as a minor either the Fano matroid or its dual.*

Immediate consequences of this are:

(3)   If M is regular any minor of M is regular.

(4)   If M is regular its dual M^* is regular.

Now suppose that M is a matroid which is representable over every field. Then M is binary and cannot contain either M (Fano) or M^* (Fano) since they are representable only over fields of characteristic 2. Thus by Theorem 1, (1) and (2) we have proved

THEOREM 2. *A matroid is regular if and only if it is representable over every field.*

Again using Tutte's theorem we can relate regular and orientable matroids.

THEOREM 3. *A matroid is regular if and only if it is orientable.*

Proof. Let M be orientable so that by Section 3 any minor of M is orientable. But the Fano matroid not its dual is orientable. Hence M cannot contain M(Fano) or M^*(Fano) as a minor and is therefore regular.

Conversely let M be regular, say $M = M(N)$ where N is a chain group over $\mathbb{Z}$. Let C be a circuit of M. Then $\exists$ a primitive chain $f \in N$, with $\|f\| = C$.

In the row of the circuit matrix of M corresponding to C assign a positive or negative sign to the entry according as f is ± 1. Orient the cocircuit matrix similarly. By Theorem 9.4.3 since M, M^* are dual, their associated chain groups are orthogonal and it follows that we have our required orientation of the circuit and cocircuit matrices. ∎

EXERCISES 4

1. Prove that the smallest matroids which are regular but not graphic are $M^*(K_{3,3})$ and $M^*(K_5)$. Similarly show that the smallest regular matroids which are not cographic are $M(K_{3,3})$ and $M(K_5)$.

2. Prove that M is a regular matroid if and only if every standard matrix representation over the rationals is *totally unimodular*, that is the determinant of every square submatrix is 1, -1 or 0. (Tutte [65])

5. CONDITIONS FOR A MATROID TO BE GRAPHIC

In this section we give different sets of conditions for a matroid to be graphic or cographic.

From Theorem 3.2 we know that a necessary condition that the matroid M is graphic is that M is orientable ($=$ regular). However this is not sufficient since $M^*(K_5)$, $M^*(K_{3,3})$ for example are both regular but not graphic. The second very deep characterization theorem of Tutte [58] shows that provided a matroid does not contain either of these minimal regular non-graphic matroids as a minor then it is graphic.

THEOREM 1. *A regular matroid M is graphic if and only if it does not contain as a minor the cocycle matroids of K_5 and $K_{3,3}$.*

The dual statement is:

THEOREM 1*. *A regular matroid M is cographic if and only if it does not contain as a minor the cycle matroids of K_5 and $K_{3,3}$.*

Thus in conjunction with Theorem 4.1 we have the following excluded minor condition for a matroid to be graphic or cographic.

(1)   A binary matroid M is graphic if and only if it does not contain as a minor any one of

$$M(\text{Fano}), \qquad M(\text{Fano})^*, \qquad M^*(K_5), \qquad M^*(K_{3,3}),$$

(2) A binary matroid M is the cycle matroid of a planar graph if and only if it does not contain as a minor any one of

$$M(Fano), \qquad M(K_5), \qquad M(K_{3,3})$$

or their dual matroids.

Theorem 1 is difficult (see Tutte [65]) and can be regarded as the complete abstract generalization of Kuratowski's theorem [30].

A much more simple-minded set of conditions for a matroid to be graphic can be obtained as the matroid analogue of a theorem about graphs proved by MacLane [37].

We know that a necessary condition that M be graphic is that it should be binary.

Let M be a binary matroid, let $\mathscr{C}$ denote its collection of circuits, $\mathscr{C}^*$ its collection of cocircuits, V the circuit space of M.

We say that a set $\mathscr{B} = \{C_1, \dots, C_q\}$ of circuits, $(q = \rho(M^*))$, form a 2-*complete basis* for the circuit space V if they form a basis in the usual sense and furthermore an element $x \in S$ belongs to at most 2 members of $\mathscr{B}$. A 2-complete basis for the cocircuit space of M is defined analogously.

The following extensions of MacLane's Theorem were proved by Graver [66] and Welsh [69].

THEOREM 2. *A matroid is graphic if and only if it is binary and has a 2-complete basis of cocircuits.*

The dual result is:

THEOREM 2*. *A matroid is cographic if and only if it is binary and has a 2-complete basis of circuits.*

Proof of Theorem 2. Let M be binary with a 2-complete family $\mathscr{B}$ of cocircuits, $\{C_i : 1 \leq i \leq p\}$. Let M_i on $S_i (1 \leq i \leq k)$, be the connected components of the matroid M. Clearly each M_i is binary and moreover since a cocircuit is contained in one matroid component, each M_i has a 2-complete family of cocircuits consisting of those members of $\mathscr{B}$ which are contained in S_i. We will here construct a graph G having M_1 as its cycle matroid. Repeating this construction for each of the components M_i proves the theorem. Hence we may assume that M is connected.

Construction. Let M be binary, connected, on S, with a 2-complete family $\mathscr{B}$ of cocircuits. If S is a single element x, then x must either be a loop of M or $\{x\}$ is a base of M. If x is a loop let G be the graph of a loop on a single vertex. If $\{x\}$ is a base let G be a single edge joining two vertices. If S has more than one element then since M is connected each pair of elements of S must be contained in a cocircuit. Hence every element belongs to at least one cocircuit. Let $\mathscr{B} = \{C_i : 1 \leqslant i \leqslant p\}$ and let G have $p + 1$ vertices $A_i (1 \leqslant i \leqslant p + 1)$. If the element $x \in S$ belongs to C_i and C_j, let the corresponding edge x' of G join the vertices A_i and A_j. If x only belongs to C_i, let x' join A_i to A_{p+1}. Then the mapping $f : S \rightarrow E(G)$ defined by $f(x) = x'$ is a bijection and we prove that under it, M is isomorphic to the cycle matroid $M(G)$.

When $|S| = 1$ it is trivially true. When $|S| > 1$, let X be dependent in M so that X contains a circuit Y. We show that $f(Y)$ is the set of edges of an Eulerian subgraph of G, and hence must contain a cycle of G and hence $f(X)$ is dependent. To see this let Z_i be those edges of $f(Y)$ which have A_i as an endpoint. Then

(3) $|Z_i| = |Y \cap C_i|$ $(1 \leqslant i \leqslant p)$

and an edge e has A_{p+1} as an endpoint if and only if

$$f^{-1}(e) \in C_1 \triangle C_2 \triangle \ldots \triangle C_p.$$

Since M is binary, $C_1 \triangle \ldots \triangle C_p$ is the union of disjoint cocircuits $C'_1, \ldots, C'_q$ of M and hence

(4) $$|Z_{p+1}| = \sum_1^q |Y \cap C'_i|$$

Again since M is binary, (3) and (4) together imply that $|Z_i|$ is even for all i and hence $f(Y)$ is an Eulerian graph. Thus we have shown that if A is independent in the cycle matroid of G, $f^{-1}(A)$ is independent in M.

To prove the converse let B be a base of M. Then B has non-null intersection with every cocircuit of M. Thus B has a non-null intersection with each C_1 $(1 \leqslant i \leqslant p + 1)$. Thus $f(B)$ is a set of edges of G which span the vertices of G. We will show that $f(B)$ is in fact the set of edges of a spanning tree of G by proving that $f(B)$ is the edge set of a connected graph on $|f(B)| + 1$ vertices.

Suppose $f(B)$ is disconnected, and let $U = \{A_i : 1 \leqslant i \leqslant k\}, (A_{p+1} \notin U)$ be the set of vertices of a connected component of $f(B)$. Consider

$$C = C_1 \triangle \ldots \triangle C_k,$$

which is the union of disjoint cocircuits and must therefore have non-null intersection with B. But $x \in C$ if and only if x is a member of exactly one of the cocircuits $C_i (1 \leqslant i \leqslant k)$. Hence $f(C)$ is a set of edges joining the set of vertices U to the vertex set $V(G) \backslash U$. Hence, since U is a connected component of the graph determined by $f(B)$, $f(C) \cap f(B) = \emptyset$ which contradicts $C \cap B \neq \emptyset$. Thus $f(B)$ is the set of edges of a spanning tree of G and is therefore a base of $M(G)$. This completes the proof. ∎

A characterization of graphic geometric lattices.

Clearly Tutte's excluded minor criterion for a matroid to be graphic can be translated into the theorem that a geometric lattice is graphic if and only if it does not contain any sublattice which is isomorphic to the lattices of $M(K_{3,3})^*$, $M(K_5)^*$, $M(Fano)^*$, $M(Fano)$, $M(U_{2,4})$. A nice refinement of this result is given in Section 17.

An alternative characterisation due to Sachs [70] is essentially the same as the cocircuit criterion of Theorem 2, though instead of talking of a 2-complete family of cocircuits it uses the identity between hyperplanes and cocircuit complements to obtain a neater lattice formulation.

THEOREM 3. *An irreducible geometric lattice $\mathscr{L}$ is isomorphic to the geometric lattice of the cycle matroid of a block graph G if and only if it satisfies the follow-ing conditions.*

(i)   *It possesses a family $\mathscr{H} = (H_i : i \in I)$ of hyperplanes such that every atom of $\mathscr{L}$ has exactly two complements in it and no two atoms have the same two complements.*

(ii) *If $J \subseteq I$,*

$$\rho\left(\bigcap_{i \in J} H_i\right) \leqslant |I \backslash J| - 1,$$

where ρ is the rank function of $\mathscr{L}$.

Recall that an irreducible geometric lattice corresponds to a connected matroid and the case of a reducible lattice can be easily treated by splitting the lattice up into components.

Proof of Theorem 3. This is very similar to that of Theorem 2. We construct from $\mathscr{L}$ a graph G whose vertices are the hyperplanes in $\mathscr{H}$, and an edge passes through a vertex if and only if the corresponding atom and hyperplane are complementary. ∎

EXERCISES 5

1. Let C_1^*, C_2^*, C_3^* be 3 distinct cocircuits of a matroid M on S. We say that C_1^* does not *separate* C_2^* and C_3^* when C_2^*/C_1^* and $C_3^* \backslash C_1^*$ are contained in the same component of $M | (S \backslash C_1^*)$. Prove that a matroid is graphic if and only if from any 3 cocircuits having a non-empty intersection there is at least one which separates the two others. (Fournier [74])

6. SIMPLICIAL MATROIDS

Let $\mathcal{T}_k(S)$ be the set of all k-sets of a set S. If $\mathcal{T} \subseteq \mathcal{T}_k(S)$ and $X_1, \ldots, X_m$ are members of $\mathcal{T}$, we say that $\{X_1, \ldots, X_m\}$ is a *circuit* if it is a minimal set such that

(1) $\mathcal{T}_{k-1}(X_1) \triangle \ldots \triangle \mathcal{T}_{k-1}(X_m) = \varnothing$.

If $\{X_1, \ldots, X_m\}$ does not contain a circuit we say that it is *independent*.

THEOREM 1. *The independent sets of $\mathcal{T}$ are the collection of independent sets of a binary matroid on $\mathcal{T}$.*

Any matroid obtained in this way we call *k-simplicial*.

Example. Let G be a simple graph, with vertex set $V(G)$. Take $S = V(G)$ and let $\mathcal{T}$ be the set of edges of G. Then the 2-simplicial matroid on $\mathcal{T}$ is just the cycle matroid of the graph G, and it is easy to see that every 2-simplicial matroid is obtained in this way.

THEOREM 2. *A matroid is 2-simplicial if and only if it is the cycle matroid of a simple graph.*

We now prove Theorem 1.

Proof. Let M be k-simplicial. Let $\mathcal{U} = \{U_i : i = 1, \ldots, p = \binom{n}{k-1}\}$ be the set of $(k-1)$ subsets of S ($|S| = n$). Then for any k-set X of S define a corresponding vector

$$\phi X = (\alpha_1(X)\, \alpha_2(X), \ldots, \alpha_p(X))$$

where $\alpha_i(X) = 1$ if $X \supseteq U_i$, and is zero otherwise. Then if M is on $\mathcal{T} \subseteq \mathcal{T}_k$ it is clear that the map ϕ: is a representation of M over the field $GF(2)$. Thus M is binary, and the assertion that the circuits as defined by (1) are the circuits of a matroid is now obvious. ∎

From the proof of Theorem 1 it becomes clear that M is just the restriction to $\mathcal{T}$ of the 'full' k-simplicial matroid on $\mathcal{T}_k(S)$ thus for many purposes it is enough to consider this class of matroid.

We denote by $G_k(S)$ the full k-simplicial matroid on the set $\mathcal{T}_k$ of all k-subsets of S.

Example. Let $|S| = 5$:

$G_2(S)$ is the cycle matroid of K_5.
$G_3(S)$ is a geometry with 5 non-trivial planes (one for each 4-element set), in rank 6-space.

$G_4(S)$ is five points in general (free) positions in rank 4 space
$G_5(S)$ is a single point.

It is not difficult to prove:

(2)   The dual of the full simplicial geometry $G_k(S)$ is the full simplicial geometry $G_{n-k}(S)$ $(n = |S|)$.
(3)   The rank of $G_k(S)$ is given by

$$\rho(G_k(S)) = \binom{|S| - 1}{k - 1}.$$

THEOREM 3. *If M is k-simplicial then all its citcuits have cardinal $\geqslant k + 1$.*

From this we note that not all binary matroids can be simplicial, in particular.

(4) $M^*(K_{3,3})$ is binary but not simplicial.

Proof. $M^*(K_{3,3})$ has a circuit of cardinal 3 and hence if it is simplicial it must be 2-simplicial—that is graphic. But this is not so.

At the moment there seems no easy way of deciding whether or not an arbitrary binary matroid is k-simplicial for some k.

EXERCISES 6

1. Prove that the property of being k-simplicial is not preserved under the following operations—taking duals, truncating, contracting, taking unions.

°2. Is every simplicial matroid regular?

NOTES ON CHAPTER 10

Theorem 1.3 is based on the papers of Lehman [64] and Minty [66], though it is also implicit in the work of Tutte [65]. The proof of Tutte's excluded minor condition given here is due to Piff [72]. Other criteria for a matroid to be binary are given by Fournier [74] who also gives new sufficient conditions for a binary matroid to be regular.

The circuit space and cocircuit space of binary matroids were introduced by Minty [66], who was also the first to study orientable matroids. Proofs of the equivalence of regular and orientable matroids were given by Minty [66] and Crapo [69].

Recently Las Vergnas [75] and Bland [75] independently developed an attractive orientation theory for general, not necessarily binary, matroids. Indeed in this theory matroids which are not even representable over any field may be orientable.

In addition to the theory developed in Section 5, Tutte [60] gives an algorithm for deciding whether or not a given binary matroid is graphic. Since there seems to be a natural correspondence between theorems characterizing planar graphs among graphs and theorems characterizing graphic matroids among matroids, it is possible that any algorithm for testing whether or not a graph is planar has an analogue which tests whether or not a matroid is graphic. For a discussion of this see Weinberg [72].

Section 6 on simplicial matroids is based on Crapo and Rota [70], [71]. Theorem 6.3 is an unpublished result of J. Davies.

CHAPTER 11

Matroids from Fields and Groups

1. ALGEBRAIC MATROIDS

In this chapter we first study the class of matroids obtained from the concept of algebraic dependence over a field (we call these algebraic matroids), and in the last section describe a new class of matroids determined by finite groups. Although entirely different classes of matroids, they both (in different ways) are relevant to the study of representability carried out in Chapter 9.

Let K be an extension field of a fixed field F. An element u of K is called *algebraically dependent* on the elements $u_1, u_2, \ldots, u_n$ of K if u is algebraic with respect to the field $F(u_1, u_2, \ldots, u_n)$, that is if u satisfies an algebraic equation

(1) $\alpha_0(u)v^m + \alpha_1(u)v^{m-1} + \ldots + \alpha_m(u) = 0$

in which the coefficients $\alpha_0(u), \ldots, \alpha_m(u)$ are polynomials in $u_1, u_2, \ldots, u_n$ with coefficients in F, and if not all of them are zero.

The following are well known properties of algebraic dependence.

(2)   Every $u_i (1 \leq i \leq n)$ is algebraically dependent on $u_1, u_2, \ldots, u_n$.

(3)   If v is algebraically dependent on $u_1, \ldots, u_n$, but v is not algebraically dependent on $u_1, \ldots, u_{n-1}$, then u_n is algebraically dependent on $u_1, \ldots, u_{n-1}, v$.

(4)   If w is algebraically dependent on $v_1, \ldots, v_s$, and every $v_j (1 \leq j \leq s)$ is algebraically dependent on $u_1, \ldots, u_n$, then w is algebraically dependent on $u_1, \ldots, u_n$.

From properties (2)–(4) it is easy to prove the following basic result.

We call a subset $X = \{x_1, \ldots, x_n\}$ of elements of K, *algebraically independent* if no element $x_i (1 \leq i \leq n)$ is algebraically dependent on the elements of $X \backslash x_i$.

THEOREM 1. *Let K be an extension of the field F, and let S be a finite subset of K. Let $\mathscr{I}$ be the set of algebraically independent subsets of S. Then $\mathscr{I}$ is the set of independent sets of a matroid M on S.*

The rank of the matroid M is the *degree of transcendence* of S with respect to F.

A characterization of independent sets is given by the following well known result.

THEOREM 2. *A subset $U = \{u_1, \ldots, u_n\}$ is algebraically independent over F if and only if the only polynomial f with coefficients in F such that*

$$f(u_1, u_2, \ldots, u_n) = 0$$

is the polynomial whose coefficients are all zero.

The following statements about the matroid M on $S \subseteq K$, are obvious.

(5)   If $x \in S \cap F$ then x is a loop of M.

(6)   If $u \in S$ and $u^k \in S$ for some integer k then u and u^k are parallel elements in M, or both are loops.

(7)   Every element of S which is algebraic over F is a loop of M.

We call an arbitrary matroid M *algebraic over a field F* if there exists some extension field K of F and a subset S of K such that M is isomorphic to the matroid induced on S by algebraic dependence over F. A matroid is *algebraic* if it is algebraic over some field. We also say that such a matroid is *algebriac in the field K*.

Very little seems to be known about algebraic matroids, and the content of this section is based on the paper of Ingleton [71] and the unpublished results of Piff [72].

We first remark that the property of being algebraic over a field F is not equivalent to the property of being linearly representable in a vector space over F.

Example 1. Let $S = \{1, 2, 3, 4\}$ and let M have as its independent sets all subsets of S of cardinality $\leqslant 2$. Then we know that M is the smallest non binary matroid and is therefore not linearly representable over $GF(2)$. However the map $\phi : S \to GF(2)(x, y)$ defined by

$$\phi(1) = x, \qquad \phi(2) = y, \qquad \phi(3) = xy, \qquad \phi(4) = x + y,$$

where x, y are indeterminates is an algebraic representation of M over $GF(2)$.

Indeed Ingleton [71] has constructed a matroid M on 11 elements which is algebraically representable over $GF(2)$ but which is not linearly representable over any field.

Example 2. Let x, y, z be independent transcendentals over $GF(2)$. Let S consist of the eleven points $\{x, y, z, yz, zx, xy, xyz, x + y, y + z, z + x, x + y + z\}$, and let M be the matroid induced on S by algebraic dependence.

Now if $S_1 = \{x, y, z, zx, xy, yz, xyz\}$ and $S_2 = \{x, y, z, x + y, y + z, z + x, x + y + z\}$ we notice that $M|S_1$ is the matroid of the Fano configuration and hence is only linearly representable over fields of characteristic 2 but that $M|S_2$ is the matroid of the configuration of Fig. 1(b) which is only linearly representable over fields of characteristic not equal to 2. It follows that M is not linearly representable over any field.

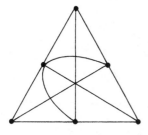
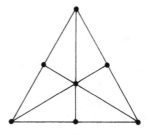

Figure 1(a) Fano matroid   *Figure 1(b)*

2. THE RELATION BETWEEN ALGEBRAIC AND LINEAR REPRESENTABILITY

We first prove that if M is (linearly) representable over a field F then M is algebraic over F. The proof given here is due to Piff [69].

THEOREM 1. *If M is linearly representable over F then M is algebraic over F.*

Proof. Let $\phi : S \to V$ be a linear representation of M in the vector space V over F. Let $\{b_1, b_2, \ldots, b_r\}$ be a base of M and let $\phi(b_i) = v_i (1 \leqslant i \leqslant r)$. Let $\theta_1, \theta_2, \ldots, \theta_r$ be independent transcendentals over F and let ψ be a map on S into $F(\theta_1, \theta_2, \ldots, \theta_r)$ defined as follows:

$$\psi(b_i) = \theta_i \qquad (1 \leqslant i \leqslant r)$$

while if $x \in S \backslash B$, and

$$\phi(x) = \sum x_i v_i \qquad (x_i \in F)$$

we let

$$\psi(x) = \sum x_i \theta_i.$$

We assert that ψ is an algebraic representation of M over F.

To see this let $\{s_1, \ldots, s_r\}$ be any base of M. The set $\{\phi(s_i): 1 \leqslant i \leqslant r\}$ is obtained from $\{\phi(b_i): 1 \leqslant i \leqslant r\}$ by some non-singular linear transformation. Thus $\{\psi(s_i): 1 \leqslant i \leqslant r\}$ is obtained from $\{\psi(b_i): 1 \leqslant i \leqslant r\}$ by the same transformation. Consequently $\{\psi(s_i): 1 \leqslant i \leqslant r\}$ must be independent transcendentals over F.

Now suppose that $\{c_1, c_2, \ldots, c_k\}$ is a circuit of M. Then there exist non zero $a_i \in F$ $(1 \leqslant i \leqslant k)$ such that

$$\sum_1^k a_i \, \phi(c_i) = 0.$$

Hence

$$\sum_1^k a_i \, \psi(c_i) = 0$$

and hence $\psi\{c_1, \ldots, c_k\}$ is an algebraically dependent set over F and thus ψ is an algebraic representation of M. $\blacksquare$

The converse is not true in general, as shown in Example 1, however when the field F has zero characteristic we have the following result from Lang [65, chapter 10, proposition 10], though the treatment there is non-matroidal.

THEOREM 2. *If M is algebraic over a field F of characteristic zero, then M is linearly representable over some transcendental extension of F.*

Standard results from the theory of derivations see for example Ingleton [71] show that:

THEOREM 3. *If M is algebraic in a field K of characteristic zero then M is linearly representable over K.*

Note that by Example 1 above, this is not true for fields of prime characteristic.

A result which emphasizes the difference between linear and algebraic representability is the following

THEOREM 4. *If M is linearly representable over the rationals it is algebraically representable over every field.*

EXERCISES 2

1, Prove that graphic matroids and cographic matroids are algebraic over every field.

○2. Is the Fano matroid algebraic over a field not of characteristic 2?

3. OPERATIONS ON ALGEBRAIC MATROIDS

Clearly any restriction of an algebraic matroid is algebraic over the same field.

THEOREM 1. *Let M be a matroid on S which is algebraic over the field F. Then for $T \subseteq S$, $M . T$ is algebraic over some transcendental extension of F.*

Proof. Let $\phi : S \to G$ be an algebraic representation of M over F.

Then G is an extension field of $F(\phi(S \setminus T)) = H$. Let $\phi' = \phi|_T$; then we assert that $\phi' : T \to G$ is an algebraic representation of $M . T$ over H. For U is independent in $M . T$ if and only if for some base B of $S \setminus T$ in M, $B \cup U$ is independent in M, which is the case if and only if $\phi(B \cup U)$ is a set of independent transcendentals over F. But this is true if and only if $\phi(U)$ is a set of independent transcendentals over $F(B)$, hence over $F(\phi(S \setminus T)) = H$. ∎

COROLLARY. *Every minor of an algebraic matroid is algebraic.*

It seems harder to prove the following result of Piff [72].

THEOREM 2. *If M is algebraic over F then any truncation of M is algebraic over some transcendental extension of F.*

However under the operation of taking functions and unions, algebraic representability is a relatively well behaved property.

THEOREM 3. *Let M on S be algebraic over F and $f : S \to T$ be any function. Then $f(M)$ is algebraic over F.*

Proof. As in the proof of Theorem 9.6.1 it is sufficient to consider a simple function $f : S \to T$.

If $\phi : S \to F$ is an algebraic representation of M we look for a representation $\psi : T \to F$ of M where

$$\psi(t_0) = \phi(s_1)^\lambda \phi(s_2)^\mu \qquad \lambda, \mu \text{ integers}$$

$$\psi(t_i) = \phi(t_i) \qquad (i = 3, \ldots, n)$$

and where $f: S = \{s_1, \ldots, s_n\} \to T = \{t_0, t_3, \ldots, t_n\}$ takes

$$f(s_1) = f(s_2) = t_0, \qquad f(s_i) = t_i \qquad 3 \leqslant i \leqslant n.$$

The remainder of the proof now follows like that of Theorem 9.6.1. ∎

In the same way as in §9.6, we get as a corollary

THEOREM 4. *If for $1 \leqslant i \leqslant t$, M_i is a matroid and each M_i is algebraically representable over a field F then $\bigvee M_i$ is algebraically representable over F.*

Since a uniform rank one matroid is algebraically representable over any field a further corollary is,

THEOREM 5. *If M is transversal, then M is algebraic over every field.*

Note that unlike the situation in linear representability there is no requirement in Theorems 3–5 that the fields have sufficiently large cardinality.

Again by making use of multiplication rather than addition we get the algebraic analogue of Theorem 9.6.2.

THEOREM 6. *If M is algebraic over F and $\Delta(M)$ is induced from M by a directed bipartite graph Δ then $\Delta(M)$ is algebraic over F.*

A non-algebraic matroid

Until recently it was not known whether there existed a matroid which was not algebraic. However Ingleton and Main [75] proved that the Vamos matroid on 8 elements shown in Section 9.2 to be not linearly representable was also non-algebraic.

Consider the Vamos matroid on

$$S = \{a_1, b_1, a_2, b_2, a_3, b_3, a_4, b_4\}$$

with bases all 4-sets except

$$\{a_1, b_1, a_2, b_2\}, \qquad \{a_1, b_1, a_3, b_3\}, \qquad \{a_1, b_1, a_4, b_4\},$$
$$\{a_2, b_2, a_3, b_3\}, \qquad \{a_2, b_2, a_4, b_4\}.$$

THEOREM 7. *The Vamos matroid is not algebraic.*

This is an immediate consequence of the following lemma.

LEMMA. *Let $S = \{a_1, b_1, a_2, b_2, a_3, b_3, a_4, b_4\}$ be a set of 8 elements of a field Ω and let F be a subfield of K such that each of the fields $F(S \backslash \{a_i, b_i\})$ ($i = 1, 2, 3, 4$) is of transcendence degree 4 over F and every 3 elements of S are algebraically* *independent over F. If each of the sets* $\{a_1, b_1, a_2, b_2\}$, $\{a_1, b_1, a_3, b_3\}$, $\{a_1, b_1, c_4, b_4\}$, $\{a_2, b_2, a_3, b_3\}$, $\{a_2, b_2, a_4, b_4\}$ *is algebraically dependent over F, then* $\{a_3, b_3, a_4, b_4\}$ *is also algebraically dependent over F.*

For details see Ingleton and Main [75].

EXERCISES 3

∘1. Is the Non-Pappus matroid of Section 9.2 algebraic over any field? We conjecture not.

∘2. Is the dual of an algebraic matroid also algebraic?

4. PARTITION MATROIDS DETERMINED BY FINITE GROUPS

In this section we sketch the main results of a recent paper by Dowling [73].

If $S = \{x_1, \ldots, x_n\}$ a *partial partition* of S is a collection $\alpha = \{A_1, \ldots, A_r\}$ of disjoint, non-empty subsets of S. The sets A_i are the *blocks* of α. The set Q_n of all partial partitions of S is partially ordered by $\alpha \leqslant \beta$ if and only if every β-block is the union of a set of α-blocks, that is if and only if for each $B_k \in \beta$ there exists a non-empty subset α_k of α such that $B_k = \bigcup_{\alpha_k} A_j$.

Ordered in this way, Q_n is isomorphic to the lattice P_{n+1} of partitions of an $(n + 1)$-set $S \cup x_0$; the isomorphism $P_{n+1} \to Q_n$ is given simply by deleting from each partition $\{A_0 \cup x_0, A_1, \ldots, A_r\}$ of $S \cup x_0$ the distinguished block $A_0 \cup x_0$. We call the block of any partition of $S \cup x_0$ which contains x_0 the *zero block* of the partition.

Formally we define the inverse map $\phi: Q_n \to P_{n+1}$ by

$$\phi(\alpha) = \{A_0 \cup x_0, A_1, \ldots, A_r\}$$

where

$$\alpha = \{A_1, \ldots, A_r\}, \qquad A_0 = S \backslash \bigcup_{j=1}^{r} A_j.$$

Now let H be a finite multiplicative group with unit element 1. We define a *partial H-partition* of the set S to be a collection α of functions into H

$$\alpha = \{a_j : A_j \to H; \quad j = 1, \ldots, r\}$$

for which the domains A_i are disjoint non-empty subsets of S.

Let $Q'_n(H)$ denote the set of all partial H-partitions of S. The map $\pi: Q'_n(H) \to Q_n$ defined by

$$\pi(\alpha) = \{A_j : j = 1, \ldots, r\}.$$

takes each partial H-partition of S into its underlying partial partition of S.

If $\alpha = \{a_j : j = 1, \ldots, r\}$ is a partial H-function over S and α_k is any non-empty subset of α, a (left)-*linear combination* of α_k is a function

$$b_k : B_k \to H \qquad (B_k = \bigcup_{\alpha_k} A_i)$$

such that the restriction of b_k to A_j is a (left) multiple $h_j a_j$ of a_j for some $h_j \in H$. That is

$$b_k(x) = h_j a_j(x) \qquad (x \in A_j)$$

and in this case we write

$$b_k = \sum_{\alpha_k} h_j a_j$$

where the summation sign is to be interpreted as the "domain-disjoint union" of the functions following it. The natural analogue in $Q'_n(H)$ of the order relation of Q_n is then the following: $\alpha \leqslant \beta$ if and only if every β-function is a linear combination of a set of α-functions; that is if and only if for each $b_k \in \beta$ there exists a subset α_k of α and elements $h_j \in H$ such that $b_k = \sum_{\alpha_k} h_j a_j$, so $B_k = \bigcup_{\alpha_k} A_j$. It follows that $\pi(\alpha) \leqslant \pi(\beta)$ in Q_n.

Now let $\mathscr{E}$ be the equivalence relation on $Q'_n(H)$ defined by $\alpha \, \mathscr{E} \, \alpha'$ if and only if $\alpha \leqslant \alpha'$, $\alpha' \leqslant \alpha$, that is if and only if there exists a bijection $a'_j \leftrightarrow a_j$ such that

$$a'_j = h_j a_j, \qquad a_j = h_j^{-1} a'_j.$$

Any member α' of an $\mathscr{E}$-class (α) is uniquely determined up to scalar multiples of its elements, and the preorder $\leqslant$ on $Q'_n(H)$ induces in the usual way a partial order, also denoted $\leqslant$ on the quotient set $Q_n(H) = Q'_n(H)/\mathscr{E}$ of $\mathscr{E}$-classes, by

$$(\alpha) \leqslant (\beta) \Leftrightarrow \alpha \leqslant \beta.$$

The situation is analogous to the identification of the points of a 1 dimensional subspace of a vector space when forming a projective geometry.

It is also easy to see that $Q_n(\{1\})$ is the usual partition lattice.

The main result of Dowling is:

THEOREM 1. *Under the ordering* $\leqslant$, $Q_n(H)$ *is a geometric lattice.*

Thus for any finite group H and integer n we have a geometric lattice which as Dowling goes on to show possesses considerable structure. These lattices seem to occupy a middle ground between the highly structured projective geometric lattices and arbitrary geometric lattices.

First observe $Q_n(H)$ has rank function ρ given by

(1) $\rho(\alpha) = n - |\alpha| \qquad \alpha \in Q_n(H)$

so that the dimension of the lattice $Q_n(H)$ depends on H only through its order m.

However the lattice $Q_n(H)$ does reflect the structure of the underlying group H since Dowling proves;

THEOREM 2. *If $n \geqslant 3$ and $Q_n(H) \simeq Q_n(H')$ then the groups H, H' are isomorphic.*

Dowling also proves the following striking representability theorem.

THEOREM 3. *Let H be a finite group of order m and let $n \geqslant 3$. If H is not cyclic the matroid determined by the lattice $Q_n(H)$ is not representable over any field. If H is cyclic, then $Q_n(H)$ is representable over*

(a)   *every field if and only if $m = 1$ (that is H is trivial)*
(b)   *the finite field of order q if and only if $m \mid (q - 1)$*
(c)   *the rational or real field if and only if $m = 1$ or 2*
(d)   *the complex field for all m.*

The key to the above result is the following theorem. If F is a field let $F^* = F \backslash 0$ denote the multiplicative group of F. Then we have:

THEOREM 4. *If $Q_n(H)$ is representable over a field F and $n \geqslant 3$ then H is isomorphic to a subgroup of F^*. Conversely if H is isomorphic to a subgroup of F^* then $Q_n(H)$ is representable over F.*

EXERCISES 4

1. Prove that any cofinite set of primes can occur as a characteristic set of a matroid.

2. Prove that for $n \geqslant 3$, $Q_n(H)$ is not transversal or graphic for any non-trivial group H.

NOTES ON CHAPTER 11

Maclane's paper [38] gives an account of the work on abstract algebraic structures done in the 1930's by Baer, Birkhoff, Menger, Teichmüller and others. Sections 1–3 are based on Ingleton [71], Ingleton and Main [75] and Piff [72]. Probably the most interesting unsolved problem in algebraic matroid theory which offers some hope of solution is deciding whether or not the dual of an algebraic matroid is also algebraic.

For an alternative treatment of Section 5 see Dowling [73a]. Dlab [66b] looks for conditions on a ring R in order that modules over R admit a "matroid like" dependence theory (necessarily infinite). For other algebraically oriented work we refer to Kertesz [60], Dlab [62a], [66b], [69].

Recently White [75] has introduced the concept of a *bracket algebra* associated with a matroid. This is an interesting idea which is based on the idea that the bracket corresponds to a determinant and hence elements in this bracket algebra must satisfy the standard identities of determinant theory.

CHAPTER 12

Block Designs

1. PROJECTIVE AND AFFINE SPACES

Projective spaces are structures of fundamental importance in combinatorial theory. Here we briefly review their properties.

Consider an ordered triple (P, L, ε) where P and L are disjoint sets and ε is a subset of $P \times L$.

The elements of P are called *points*, the elements of L are called *lines* and the incidence relation ε is interpreted by: if for a point x and line l, (x, l) belongs to the set ε, then we say that the point x *is on* the line l or that the line l *contains* or *passes through* x. If a line l contains two points x, y we call it the line xy.

If distinct points x, y, z are on a common line l we say they are *collinear*.

Definition. A *projective space* is a triple (P, L, ε) as described above, satisfying the following axioms:

(P1) Any two distinct points are on exactly one line.
(P2) If x, y, z, w are four distinct points, no three collinear, and if xy intersects zw then xz intersects yw.
(P3) Each line contains at least three points.

The axiom (P2) is known as the Veblen–Young axiom. The axiom (P3) serves to rule out certain degenerate cases.

Definition. A *projective plane* is a triple (P, L, ε) as described in the opening paragraph satisfying the axioms:

(PP1) Any two distinct points are on exactly one line.
(PP2) Any two distinct lines have exactly one point in common.
(PP3) There exist four points, no three of which are collinear.

The simplest non-trivial example of a projective plane is the set of points and lines of the Fano configuration (see Fig. 1.11.2).

Now if (P, L, ε) is a projective space and x, l are respectively a point and line with $x \notin l$, the *plane* (x, l) is the ordered triple (P', L', ε') with

$P' = \{y \in P: y = x \text{ or } yx \text{ intersects } l\}$
$L' = \{l \in L: l \text{ passes through two distinct points of } P'\}$
$\varepsilon' = \varepsilon \text{ restricted to } P' \text{ and } L'.$

It is routine to check the following result.

THEOREM 1. *Every plane in a projective space is a projective plane.*

We can now recursively define the projective subspaces of a projective space.

Let $Q = (P, L, \varepsilon)$ be a projective space. A *2-dimensional subspace* of Q is a plane as defined above.

If $n > 2$, an *n-dimensional subspace* $Q_n = Q_n(x, Q_{n-1})$ is defined from an $(n - 1)$-dimensional subspace Q_{n-1} by taking a point x of P not in Q_{n-1} and letting $Q_n(x, Q_{n-1})$ be the ordered triple $(P_n, L_n, \varepsilon_n)$ where

$$P_n = \{y \in P: y = x \text{ or } yx \text{ intersects } Q_{n-1}\}$$
$$L_n = \{l \in L: l \text{ contains two points of } P_n\}$$
$$\varepsilon_n \text{ is the restriction of } \varepsilon \text{ to } P_n \times L_n.$$

Thus we can define the *dimension* of a projective space Q to be

-1 if $P = \varnothing$,
0 if $|P| = 1$,
1 if $|L| = 1$,
n if Q contains an n-dimensional subspace with point set P,
∞ if Q contains an n-dimensional subspace for all n.

Exercise. Prove that a projective plane is a projective space of dimension 2, and that every projective space of dimension 2 is a projective plane.

We can intuitively picture the subspaces of a finite dimensional projective space as the flats of a matroid, and in fact it is not difficult to prove.

THEOREM 2. *If $Q = (P, L, \varepsilon)$ is a finite projective space of dimension n, the set of subspaces of Q ordered by inclusion forms a geometric lattice of rank $n + 1$.*

Conversely we can give an alternative and in many ways simpler description of a finite projective space.

THEOREM 3. *If M is a simple matroid on S with the following properties:*

(P1)′ $|L| \geqslant 3$ *for every 2-flat L of M;*

(P2)′ $L \cap H \neq \emptyset$ *for every 2-flat L and hyperplane H of M;*

then the points and 2-flats of M with the obvious incidence relation form a projective space, and every projective space can be got from a simple matroid in this way.

Proof. Routine checking of axioms. ∎

It is well known and easy to prove that the lattice of flats of a projective space is modular. We close this section by briefly presenting an alternative lattice theoretic characterization of projective spaces.

First notice that there exist modular geometric lattices which are not the lattice of flats of a projective space, for example a Boolean algebra.

Also it is easy to prove that the direct product of modular lattices is modular, so taking $\mathscr{L}_1$ a Boolean algebra, $\mathscr{L}_2$ any projective space, the product $\mathscr{L}_1 \times \mathscr{L}_2$ is a geometric modular lattice which is neither a projective space nor a Boolean algebra. However these are essentially the only possibilities.

THEOREM 4. *Any modular geometric lattice is the direct product of a Boolean algebra with projective spaces.*

Proof. See Birkhoff [67]. ∎

Expressed in matroid language we have:

COROLLARY. *If $\mathscr{L}$ is a modular geometric lattice whose underlying simple matroid M is connected, then $\mathscr{L}$ is either a Boolean algebra or is the lattice of flats of a projective space.*

For further lattice theoretic properties of projective spaces we refer to Birkhoff's book [67, Chapter 4, Section 7].

The finite projective geometry PG(n, q)

Let F be a finite field and n be a non-negative integer. The *projective geometry* of dimension n over F, denoted by $PG(n, F)$, is constructed as follows.

Let $\mathscr{L}$ be the lattice of subspaces of the vector space $V(n + 1, F)$. Let P be the set of atoms of this lattice and L be the set of 2-flats. Then $PG(n, F)$ is (P, L, ε) where two atoms x, y belong to a line $l \varepsilon L$ if and only if the flat corresponding to l contains x, y.

The fundamental theorem of projective spaces is the following coordinatization result.

THEOREM 5. *The only finite projective spaces of dimension $\geqslant 3$ are the projective geometries over finite fields.*

Thus for projective spaces of dimension not less than 3 the existence problem is completely settled.

There exists a projective space $PG(n, q) = PG(n, GF(q))$ for each prime power q, and up to isomorphism these spaces are unique.

For projective spaces of dimension 2, that is projective planes, the existence problem is one of the classic unsolved problems of combinatorial theory. First we state a basic property of projective planes.

THEOREM 6. *Corresponding to each projective plane there is a cardinal number n, called its order, such that:*

(1)   *each line in the plane contains $n + 1$ points;*
(2)   *through each point of the plane pass $n + 1$ lines;*
(3)   *the plane contains $n^2 + n + 1$ points and $n^2 + n + 1$ lines.*

Now $PG(2, q)$ is a projective plane with $q + 1$ points on every line and hence we know:

(4)   There exists a projective plane of any prime power order.

The big difference between projective planes and projective spaces of higher dimension is contained in the following two statements.

(5)   There exist projective planes of prime power order q which are not isomorphic to $PG(2, q)$; in other words are not coordinatizable over a field, see for example Hall [67].

(6)   It is not known whther or not there exists a plane of order not a prime power.

In 1949 Bruck and Ryser proved:

THEOREM 7. *If $n \equiv 1 \pmod 4$ or $n \equiv 2 \pmod 4$ and if n is not the sum of two squares, then there is no projective plane of order n.*

The smallest $n > 1$ that are not prime powers are $n = 6$, $n = 10$. By the Bruck–Ryser theorem, 6 cannot be the order of a projective plane. However $10 = 3^2 + 1^2$ is not excluded by this theorem. A massive amount of work has gone into trying to settle whether or not there exists a plane of order 10 but at the moment the problem is still open.

Affine geometries and affine planes

If we use the basic incidence properties of the Euclidean plane to construct an abstract geometrical configuration we are led to the following definition.

An *affine plane* is a set P of *points* and a collection L of subsets of P called *lines* such that:

(A1)   Two distinct points are on exactly one line.

(A2)   If l is a line and x is a point not on l, there is exactly one line through x which does not intersect l.

(A3)   There are four points, no three collinear.

The fundamental relationship between projective and affine planes is summarized in the following theorem.

THEOREM 8. (a) *If one line and all the points on it are removed from a projective plane the remaining incidence structure is an affine plane.*
(b) *Given any affine plane there exists a projective plane which determines it by the construction outlined in* (a).

Hence from Theorem 6 we have

COROLLARY. *Associated with any affine plane is a cardinal n called its order such that the following statements hold.*

(7)   *Each line of the plane contains n points.*

(8)   *Each point of the plane is on $n + 1$ lines.*

(9)   *The plane contains n^2 points.*

(10)   *The plane contains $n^2 + n$ lines.*

Now we could (see Bumcrot [69]) define an affine space in a way similar to our original definition of projective space. However with our geometric lattice approach it serves the purposes of this chapter to proceed as follows.

Let M be the simple matroid on the set S of atoms of the geometric lattice $PG(n, q)$. Let H be any hyperplane of M. Let M' be the matroid $M|(S \backslash H)$. Let $\mathscr{L}'$ be the geometric lattice of M'.

The affine geometry $AG(n, q)$ has the set of points S' and lines the 2-flats of $\mathscr{L}'$ with incidence defined by the inclusion relation.

Note by symmetry this definition is (up to isomorphism) independent of the choice of hyperplane H. Moreover it is the natural analogue of the construction of affine planes from projective planes.

Henceforth we shall be more concerned with the various subspaces of projective and affine space and shall often identify $PG(n, q)$, $AG(n, q)$ with their geometric lattices of flats.

EXERCISES 1

1. Find some modular non-geometric lattices.

2. Prove that the dimension of a projective space is one less than its rank as a matroid.

3. Prove that $AG(n, q)$ has q^n points and each hyperplane of $AG(n, q)$ contains q^{n-1} points.

4. Two hyperplanes of $AG(n - q)$ are *parallel* if they do not intersect. Prove that two non-parallel hyperplanes intersect in the same number of elements.

5. Prove that the flats of a given rank in $AG(n, q)$ have a common cardinality.

$^\circ$6. Is it true that any geometric lattice of rank 3 can be embedded as a sublattice in the lattice of flats of some projective plane?

2. BLOCK DESIGNS AND STEINER SYSTEMS

A *balanced incomplete block design* (BIBD) on a set S with $|S| = v$, is a family of subsets $(B_1, \ldots, B_b)$ of S called *blocks* such that:

(1) $|B_i| = k \quad i \leqslant i \leqslant b$.

(2)   If $x \in S$ then x belongs to exactly r of the blocks B_i.

(3)   If x, y are distinct elements of S then $\{x, y\}$ is contained in exactly λ blocks.

We shall denote such a block design by $D = D(b, v, r, k, \lambda)$ and the integers b, v, r, k, λ are called the *parameters* of the design. These parameters are clearly not independent of each other.

Elementary counting arguments show that

(4) $v = bk$,

(5) $(v - 1)\lambda = (k - 1)r$.

Fisher's inequality (see Hall [67]) shows

(6) $b \geqslant v$,

(7) $b + k \geqslant v + r$.

A design with $b = v$ is called *symmetric*. In such a design by (4), $r = k$ and hence such structures have only three parameters and are often called (v, k, λ)-designs. Comparison of definitions shows there is an obvious isomorphism between projective planes and symmetric designs with $\lambda = 1$.

(8)   A projective plane is a $(v, k, 1)$ design when its lines are taken as blocks and conversely.

More generally we have:

(9)   The points and hyperplanes of projective space $PG(n, q)$ are the points and blocks of a symmetric design with parameters

$$v = \frac{q^{n+1} - 1}{q - 1}, \qquad k = \frac{q^n - 1}{q - 1}, \qquad \lambda = \frac{q^{n-1} - 1}{q - 1}.$$

An important characteristic of a symmetric (v, k, λ) design is:

(10)   Distinct blocks B_i, B_j have $|B_i \cap B_j| = \lambda$.

Many other symmetric designs exist apart from those given by (9), see Hall [67].

The interest in symmetric designs is partly intrinsic because of their tighter structure and also because when they exist they "contain" other non-symmetric block designs.

THEOREM 1. *If $D(v, k, \lambda)$ is a symmetric block design on S the following constructions give BIBDs.*

(a)   *The derived design on B_i has as blocks the sets $B_i \cap B_j (j \neq i)$ and has parameters $(v - 1, k, k - 1, \lambda, \lambda - 1)$.*

(b)   *The residual design on $S \backslash B_i$ with blocks $B_j \backslash B_i (j \neq i)$ with parameters $(v - 1, v - k, k, k - \lambda, \lambda)$*

Thus from the symmetric design (9)—the points and hyperplanes of $PG(n, q)$ the residual design is the points and hyperplanes of $AG(n, q)$.

On arbitrary BIBDs we may perform the following constructions to produce other designs.

(11)   If $D(b, v, r, k, \lambda)$ is a design on S with blocks $B_i (1 \leqslant i \leqslant b)$ the *complementary design* on S has as blocks the sets $S \backslash B_i (1 \leqslant i \leqslant b)$ and has parameters $(b, v, b - r, v - k, b - 2r + \lambda)$.

(12)   If the blocks of $D = D(b, v, r, k, \lambda)$ are distinct as sets the *subtract design* has as blocks all k-sets of S which are not blocks of D. This has parameters

$$\left[\binom{v}{k} - b, \; v, \; \binom{v - 1}{k - 1} - r, \; k, \; \binom{v - 2}{k - 2} - \lambda \right].$$

(13)   If $D_1(b, v, r, k, \lambda)$ and $D_2(b', k, r', k', \lambda')$ are designs then there exists a design with parameters $(bb', v, rr', k', \lambda\lambda')$, called the *composition* of D_1 and D_2.

Proof. Replace each block B of D_1 by a design D_B isomorphic to D_2 on the set of elements of B.

The problem of deciding for what values of (v, k, λ) there exist symmetric designs is, since it is a generalization of the projective plane problem, open. A table of parameter values for which designs are known to exist, not exist, or still undecided can be found at the back of Hall's book [67].

Steiner systems and t-designs

For integer $t > 1$, a *t-design* D on a set S ($|S| = v$) is a collection of b subsets of S called blocks with the property that:

(14)   Each block has cardinal k.
(15)   Each t-subset of S is contained in exactly λ blocks.

Such a design is called a t-(v, k, λ) design.
It is clear that:

(16)   Any t-(v, k, λ) design is an s-(v, k, λ_s) design for all $s \leqslant t$, where

$$\lambda_s = \lambda \binom{v-s}{t-s} \bigg/ \binom{k-s}{t-s}.$$

(17) $\lambda \dbinom{v}{t} = b \dbinom{k}{t}.$

Thus a BIBD, $D(b, v, r, k, \lambda)$ is a 2-(v, k, λ) design and by (16) the blocks of any t-design are also the blocks of a BIBD on the same set.

A t-design on S is *symmetric* if it has $|S|$ blocks.

THEOREM 2 (Mendelsohn [71]). *There is no non-trivial symmetric t-design with* $t \geqslant 3$.

A *Steiner system* $S(d, k, n)$ is a d-$(n, k, 1)$ design.
We list the following properties of Steiner systems.

(18) $S(d, k, n)$ is a BIBD with parameters

$$b = \binom{v}{d} \bigg/ \binom{k}{d}, \quad v = n, \quad r = \binom{v-1}{d-1} \bigg/ \binom{k-1}{d-1}, \quad \lambda = \binom{v-2}{d-2} \bigg/ \binom{k-2}{d-2}.$$

(19)   If $S(d, k, n)$ exists then so does $S(d - 1, k - 1, n - 1)$.

Proof. Take the blocks $S(d, k, n)$ which contain a given element x. These blocks then form the required $S(d - 1, k - 1, n - 1)$ on deleting x from each block and the underlying set.

A Steiner system of type $S(2, 3, n)$ is known as a *Steiner triple system*. The existence problem has been settled in the following result proved independently by Reiss and Kirkman (see Hall [67]).

THEOREM 3. *An $S(2, 3, n)$ exists if and only if $n = 1, 3 \pmod 6$.*

A *quadruple system* is an $S(3, 4, n)$. Hanani [60] has proved

THEOREM 4. *An $S(3, 4, n)$ exists if and only if $n \geqslant 4$ and $n \equiv 2$ or $4 \bmod 6$.*

Very few non-trivial t-designs are known to exist for $t > 3$. Until recently the only known Steiner systems with $d > 3$ were

$$S(4, 5, 11), \qquad S(5, 6, 12), \qquad S(4, 7, 23), \qquad S(5, 8, 24).$$

However Denniston [75] has very recently found Steiner systems $S(5, 6, 24)$ $S(5, 7, 28)$, $S(5, 6, 48)$ and $S(5, 6, 84)$. Moreover unlike all other Steiner systems with $d = 5$ there are two non-isomorphic ones of type $(5, 6, 24)$. The Steiner system $S(5, 8, 24)$ has many interesting properties, we study it in more detail in Section 6.

EXERCISES 2

1. Show that the t-flats of $PG(n, q)$ are the blocks of a BIBD on the points of $PG(n, q)$ and find its parameters.

2. Is every BIBD with $r = k + 1$ and $\lambda = 1$ an affine plane?

3. Prove that if there exists a (b, v, r, k, λ) design there exists a $(tb, v, tr, k, t\lambda)$ design.

4. Prove that if a BIBD has the property that any two distinct blocks intersect in a set of size λ then the design must be symmetric.

5. Prove that $S(3, q + 1, q^2 + 1)$ exists for all prime powers q. (It is the *inversive plane* $IP(q)$).

°6. Prove there is no non-trivial t-design with $t \geqslant 6$.

3. MATROIDS AND BLOCK DESIGNS

Consider the following classes of block designs:

(1) the points of a projective geometry with blocks its hyperplanes;
(2) the points of an affine geometry with blocks the hyperplanes;
(3) the points of a projective or affine geometry with blocks the k-flats for any k.

In each case the blocks of the associated design are the hyperplanes of a geometric lattice (in (3) we just truncate the lattice $\mathscr{L}(PG(n, q))$ or $\mathscr{L}(AG(n, q))$).

Consider a Steiner system $S(d, k, n)$, on a set S. Since every d-subset of S is contained in a unique block, by Theorem 2.3.1. we see:

(4) The blocks of an $S(d, k, n)$ on S are the hyperplanes of a paving matroid on S.

We define a *matroid design* to be a matroid on S whose hyperplanes are the set of blocks of a balanced incomplete design.

A set of parameters (b, v, r, k, λ) is called *matroidal* if there exists a matroid design with those parameters.

Thus a design $D(b, v, r, k, \lambda)$ has its blocks the hyperplanes of a matroid design if its blocks are distinct subsets which satisfy the hyperplane axioms of Theorem 2.2.3.

We will often use the term *matroid design* to refer to the overall structure regardless of whether we are regarding it as a matroid or as a design.

For non-symmetric designs the problem of deciding which parameter sets are matroidal is still unsettled.

THEOREM 1. *Any* $D(b, v, r, k, 1)$ *is a matroid design of rank* 3.

Proof. Since each 2-subset belongs to exactly one block the blocks form the hyperplanes of a paving matroid. ∎

THEOREM 2. *The parameter set* $(b, v, r, k, 2)$ *is always non-matroidal except for trivial designs.*

Proof. Let x, y be distinct elements. The set $\{x, y\}$ is contained in precisely two blocks B_1, B_2. Hence if $z \notin B_1 \cup B_2$, $z \cup (B_1 \cap B_2)$ contains $\{z, x, y\}$ and there does not exist a third block containing $\{x, y\}$. Thus the hyperplane axiom cannot be satisfied. ∎

However for $\lambda \geqslant 3$ little seems to be known.

THEOREM 3. *For $\lambda > 2$, necessary conditions that (b, v, r, k, λ) be matroidal are that*

(5) (a) $\lambda(k - 2) + 2 \geqslant v$,

    (b) $b\dbinom{k}{3} \geqslant \dbinom{v}{3}$.

There exists an integer i, $2 \leqslant i \leqslant k - 2$ with

    (c) $b\dbinom{k}{i + 1} \leqslant \dbinom{v}{i + 1}$,

    (d) $(v - i)/(k - i)$ *is an integer p not greater than λ,*

    (e) $k \leqslant 1 + p(i - 1)$.

The proof of Theorem 3 is not difficult and we leave it to the reader as an exercise.

In the next section we shall completely characterize symmetric matroid designs.

We close this section with a brief discussion of other ways of relating matroids and designs. Classifying matroids whose circuits are the blocks of a design is clearly just the dual problem to finding matroid designs.

More interesting is the problem of deciding which matroids have a set of bases which are the blocks of a design. We describe the main results of Main and Welsh [75] where proofs of all assertions may be found.

A block design $D(b, v, r, k, \lambda)$ on S is said to be a *base design* or *basic* if it has distinct blocks and these blocks are the bases of a matroid on S. The associated matroid is denoted by $M(D)$ where appropriate.

Similarly a matroid is a base design if its bases are the blocks of a BIBD.

An example of a base design is the k-uniform matroid for any k. We shall call such a base design *trivial*.

Non-trivial examples of base designs are given in the following assertions.

(6) The bases of $PG(n, q)$ are the blocks of a design.

(7) The paving matroid of a Steiner system is a base design.

It is immediate that:

(8) If D is a base design, the complemetary design D is also a base design and is the set of bases of $[M(D)]^*$.

Our next result shows there are no non-trivial symmetric base designs.

(9) If $D(v, k, \lambda)$ is a symmetric base design then $v = k + 1$ and $M(D)$ is k-uniform.

Proof. Let B_1, B_2 be distinct blocks of D and let $x \in B_1 \backslash B_2$. Then there exists $y \in B_2 \backslash B_1$ such that $B_3 = (B_1 \backslash x) \cup y \in D$. But then $B_1 \cap B_3 = B_1 \backslash x$, so $|B_1 \cap B_3| = |B_1 \backslash x| = k - 1$. But since D is symmetric any two distinct blocks have intersection of cardinal λ, so $\lambda = k - 1$ and thus since $(v - 1) = k(k - 1)$ we get $v = k + 1$. Thus the blocks of D must be all k-sets of S and $M(D)$ is k-uniform.

As one might expect the best we can do for non-symmetric designs is to give necessary conditions on the parameter set in order that a design with those parameters be basic.

(10)   Necessary conditions for a BIBD $D(b, v, r, k, \lambda)$ to be a base design are:

   (a) $\lambda \geqslant k - 1$, or equivalently $r \geqslant v - 1$, with equality in each case if and only if $M(D)$ is trivial.
   (b) $(r - 1)(v - k) \geqslant rk/(k - 1)$ unless $M(D)$ is trivial.
   (c) $b - r \geqslant v - 1$, with equality if and only if $M(D)$ is trivial.

However base designs seem to be far more plentiful than matroid designs. A large class can be constructed by the following theorem.

THEOREM 4. *Let $D(b, v, r, k, \lambda)$ be a block design with the property that no two blocks intersect in more than $k - 2$ elements. Then the subtract design $-D$ is basic.*

Proof. Let B_1, B_2 be distinct blocks of $-D$, let $e \in B_1 \backslash B_2$. If there is no $f \in B_2 \backslash B_1$ for which $(B_1 \backslash e) \cup f$ is a block of $-D$, then $(B_1 \backslash e) \cup f$ is a block of D for each $f \in B_2 \backslash B_1$. But if $B_2 \backslash B_1 \supseteq \{f_1, f_2\}$ then

$$|(B_1 \backslash e) \cup f_1| \cap |(B_1 \backslash e) \cup f_2| = B_1 \backslash e$$

has cardinal $k - 1$, giving a contradiction. Hence $-D$ is a base design. ∎

EXERCISES 3

1. Prove that a matroid design is a connected matroid.

2. Call a matroid *equicardinal* if all its hyperplanes are the same cardinality. Find an equicardinal matroid which is not a matroid design.

3. Find the graphic equicardinal matroids. (Murty [71a])

4. Prove that if M is a non-trvial base design it is a connected matroid. (Main and Welsh [75])

5. Let $D(b, v, r, k, 2)$ be a BIBD whose blocks are the circuits of a matroid. Prove that $k = 3$ or 4 and that if $k = 3$, $b = 0 \bmod 4$. (Main [73])

$\circ$6. Do the bases of a matroid design form a base design? (Main and Welsh [75])

4. THEOREMS OF DEMBOWSKI, WAGNER AND KANTOR

In this section we shall completely characterize those matroid designs which are symmetric. The key result is an extension by Kantor [69] of a classic theorem by Dembowski and Wagner [60].

THEOREM 1 (Dembowski–Wagner). *Let D be a symmetric (v, k, λ)-design with $\lambda > 1$, and $n = k - \lambda > 1$. Then each of the following properties is necessary and sufficient for D to be a finite projective space, with blocks as hyperplanes.*

(a)   *Every line meets every block.*

(b)   *Every line contains exactly $2 + (n - 1)/\lambda$ points.*

(c)   *Every plane is contained in exactly $\lambda(\lambda - 1)/(n - \lambda - 1)$ blocks.*

Lines, triangles, and planes are defined below.

We first need some preliminary lemmas.

Let D be a (v, k, λ)-symmetric design with $\lambda > 1$ and $n = k - \lambda > 1$.

The *line* joining two points x, y and denoted by xy is the intersection of all blocks containing x, y.

LEMMA 1. *If $x, y \in pq$ and $x \neq y$, then $xy = pq$.*

Proof. Let $B_1, \ldots, B_\lambda$ be the blocks through p and q. Then if $x, y \in pq$ each $B_i (1 \leq i \leq \lambda)$ contains x and y. But there can be no other blocks containing x and y. Hence $xy = pq$.

Now if p, q, r are three points and $p \notin qr$ then by Lemma 1, $y \notin pr$, $r \notin pq$, and we say that p, q, r form a *triangle*. The *plane spanned by the triangle p, q, r* denoted by pqr is defined to be the intersection of all blocks through p, q, and r.

LEMMA 2. *If π is a plane and $p, q \in \pi$, $p \neq q$, then $pq \subset \pi$.*

Proof. An immediate consequence of the definitions of lines and planes.

If the triangle p, q, r is contained in the plane π it may well be that the plane pqr is a proper subset of π. We now make the following definition: A design D satisfies *condition* (*) if every plane is contained in the same number of blocks.

Analogously to Lemma 1 we obtain:

LEMMA 3. *If D satisfies (*) and if the triangle p, q, r is contained in the plane π then pqr = π.*

We also deduce

LEMMA 4. *Let D satisfy (*). Further let B and π be a block and plane of D with π not contained in B. Then if B and π have two distinct points p, q in common, pq = B $\cap$ π.*

Proof. From Lemma 2, $pq \subseteq B \cap \pi$. If pq were different from $B \cap \pi$ there would exist $r \in B \cap \pi$ with $r \notin pq$. But then it follows from Lemma 3 that $\pi = pqr \subset B$ which contradicts our initial assumption.

LEMMA 5. *If every plane of D is contained in u blocks, then every line of D contains exactly $(\lambda k - uv)/(\lambda - u)$ points.*

Proof. Let h be an arbitrary line. Suppose there are d blocks not intersecting h, and $d(p)$ of these pass through the point $p \notin h$. Then

$$(1) \qquad \sum_{p \,\in\, h} d(p) = kd.$$

Since D satisfies condition (*), it follows from Lemmas 2 and 3 that there are exactly r blocks containing h and any point p not on h. If x is a point of h, there are λ blocks containing p and x, and $\lambda - u$ of these have only x in common with h. Hence, if h contains $m + 1$ points:

$$(2) \qquad k = d(p) + u + (\lambda - u)(m + 1) \qquad (p \notin h).$$

Next, there are λ blocks containing h, and hence the n blocks through a point $x \in h$ which do not contain h have no point other than x in common with h. This proves.

$$(3) \qquad v = d + \lambda + n(m + 1).$$

Combining (1), (2) and (3) we obtain

$$k[v - \lambda - n(m + 1)] = kd \qquad = \sum_{p \notin h} d(p)$$
$$= [k - u - (\lambda - u)(m + 1)](v - m + 1),$$

or on rearrangement

$$(4) \quad (\lambda - u)m^2 - [(\lambda - u)(v - 1) - n(k - 1)]m = k(v - k) - n(v - 1).$$

But since $\lambda(v - 1) = k(k - 1)$ the right-hand side of (4) is zero. Since $m \neq 0$ an easy calculation proves the lemma.

Proof of Theorem 1

Assume (a) holds.

Consider any line h of D. There are exactly λ blocks containing h. The remaining $v - \lambda$ blocks meet h in just one point.

Through every point of h there are exactly n blocks not containing h. Hence there must be $(v - \lambda)/n$ points on h. But $(v - \lambda) = n(2 + (n - 1)/\lambda)$ and (b) holds.

Similarly if (b) holds, a similar consideration shows that (a) must hold. Thus (a) and (b) are equivalent.

Next we show that (a) and (b) together imply (c). Let p, q, r be a triangle in D and let u be the number of blocks containing the plane pqr. We wish to show that u is independent of p, q, r. The number of blocks through p not containing the line $h = qr$ is $n + \lambda - u$. Let x be any point on h. There are λ blocks through p and x, and of these r contain the line h, hence the remaining $\lambda - u$ blocks meet h only in x. Since there are $2 + (n - 1)/\lambda$ points on h and since h meets every block it follows that the number of blocks through p which do not contain h is also $(\lambda - u)(2 + (n - 1)/\lambda)$. Hence we have shown that

$$n + \lambda - u = (\lambda - u)(2 + (n - 1)/\lambda)$$

which on rearrangement gives $u = \lambda(\lambda - 1)/(n + \lambda - 1)$, proving (c).

From Lemma 5 and the identity $\lambda(v - 1) = k(k - 1)$ it follows by easy calculation that (c) implies (b) so that we know the equivalence of (a), (b) and (c). Since condition (*) follows from (c) it will suffice for the remainder of the proof to show that properties (a) and (b) together with (*) imply that D is a projective space.

Let h_1, h_2 be two different lines through a point $s \in D$. Let p_i and q_i be different points $\neq s$ on h_i, $i = 1, 2$. (It follows from (b) that we can find such points). It will be sufficient to show that the lines $p = p_1 p_2$ and $q = q_1 q_2$ have a point in common. Let π be the plane $sp_1 p_2$. It follows by repeated application of Lemma 2 that $\pi = sq_1 q_2$ and that p and q are contained in π.

Now let B be a block containing p but not q. Such a block exists, since if every block through p contained q we would have $p = \pi$ which is absurd. By Lemma 4 we have $\pi \cap B = p$, and since the line q which is contained in π, meets B by property (a), it follows that the lines p and q have non-empty intersection, as we needed to show.

To complete the proof it only remains to show that the hyperplanes of the projective space determined by the points and lines of D are in fact the blocks of D.

Since every block contains the line joining any two of its points the blocks are linear subspaces. The hyperplanes are the only subspaces which meet every line, hence it follows from (a) that the blocks are hyperplanes. Since the number of blocks equals the number of points it is therefore equal to the number of hyperplanes and the blocks must be all the hyperplanes. This completes the proof of the theorem. ■

We now state an extension of the Dembowski–Wagner theorem due to Kantor [69].

THEOREM 2. *Let* $D(v, k, \lambda)$ *be a symmetric design. Then the blocks of D are the hyperplanes of a matroid if and only if either*

(a) $\lambda = 1$, *in which case the design is a projective plane; or*
(b) $\lambda > 1$ *and the blocks are the hyperplanes of a projective space.*

Kantor in fact proves much more than this, he characterizes affine spaces in a similar fashion.

We leave it as a (not-straightforward) exercise to prove that if the blocks of a symmetric design satisfy the hyperplane axioms of Theorem 2.2.3 then each of (a), (b), (c) of the Dembowski–Wagner theorem must hold. (Kantor's proof uses induction on λ and I know of no easier proof).

As a consequence of Theorem 2 we get:

COROLLARY. *The parameter set* (v, k, λ) *is matroidal if and only if either* $\lambda = 1$ *or*

$$(5) \quad v = (q^{n+1} - 1)/(q - 1), k = (q^n - 1)/(q - 1), \lambda = (q^{n-1} - 1)/(q - 1),$$

for some prime power q.

Note that Theorem 2 does not imply that every design with parameters (5) is a $PG(n, q)$. In another paper Kantor [74] proves that necessary and sufficient conditions that a non-trivial symmetric design is isomorphic to a projective space are that it has an automorphism group which is 2-transitive on its points.

We close this section by stating without proof the main theorem of Kantor [69].

THEOREM 3. *If a matroid design on v points with blocks of cardinal k has the following properties:*

(a)   *any two distinct intersecting blocks have* $\mu \geqslant 1$ *points in common*
(b)   *given distinct points x, y there is a block containing x but not y*
(c) $v - 2 \geqslant k \geqslant \mu$
(d) $(\mu - 1)(v - k) \neq (k - \mu)^2$
(e)   *its matroid rank is at least 5*

then D is a projective or affine space.

As Kantor pointed out the Steiner system $S(3, 6, 22)$ satisfies (a)–(d) above so that the condition (e) is essential.

EXERCISES 4

1. Prove that a necessary and sufficient condition that a (v, k, λ)-design with $\lambda > 1$ is a projective space is that it has an automorphism group transitive on non-incident point-line pairs. (Dembowski and Wagner [60])

2. Prove that a matroid design is isomorphic to a projective space if and only if it satisfies (a)–(d) of Theorem 3 and some block meets every other block. (Kantor [69])

3. Prove that a matroid design is isomorphic to a finite affine space if and only if it satisfies (a)–(d) of Theorem 3 and for some non-incident point-block pair there is precisely one block not meeting the given block and incident with the given point. (Kantor [69])

5. PERFECT MATROID DESIGNS

A *perfect matroid design* or PMD is a matroid M in which each k-flat has a common cardinal α_k, $1 \leqslant k \leqslant \rho(M)$. These were first studied by Young, Murty and Edmonds [70]. Examples of PMDs are:

(1)   uniform matroids,
(2)   projective geometries,
(3)   affine geometries,
(4)   the matroid designs of Steiner systems.

Now consider the geometric lattice associated with any PMD. The following remark is obvious.

(5)   If M is a PMD any truncation of M is a PMD.

THEOREM 1. *Let M be a PMD of rank n on S. Then for any i, j, k, and any i-flat F_i and k-flat F_k with $F_i \subseteq F_k$, the number $t(i, j, k)$ of j-flats F_j such that $F_i \subseteq F_j \subseteq F_k$ is independent of the choice of F_i and F_k.*

Proof. We use the following lemma.

LEMMA. *$F_k \backslash F_i$ is partitioned by sets of the form $F_{i+1} \backslash F_i$ where F_{i+1} is an $i + 1$ flat such that $F_i \subseteq F_{i+1} \subseteq F_k$.*

Proof. The sets $F_{i+1} \backslash F_i$ are just the rank 1 flats of the minor on $F_k \backslash F_i$ corresponding to the geometric interval lattice $[F_i, F_k]$.

Hence a simple counting argument shows that the number of $(i + 1)$-flats F_{i+1} such that $F_i \subseteq F_{i+1} \subseteq F_k$ is given by

$$(6) \qquad t(i, i + 1, k) = \frac{|F_k| - |F_i|}{|F_{i+1}| - |F_i|} = \frac{\alpha_k - a_i}{\alpha_{i+1} - \alpha_i}$$

and thus $t(i, i + 1, k)$ is independent of the choice of F_i, F_k.

Now use induction on j, we have proved it true for $j = i + 1$. Let $\mathscr{F}^j$ denote the set of j-flats $\{A_j\}$ such that $F_i \subseteq A_j \subseteq F_k$. Let $\mathscr{F}^{j+1}$ denote the set of $(j + 1)$-flats $\{A_{j+1}\}$ such that $F_i \subseteq A_{j+1} \subseteq F_k$. By the induction hypothesis $|\mathscr{F}^j| = t(i, j, k)$. But by (6) each member of $\mathscr{F}^j$ is contained in $t(j, j + 1, k)$ members of $\mathscr{F}^{j+1}$. By the induction hypothesis each member of $\mathscr{F}^{j+1}$ contains $t(i, j, j + 1)$ members of $\mathscr{F}^j$. Hence

$$|\mathscr{F}^{j+1}| = \frac{t(i, j, k)\, t(j, j + 1, k)}{t(i, j, j + 1)}$$

independently of the choice of F_i and F_k, which completes the proof of Theorem 1. ∎

COROLLARY 1. *If M is a PMD on S and $F \subseteq G$ are flats of M the matroid $M \cdot (S\backslash F)|(G\backslash F)$ associated with the interval $[F, G]$ of $\mathscr{L}(M)$ is also a PMD.*

We call $t(., ., .)$ the *t-function* of the perfect matroid design.

Call an integer valued function $t(i, j, k)$ of integers $0 \leqslant i \leqslant j \leqslant k \leqslant n$ *T-consistent* if it satisfies the following relations:

(T0) (i) $t(i, i, k) = t(i, k, k) = 1$ for all integers $0 \leqslant i \leqslant k \leqslant n$.
        (ii) $t(0, 1, i + 1) > t(0, 1, i)$ $\qquad\qquad 0 \leqslant i \leqslant n - 1$.

(T1) $t(i, i + 1, k) = \dfrac{t(0, 1, k) - t(0, 1, i)}{t(0, 1, i + 1) - t(0, 1, i)}$ $\qquad 0 \leqslant i \leqslant k \leqslant n.$

(T2) $t(i, j, k) = \dfrac{t(i, l, k)\, t(l, j, k)}{t(i, l, j)}$ $\qquad 0 \leqslant i \leqslant l \leqslant j \leqslant k \leqslant n.$

(T3) $t(i, i + 1, k) \leqslant t(i, j, k).$

THEOREM 2. *If $t(i, j, k)$ is the t-function of a rank n PMD then it is T-consistent.*

Proof. (T0) is obvious, and (T1) has been proved above (6). Similarly (T2) is proved in the proof of Theorem 1. Finally (T3) is a special case of the more general result which says that in *any* matroid there are at least as many hyperplanes as there are points, and which we will prove in Chapter 16. ∎

Another consequence of Theorem 2 is the following.

THEOREM 3. *If M is a* PMD *on S the hyperplanes of M are the blocks of a* BIBD *on the set of* 1-*flats of M.*

Proof. Consider any pair of distinct 1-flats of M, where M is assumed to have rank r. They are in a unique 2-flat of M, say F_2 and this in turn is contained in $t(2, r - 1, r)$ hyperplanes of M. Similarly any 1-flat of M is contained in exactly $t(1, 1, r - 1)$ hyperplanes of M. Thus we have a block design with parameters

$$b = t(0, r - 1, r), \qquad v = t(0, 1, r), \qquad r = t(1, 1, r - 1), \qquad k = t(0, 1, r - 1),$$
$$\lambda = t(2, r - 1, r).$$
∎

COROLLARY. *If M is a* PMD *it is a matroid design.*

The relations (T0)–(T3) can be simplified by the following device. Let d_i be defined for $1 \leqslant i \leqslant \rho(M)$ by

$$d_i = t(0, 1, i) - t(0, 1, i - 1)$$

where $t(0, 1, 0) = 0$. Then the sequence $(d_\rho, d_{\rho - 1}, \ldots, d_1)$ is called the *d-sequence* of the PMD.

Relations (T0) to (T3) can now be rewritten in the simpler form

(D1) $d_1 = 1, \quad d_i \geqslant 1, \quad 1 \leqslant i \leqslant \rho.$

(D2) $\dfrac{\prod\limits_{l=i+1}^{j} \sum\limits_{m=l}^{k} d_m}{\prod\limits_{l=i+1}^{j} \sum\limits_{m=l}^{j} d_m}$ is integral $0 \leqslant i \leqslant j \leqslant k \leqslant \rho.$

(D3) $d_{i-1} \mid d_i, \quad 2 \leqslant i \leqslant \rho.$

(D4) $d_{i+1} d_{i-1} \geqslant d_i^2, \quad 2 \leqslant i \leqslant \rho - 1.$

The four classes of PMD given in Examples 1 to 4 are characterized by their *d*-sequences as follows.

THEOREM 4. *Let M be a perfect matroid design.*

(1) *M is a uniform matroid of rank k on n elements if and only if it has d-sequence* $(n - k + 1, 1, 1, \ldots, 1)$.

(2) *M is the projective space* $PG(n, q)$ *if and only if it has d-sequence* $(q^n, q^{n-1}, \ldots, q, 1)$.

(3) *M is the affine space* $AG(n, q)$ *if and only if it has d-sequence* $(q^{n-1}(q - 1), q^{n-2}(q - 1), \ldots, q - 1, 1)$.

(4) *M is a Steiner system S(d, k, n) if and only if it has d-sequence (n − k, k − d + 1, 1, ..., 1).*

A natural question to ask is, given a sequence $(d_r, d_{r-1}, ..., d_1)$ satisfying the conditions (D1)–(D4) does there exist a PMD having this sequence as its d-sequence? In other words, as well as being necessary, are conditions (D1) to (D4) sufficient for the existence of a PMD? The answer is no—the sequence (36, 4, 2, 1) satisfies (D1) to (D4) but geometrical arguments show that it is not realisable as the d-sequence of a PMD.

Three further properties of PMDS which illustrate their high degree of structure are the following:

THEOREM 5
(a) *If M is a PMD and $\mathscr{I}_j$ is the set of independent sets of cardinal j, then $\mathscr{I}_j$ is the set of blocks of a BIBD.*

(b) *If M is a PMD and $\mathscr{C}_j$ is the set of circuits of cardinal j then $\mathscr{C}_j$ is the (possibly vacuous) set of blocks of a BIBD.*

In some cases these appear to give new balanced incomplete block designs, however because of the high parameters which they turn out to have this is not very easy to check.

We close this section by stating the following very useful theorem.

THEOREM 6. *If M is a perfect matroid design on a set of n-elements then unless it is a uniform matroid and hence a trivial design its hyperplanes must have cardinal $\leqslant n/2$.*

For proofs see Young, Murty, Edmonds [70]

This result is best possible, as can be seen from the Steiner systems $S(3, 4, 8)$ or $S(5, 6, 12)$.

EXERCISES 5

1. If $D(d_n, d_{n-1}, ..., d_2, 1)$ is the d-sequence of a PMD such that $n \geqslant 3$ and $d_3 = d_2^2$, $d_2 \geqslant 2$ then

$$\sum_{i=1}^{n} d_i = (d_2^h - 1)/(d_2 - 1)$$

for some integer h.

Deduce that (36, 4, 2, 1) is d-consistent in that it satisfies (D1–D4) but is not realizable as the d-sequence of a PMD. (R. M. Wilson)

2. Prove that if M is a PMD and $c(M)$ denotes the minimum circuit cardinality in M then

(a) $c(M) \leqslant c(M^*)$,

(b) $\rho(M) \leqslant c(M^*)$.

(Young, Murty and Edmonds [70])

3. Prove that if a M is a perfect matroid design and W_k is the number of k-flats of M then the sequence $(W_k : 1 \leqslant k \leqslant \rho(M))$ is a unimodal sequence.

6. THE STEINER SYSTEM $S(5, 8, 24)$ AND ITS SUBSYSTEMS

We now study in more detail a remarkable combinatorial structure, the Steiner system $S(5, 8, 24)$ and its related Steiner systems.

The interesting features of $S(5, 8, 24)$ cause it to occur in several branches of mathematics, we list a few:

(1)   It is one of the largest known Steiner systems with $d = 5$.
(2)   Its automorphism group is 5-transitive. Apart from the symmetric and alternating groups no group of higher transitivity is known.
(3)   The incidence vectors of its blocks against its points regarded as vectors over $GF(2)$ generate the set of code words of the unique perfect Golay code. (see Van Lint [71]).

For a description of a method of obtaining $S(5, 8, 24)$ we refer the reader to Biggs [72, p. 68], or Todd [59].

We will denote $S(5, 8, 24)$ by D_{24}.

If x is a point of a design D on the set S, we let D_x denote the design on $S \backslash x$ together with the sets $B \backslash x$, where B is a block containing x.

Thus $(D_{24})_x$ is the Steiner system $S(4, 7, 23)$ denoted by D_{23} and $(D_{23})_y$ is the Steiner system $S(3, 6, 22)$ denoted by D_{22}.

Regarded as a paving matroid D_{24} has the following easily checked properties.

(4)   The rank of D_{24} is 6 and its bases are those 6-sets not contained in any block.
(5)   All sets of cardinality $\leqslant 5$ are independent and its circuits all have cardinality 6.
(6)   If $\mathscr{L}(D_{24})$ is the geometric lattice obtained from D_{24} then if x is any atom, xy any 2-flat, xyz any 3-flat the interval sublattice satisfy
$[x, S] \simeq \mathscr{L}(D_{23})$
$[xy, S] \simeq \mathscr{L}(D_{22})$
$[xyz, S] \simeq PG(2, 4)$,
where S is the top element of $\mathscr{L}(D_{24})$.

In other words, these other Steiner systems are all "hanging from the top" of $\mathscr{L}(D_{24})$.

(7) Any two blocks (hyperplanes) B_1, B_2 of D_{24} meet in 0, 2 or 4 points. Also if $|B_1 \cap B_2| = 4$ then $B_1 \bigtriangleup B_2$ is a block.

From D_{24} we can derive the Steiner system $S(5, 6, 12)$, to be denoted by D_{12} as follows.

Take B_1, B_2 blocks of D_{24} such that $|B_1 \cap B_2| = 4$. Let $T = B_1 \bigtriangleup B_2$ and D_{12} is the design on T which has as its blocks precisely the intersections of size 6 of T with blocks of D_{24}.

Clearly $(D_{12})_x = S(4, 5, 11) = D_{11}$.

We now exhibit some interesting matroid properties of D_{24}.

Consider the collection of circuits of D_{24}. Since D_{24} is a perfect matroid design these circuits are the blocks of a BIBD, in fact the blocks of a 5-(3, 6, 24) design.

Hence since the complement of a circuit is a hyperplane of the dual matroid we see that the complementary design $\bar{D}_{24}$ of D_{24} has its blocks the hyperplanes of a matroid and hence is certainly a matroid design. However, since these hyperplanes are of cardinal 18, by Theorem 5.6 the structure $\bar{D}_{24}$ cannot be a perfect matroid design.

Doyen and Hubaut [71] studied the class of geometric lattices $\mathscr{L}$ which had the following locally projective property.

Let $\mathscr{L}$ be the geometric lattice of the simple matroid M on S, we say $\mathscr{L}$ is *locally projective* if for any point p of S, the interval sublattice $[p, S]$ is the geometric lattice of a projective space, and all its 2-flats have the same cardinal.

In other words a simple matroid is *locally projective* if its 2-flats have the same cardinal and the contraction of any point is (after the removal of all parallel elements) a projective space.

Obvious examples of locally projective matroids are the projective and affine geometries. Doyen and Hubaut were able to prove the following very interesting results.

THEOREM 1. *A finite locally projective space of dimension* > 3 *is either projective space or affine space.*

However for dimension 3 the situation is different inasmuch as the perfect matroid design associated with the Steiner system $S(3, 6, 22)$ is locally projective, for any atom p, the interval lattice $[p, S]$ is isomorphic to $PG(2, 4)$.

More precisely they proved:

THEOREM 2. *Let* $\mathscr{L}$ *be a geometric lattice of rank* 4. *Suppose that for any atom p, the interval* $[p, I]$ *is a finite projective plane. Then all lines have the same cardinal m, all projective planes* $[p, I]$ *have the same order q and one of the following is true:*

(i) $m = q + 1$, $\mathcal{L} \simeq PG(3, q)$
(ii) $m = q$, $\mathcal{L} \simeq AG(3, q)$
(iii) $q = m^2$
(iv) $q = m(m^2 + 1)$.

Now $S(3, 6, 22)$ is the case $m = 2$ of (iii). No other example of (iii) or (iv) is known.

The reader will also recall that $S(3, 6, 22)$ is an example of a matroid design satisfying all but the dimensionality restriction of Kantor's Theorem 4.3 characterizing projective and affine spaces.

Finally we mention another recent result of Kantor [74a].

THEOREM 3. *The paving matroids of the Steiner systems* $S(3, 6, 22)$, $S(4, 7, 23)$ *and* $S(5, 8, 24)$ *are not representable over any field.*

Before proving this theorem we note that this result does not contradict an embedding result of Todd [59] in which the lattice of D_{24} is represented inside the lattice of flats of $PG(11, 2)$. In this "representation" its points, 2-flats, 3-flats, 4-flats, 5-flats are represented by points, 2-flats, 3-flats, 4-flats and 7-flats respectively.

Proof. Since D_{24} and D_{23} have D_{22} as a minor it is clearly sufficient to prove that D_{24} is not representable.

Suppose there is a representation $\phi: D_{22} \rightarrow V$ where V is some vector space. Then there is an obvious extension of $\phi: \mathcal{L}(D_{22}) \rightarrow \mathcal{L}_0$ where $\mathcal{L}_0$ is the modular lattice of flats of V, and we may assume ϕ is the inclusion map.

Let $\wedge$, $\vee$ be the operations in $\mathcal{L}_0$. Fix a 2-flat F of D_{22}. There are 2-flats F_1, F_2, K of D_{22} such that F, F_1, F_2 are in a common 3-flat and no two have a common point, while $F \vee K$, $F_1 \vee K$, and $F_2 \vee K$ are 3-flats $\neq F \vee F_1$. Now each of the 6 points of D_{22} which are in the 3-flat $F \vee F_1$ is on one of F, F_1, F_2. Since

$$\rho(F \vee F_1 \vee F_2 \vee K) = 4$$

there is a point α of $\mathcal{L}_0$ such that

$$\alpha = F \wedge F_1 = F \wedge (F_1 \vee K) = F \wedge K.$$

By symmetry $\alpha = F \wedge F_2$. Each of the 3-flats of D_{22} other than $F \vee F_1$ which contains F has such a K. Thus there are $11(= 22/2)$ 2-flats $\alpha \vee x$,

$x \in D_{22}$. Each such 2-flat is in 5 3-flats of D_{22}. Each 3-flat of D_{22} which contains α (such as $F \vee F_1 \vee F_2$) contains 3 such lines. Thus α is on 11. 5/3 3-flats of D_{22} which is ridiculous. ∎

EXERCISES 6

1. Prove that any two blocks of D_{23} meet in 1 or 3 points, and any two blocks of D_{22} meet in 0 or 2 points.

2. In D_{22} prove that through 2 distinct points there are 5 blocks and that there are 60 blocks meeting a fixed block in 2 points and 16 blocks avoiding it completely.

3. Consider D_{24}; prove that a fixed block intersects 280 blocks in 4 points, 448 blocks in 2 points and 30 blocks in no points and that as a matroid it has 113,344 bases.

4. Prove that as a matroid D_{12} is self dual.

⁾5. Can $S(6, 8, n)$ exist for any integer n. Simple counting arguments show that if such an n exists then $n \geqslant 29$. Can this bound be improved?

NOTES ON CHAPTER 12

Sections 1 and 2 are well known and are based on the books by Bumcrot [69] and Hall [67].

Interest in matroids and block designs seems to have originated in the work of Murty [70] through his work on binary matroids all of whose circuits had the same cardinality. They also arise in the work on *identically self dual matroids* by Bondy and Welsh [71] and Graver [73]. Section 3 is based on Welsh [71c], and Main and Welsh [75]. Section 4 is based on the papers of Dembowski and Wagner [60] and Kantor [69]. Section 5 is a summary of the main results of the paper by Young, Murty and Edmonds [70], for further results we refer to Young and Edmonds [72].

Theorem 6.3 is due to Kantor [74a], this paper and the review paper Kantor [74] contain a wealth of interesting material on embedding problems and representations of Steiner systems in modular lattices.

For a survey of balanced incomplete block designs and parameter systems for which they exist see Hanani [75], and Hall [67].

CHAPTER 13

Menger's Theorem and Linkings in Graphs

1. THE BASIC LINKAGE LEMMA

This chapter makes precise the rather fuzzy notion that Hall's theorem is closely related to the theorem of Menger [27] about graphs. We shall show that in a very natural sense Hall's theorem is the "matroid dual" of Menger's theorem for directed graphs.

Let G denote a digraph with vertex set V and edge set E. A *path* in G is a sequence $P = (v_0, v_1, \ldots, v_k)$ of distinct vertices of G such that $k \geqslant 0$ and (v_{i-1}, v_i) is an edge of E, $(i = 1, 2, \ldots, k)$. P has *initial* vertex v_0 and *terminal* vertex v_k. Two paths are said to be *disjoint* if their vertex sets are disjoint.

If A, B are subsets of V we say that there exists a *linking* of A *onto* B if for some bijection $\alpha \colon A \to B$ we can find disjoint paths $(P_v \colon v \in A)$ in G such that P_v has initial vertex v and terminal vertex $\alpha(v)$. Note that A can always be linked onto itself. A can be *linked into* B if α is an injection.

If $A, B, C \subseteq V$, C is said to *separate* A from B if every path with initial vertex in A and terminal vertex in B has nonempty intersection with C. We denote this by $C \in S(A, B)$.

We define two pairs of functions on 2^V as follows. For $Z \subseteq V$,

$$\tilde{Z} = Z \cup \{v \in V \colon (z, v) \in E \text{ for some } z \in Z\}$$
$$\partial Z = \tilde{Z} \backslash Z.$$

We now prove a fundamental lemma about linkings due to Ingleton and Piff [73].

For each vertex v define

$$A_v = \{\tilde{v}\}.$$

Let $\mathscr{A}$ be the family of sets $(A_v \colon v \in V)$ and for $X \subseteq V$ let $\mathscr{A}_X$ denote the subfamily $(A_v \colon v \in X)$.

THE LINKAGE LEMMA. *Let X, Y be subsets of V. Then X can be linked onto Y in the digraph G if and only if $V \backslash X$ is a transversal of the family of sets $\mathscr{A}_{V \backslash Y}$.*

Proof. First suppose that X is linked onto Y in G by disjoint paths $(P_v: v \in X)$. We define a function $\alpha: V\backslash X \to V\backslash Y$ by

$$\alpha u = \begin{cases} v & \text{if } (v, u) \in P_x \text{ for some } x \in X, \\ u & \text{otherwise.} \end{cases}$$

Then α is well defined, since the paths are disjoint, and is an injection. Clearly $u \in A_{\alpha(u)}$ for all $u \in V\backslash X$ and $V\backslash X$ is a transversal of the family $\mathscr{A}_{V\backslash Y}$.

Conversely, suppose we have a bijection $\alpha: V\backslash X \to V\backslash Y$ such that $u \in A_{\alpha(u)}$ for all $u \in V\backslash X$. Take $v \in Y\backslash X$. Then $(\alpha(v), v) \in E$. Either $\alpha(v) \in X$ or $(\alpha^2(v), \alpha(v)) \in E$. Either $\alpha^k(v) \in X$ for some k or we obtain an infinite sequence $(\alpha^r(v): 1 \leqslant r \leqslant \infty)$. But since G is finite we must then have $\alpha^r(v) = \alpha^s(v)$ for some $r < s$. Choosing r minimal, we then have $\alpha(\alpha^{r-1}(v)) = \alpha(\alpha^{s-1}(v))$. But then, since α is bijective, we have $\alpha^{r-1}(v) = \alpha^{s-1}(v)$, a contradiction of the minimality of r. Thus we obtain a path $\{\alpha^k(v), \ldots, \alpha(v), v\}$ from X to Y. Similarly we obtain paths for every $v \in Y\backslash X$. These paths are disjoint since α is bijective, and by adjoining the trivial paths (v) for $v \in X\backslash Y$ we obtain a linking of X onto Y. ■

We can now prove Menger's theorem.

THEOREM 1. *Let $G = (V, E)$ be a digraph. If X, Y are subsets of V, X can be linked into Y in G if and only if there is no set $C \subseteq V$ which separates X from Y, with $|C| < |X|$.*

The reader will easily verify that Menger's theorem can be restated:

THEOREM 1'. *If u, v are two vertices of a digraph G and every set which separates u from v has cardinal at least k then there exist at least k paths in G from u to v which are pairwise disjoint apart from their end points.*

Proof of Theorem 1.

By the linkage lemma, X can be linked into Y if and only if $V\backslash X$ contains a transversal of $\mathscr{A}_{V\backslash Y}$. By Hall's theorem this happens if and only if for all $Z \subseteq V\backslash Y$,

$$|A(Z)\backslash X| \geqslant |Z|.$$

We show the equivalence of these conditions to Menger's conditions.

First, suppose that for some $C \subseteq V$ with $|C| < |X|$, C separates X from Y. Put

$$W = \{v \in V: C \text{ separates } \{v\} \text{ from } Y\}.$$

Then $X \cup C \subseteq W$. Put $Z = W \backslash C \subseteq V \backslash Y$. Then $A(Z) \subseteq W$ so that

$$|A(Z) \backslash X| \leqslant |W \backslash X| = |C| + |Z| - |X| < |Z|.$$

Conversely suppose that for some $Z \subseteq V \backslash Y$,

$$|A(Z) \backslash X| < |Z|.$$

Since $\mathscr{A}_{V \backslash Y}$ possesses a transversal, we must have, putting $W = A(Z)$ that $T = W \cap X \neq \varnothing$ and our condition is that $|W| - |T| < |Z|$.

Put $D = W \backslash Z$, so that $|D| < |T|$. Then D separates W from Y, and hence separates T from Y. Thus the set $C = D \cup (X \backslash T)$ clearly separates X from Y and $|C| < |X|$ which completes the proof. ∎

EXERCISES 1

1. Let G be a digraph with vertex set V and let X, Y be subsets of V. Prove that X can be linked to Y if and only if for all $U \subseteq V \backslash Y$,

$$|X \backslash \tilde{U}| \leqslant |\partial U|.$$

(McDiarmid [72])

2. Let G be a digraph, and let A, B be subsets of $V(G)$. Prove that $S(A, B)$, the collection of sets which separate A from B is the collection of sets of the form

$$(A \backslash \tilde{Y}) \cup \partial Y$$

for $Y \subseteq V \backslash B$. (McDiarmid [72]).

2. GAMMOIDS

We now use the linkage lemma to obtain a new class of matroids which we shall call gammoids. This class was discovered by Perfect [68]. At first gammoids appeared to be a very intractable class of matroids; for example in 1970 it was still not easy to decide whether or not an arbitrary matroid was a gammoid (see for example problem 31 of Welsh [71b]).

Let $G = (V, E)$ be a digraph. Choose a fixed subset B of V and let $L(G, B)$ denote the collection of subsets of V which can be linked into B. That is $X \in L(G, B)$ if $\exists Y \subseteq B$ such that there is a linking of X onto Y.

THEOREM 1. *$L(G, B)$ is the collection of independent sets of a matroid on V.*

Any matroid which can be obtained from a digraph G and some subset B of $V(G)$ in this way is called a *strict gammoid*. A *gammoid* is a matroid which is obtained by restricting a strict gammoid $L(G, B)$ to some subset of $V(G)$.

We shall use $L(G, B)$ to denote both the strict gammoid and its collection of independent sets.

Example. Let G be the digraph of Fig. 1. Then $L(G, \{6\})$ has bases all singleton sets of $\{1, \ldots, 6\}$. $L(G, \{3, 6\})$ has an bases all $\{x, y\}: x \leqslant 3, y \leqslant 6$.

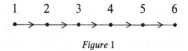

1   2   3   4   5   6

Figure 1

Proof of Theorem 1.

If $X \subseteq V(G)$, X can be linked into B if and only if X can be linked onto $B' \subseteq B$ and by the linkage lemma, this is equivalent to the condition that $V\backslash X$ is a transversal of the family $\mathscr{A}_{V'B'}$.

Thus $X \in L(G, B)$ if and only if $V\backslash X$ contains a transversal of $\mathscr{A}_{V\backslash B}$. In other words $X \in L(G, B)$ if and only if $V\backslash X$ is spanning in the transversal matroid $M[\mathscr{A}_{V\backslash B}]$. But the complement of a spanning set of a matroid M is an independent set of the dual matroid M^*. Thus we have proved the theorem and incidentally proved the following more interesting result. ■

THEOREM 2. *A matroid M is a strict gammoid if and only if its dual matroid M^* is transversal.*

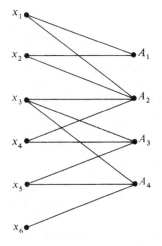

Figure 2

By following the proof of Theorem 1 we see how to establish the one-one correspondence between families $\mathscr{A}$ of subsets and directed graphs, and hence for any transversal matroid M with presentation $\mathscr{A}$ get a representation of its dual M^* as a strict gammoid, and conversely. The construction is as follows:

Given $L(G, B)$ just take $\mathscr{A}$ to be the family $(A_v : v \notin B)$ and $L(G, B)^* = M[\mathscr{A}]$

Given $M[\mathscr{A}]$ on S, select any base B of $M[\mathscr{A}]$ say $B = \{x_1, \ldots, x_r\}$ where x_i represents A_r. Then join x_i to all $y \in A_i \backslash x_i$ by a directed edge, (x_i, y_i). This gives a digraph G with $V(G) = S$ and $L(G, S \backslash B) = (M[\mathscr{A}])^*$

Example 1. Suppose $M = M[A_1, A_2, A_3, A_4]$ is the transversal matroid on $\{x_1, \ldots, x_6\}$ having associated bipartite graph shown in Figure 2.
Take any base of M say $\{x_1, x_4, x_3, x_6\}$ where $x_1 \in A_1$, $x_4 \in A_2$, $x_3 \in A_3$, $x_6 \in A_4$.
The corresponding cobase $B_0 = \{x_2, x_5\}$, and $(M[\mathscr{A}])^*$ is the strict gammoid $L(G, B_0)$ where G is the directed graph of Fig. 3.

G:

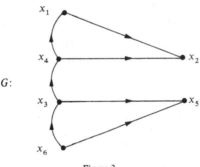

Figure 3

Example 2. Suppose M is the strict gammoid given by $L(G, B)$ where G is the digraph of Fig. 4 and $B = \{1, 2\}$.
Then M^* is the transversal matroid $M[A_3, A_4, A_5]$ where $A_3 = \{3, 1\}$, $A_4 = \{4, 1, 2, 3\}$, $A_5 = \{5, 3, 4\}$.

An alternative proof of Theorem 2 has been given by McDiarmid [72]; a byproduct of his proof is the following result.

(1)   The rank function ρ of the strict gammoid $L(G, B)$ is given by,

$$\rho X = \min_{Z \subseteq V \backslash B} \{|X \backslash \tilde{Z}| + |\partial Z|\} \qquad (X \subseteq V(G)).$$

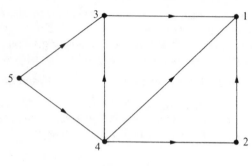

Figure 4

This is a special case of Theorem 3.1 which we prove later.

Now suppose that M is a transversal matroid on S, say with presentation $\mathscr{A} = (A_i = i \in I)$. Let $\Delta = (S, I; E)$ be its associated bipartite graph and suppose we direct the edges of Δ so that they all have initial vertex in S and terminal vertex in I. Let Δ' be the resulting digraph. Then it is clear that

$$M = L(\Delta', I)|S$$

In other words we have proved:

(2)   Every transversal matroid is a gammoid.

However it is easy to find examples to show that the converse is false.

An important characteristic of gammoids, which is not shared by transversal matroids or strict gammoids is that they form a closed set under the operations of restriction and contraction. That is:

THEOREM 3. *Any minor of a gammoid is a gammoid.*

Proof. If M is a gammoid on S there exists a digraph G, a subset B of $V(G)$ and a subset S of $V(G)$, such that $M = L(G, B)|S$. Thus for $T \subseteq S$,

$$M|T = L(G, B)|T$$

which shows that a restriction of a gammoid is a gammoid.

A contraction minor $M \cdot T$ can be represented in the form

$$M \cdot T = (N|S) \cdot T$$

where N is a strict gammoid on some set S' containing S. Now since the operations of deleting elements and contracting elements commute, we can write

$$M \cdot T = (N \cdot T')|T$$

where $T' = S' \setminus (S \setminus T)$. But

$$N \cdot T' = (N \cdot T')^{**} = (N^* | T')^*$$

and N^* is transversal so that $N^* | T'$ is transversal and hence its dual $N \cdot T'$ is a strict gammoid. Thus its restriction to T, $M \cdot T$ is a gammoid and the proof is complete. ∎

COROLLARY. *The dual of a gammoid is a gammoid.*

Proof. Let M be a gammoid, $M = N | S$ where N is a strict gammoid.

Then $M^* = N^* \cdot S$ and N^* is transversal and hence a gammoid. Thus by Theorem 3 the contraction minor $N^* \cdot S$ is a gammoid. ∎

Since a transversal matroid is linearly representable over fields of every characteristic, and representability is preserved under duality we see that strict gammoids are linearly representable over fields of every characteristic. Taking restriction minors we get:

(3)   A gammoid is linearly representable over every sufficiently large field.
(4)   The characteristic set of a gammoid is the full set of primes.

The only other known criterion for a matroid to be a strict gammoid is that of Mason [72]. It is rather unwieldy, and for large matroids difficult to apply.

THEOREM 4. *Given a matroid M on a set S, for each subset X of S define α recursively by*

$$\alpha X = |X| - \rho X - \sum \alpha(F)$$

where the summation is over all flats F of M which are properly contained in X. Then M is a strict gammoid if and only if $\alpha X \geqslant 0$ for all $X \subseteq S$.

Undirected gammoids

With the natural definitions of linking the above theory can be extended to undirected graphs, edge disjoint paths and so on. However there is an interesting example to show that linking in directed graphs gives a larger class of matroids than linking in undirected graphs.

Call a matroid M a *strict undirected gammoid* if there exists an undirected graph G and a subset B of $V(G)$ such that $M \simeq L(G, B)$. Clearly if M is a strict undirected gammoid it is a strict gammoid (we can replace G by a digraph G' with the same vertex set, and with a pair of directed edges replacing each undirected edge of G).

However there exist strict gammoids which are *not* undirected gammoids.

Example. Let G_0 be the digraph of Figure 5.

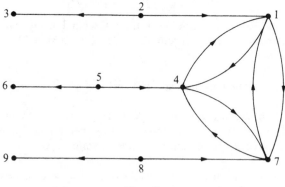

Figure 5

Consider the strict gammoid $L(G_0, \{1, 3, 4, 6, 9\})$. Woodall [75] shows that this is not an undirected gammoid, strict or otherwise. For further details on undirected linkings we refer to his paper.

EXERCISES 2

1. Find the smallest gammoid which is (a) not a strict gammoid and is (b) not transversal.

2. Prove that the union of gammoids is a gammoid. (Narayanan and Vartak [73])

3. Prove that a matroid is a gammoid if and only if it is the contraction of a transversal matroid. (Ingleton and Piff [73])

4. Prove that if M is a strict gammoid and B is any base of M there exists a digraph G such that $M \simeq L(G, B)$. (Mason [72])

5. Prove that if M is a gammoid and F is any field then M is algebraic over some transcendental extension of F. (Piff [72])

6. Prove that if M is a gammoid on a set of n elements then M is linearly representable over any field of at least 2^n elements. (Lindström [73])

3. MATROIDS INDUCED BY LINKINGS

We now show how the linkage lemma enables us to extend the idea of inducement of matroids by bipartite graphs developed in chapter 8 to arbitrary digraphs.

Let $G = (V, E)$ be a digraph and let M_0 be a matroid on V. Let $L(G, M_0)$ denote the collection of subsets of V which can be linked in G onto an independent subset of M_0.

THEOREM 1. *The collection $L(G, M_0)$ is the set of independent sets of a matroid on $V(G)$. Its rank function ρ is given in terms of the rank function ρ_0 of M_0 by*

$$\rho X = \min_{Z \subseteq V} (|X \backslash \tilde{Z}| + \rho_0 Z + |\partial Z|) \qquad (X \subseteq V)$$

Proof (Sketch). If $A_v = \{\tilde{v}\}(v \in V)$, and $\mathscr{A} = (A_v : v \in V)$, by the linkage lemma, X is linked onto some $Y \in \mathscr{I}(M_0)$ if and only if $V \backslash X$ contains a transversal of the family $(A_v : v \in V \backslash Y)$. In other words, in the terminology of Section 8.2 $L(G, M_0)$ is the set of independent sets of the dual of the matroid $Q(M_0^*, \mathscr{A})$. The rank function formula follows by applying Theorem 8.1.2 with the appropriate submodular function (see McDiarmid [75]). ∎

We use $L(G, M_0)$ to denote both the collection of independent sets and the matroid itself and call it the *matroid induced from M_0 by G*.

COROLLARY. *The dual of $L(G, M_0)$ is the bipartite induced matroid $Q(M_0^*, \mathscr{A})$.*

When M_0 is the trivial matroid on V with a single base B, we see that $L(G, M_0)$ is the strict gammoid $L(G, B)$. Thus the rank function of $L(G, B)$ is given by the formula

$$\rho X = \min_{Z \subseteq V} (|X \backslash \tilde{Z}| + |Z \cap B| + |\partial Z|) \qquad (X \subseteq V).$$

Elementary graph theory shows that this reduces to the form (2.1).

A matroid extension of Menger's theorem.

We next show how Menger's theorem can be extended in a natural way to give necessary and sufficient conditions for there to exist a linking between the independent sets of two matroids defined on the vertex set of a digraph. The theorems were first noted by Brualdi [71a] [71c] but the simple statements and treatment given here is essentially due to McDiarmid [75].

Let G be a digraph and let M_1, M_2 be two matroids on the vertex set and suppose $\rho M_1 = \rho M_2 = r$. A *linking* of M_1 onto M_2 in G is said to exist if we can link a base of M_1 onto a base of M_2 in G.

THEOREM 2. *If M_1, M_2 are matroids on $V(G)$ with $\rho M_1 = \rho M_2$ there exists a linking of M_1 onto M_2 if and only if for all $X \subseteq V$*

$$\rho_1^*(\tilde{X}) + \rho_2(X) \geqslant |X|.$$

where ρ_1, ρ_2 are the rank functions of M_1, M_2 respectively.

We can restate Theorem 2 in the form:

THEOREM 2'. *If* M_1, M_2 *are matroids of equal rank on* $V(G)$ *and* $\max(M_1, M_2)$ *denotes the maximum cardinal of a set independent in* M_1 *which is linked in* G *onto a set independent in* M_2 *then*

(1) $$\max(M_1, M_2) = \min_{X \subseteq V}\{\rho_1(V\backslash\tilde{X}) + |\partial X| + \rho_2 X\}.$$

Brualdi's original version of Theorem 2 says that

(2) $$\max(M_1, M_2) = \min\{\rho_1 Z_1 + |Z_0| + \rho_2 Z_2\}$$

where the minimum is taken over all triples (Z_0, Z_1, Z_2) where Z_1, Z_2 are subsets of V and $Z_0 \in S(V_1\backslash Z_1, V\backslash Z_2)$.

To see that (1) is equivalent to (2) note that for $Z \subseteq V$, ∂Z separates $Z \cup \partial Z$ and $V\backslash Z$. Now let $Z_1, Z_2 \subseteq V$ and let Z_0 separate $V\backslash Z_1$ from $V\backslash Z_2$ and let $Z = \{v \in V : Z_0 \notin S(V\backslash Z_1, \{v\})\}$. Then it is easily checked that

$$\rho_1 Z_1 + |Z_0| + \rho_2 Z_2 \geqslant \rho_1(V\backslash\tilde{Z}) + \partial Z + \rho_2 Z.$$

Hence the right-hand sides of (1) and (2) are equal and we have proved the equivalence of the two versions of the theorem.

Proof of Theorem 2'. Let ρ_3 be the rank function of the induced matroid $L(G, M_2)$. Then, since $\max(M_1, M_2)$ is the maximum cardinal of a set independent in M_1 and $L(G, M_2)$ we have by Edmonds' Intersection Theorem 8.5.1 that

$$\max(M_1, M_2) = \min\{\rho_1(V\backslash A) + \rho_3(A) : A \subseteq V\}.$$

Hence by Theorem 1

(3) $$\max(M_1, M_2) = \min\{\rho_1(V\backslash A) + |A\backslash\tilde{B}| + \rho_2 B + |\partial B|\}$$

where the minimum is taken over all subsets A, B of V. But for $A, B \subseteq V$,

$$\rho_1(V\backslash A) + |A\backslash\tilde{B}| \geqslant \rho_1(V\backslash B).$$

Hence the minimum in (3) is achieved with $A = \tilde{B}$ and thus

$$\max(M_1, M_2) = \min\{\rho_1(V\backslash B) + \rho_2 B + |\partial B| : B \subseteq V\}. \blacksquare$$

It is interesting to see how Theorem 2 can be applied when G is a directed bipartite graph $\Delta = (S, T; E)$ in which each edge is directed from T to S. Let M_1 be a matroid on S, M_2 be a matroid on T. Then $\max(M_1, M_2)$ is the maximum cardinality of a set of edges in Δ whose endpoints are independent sets of the two matroids.

But by Brualdi's Theorem 8.5.2

$$(4) \qquad \max(M_1, M_2) = \min_{B \subseteq T} (\rho_2 B + \rho_1(\partial(T \backslash B))).$$

It seems to be a non-trivial exercise to recover the form (4) from our main theorem.

We mention finally that much of this theory can be extended to linkings by edge disjoint paths in a directed or undirected graph G and with matroids on $V(G) \cup E(G)$. The interested reader is referred to Woodall [75].

EXERCISES 3

1. Prove that if M is linearly representable over a field F then for any digraph $G, L(G, M)$ is linearly representable over some extension of F.

2. Let G be a digraph and let M be a matroid on $V(G)$ which is algebraic over a field F. Prove that the induced matroid $L(G, M)$ is algebraic over some transcendental extension of F. (Piff [72])

3. Use both Theorems 1 and 2 to show that necessary and sufficient conditions that there exist a linking in the directed graph G of a subset D of V onto a subset C of V where $|C| = |D|$ is that for all $X \subseteq V$,

$$|C \cap X| + |\partial X| \geqslant |D \cap X| + |D \cap \partial X|.$$

4. A NEW CLASS OF MATROIDS FROM GRAPHS

Recently McDiarmid [74] has discovered a new class of matroids on the edge sets of graphs and digraphs. We sketch the main points in this section; concentrating only on the case where G is undirected and omitting all proofs.

Let $G' = (V, E)$ be a graph. A *trail* in G is a sequence $(v_1, v_2, \ldots, v_m)$ of vertices of G such that the edges (v_i, v_{i+1}) exist $1 \leqslant i \leqslant m - 1$. Note we do not insist that the vertices v_i are distinct.

v_1 is called the *initial vertex*, v_m the *terminal vertex* and v_1, v_m are the *end vertices*.

A well known variant of Euler's famous theorem, see for example Ore [62] is the following:

LEMMA. *If a connected graph $G = (V, E)$ has exactly $2k > 0$ vertices of odd degree there exist trails $P_1, \ldots, P_k$ in G such that the edge sets of these trails partition E. Further, at each odd vertex some trail P_i must have an end vertex and so we cannot partition E into fewer than k trails.*

Suppose now that $G = (V, E)$ is given. We say that any set $\mathscr{P} = \{P_1, \ldots, P_n\}$ of trails of G such that $\{E(P_1), \ldots, E(P_n)\}$ is a partition of E is a *trail partition* of G.

If $\mathscr{P}$ has the least possible number of trails, that is, if n is the sum of half the number of odd vertices and the number of non-trivial Eulerian components we say that $\mathscr{P}$ is a *minimum trail partition of G*.

Given $X \subseteq E$ we say that X is *minimum trail separated* in G if there exists a minimum trail partition $\mathscr{P}$ of G such that no two edges of X are in the same trail in $\mathscr{P}$.

THEOREM 1. *For any graph $G = (V, E)$ the minimum trail separated subsets of E are the independent sets of a matroid on E.*

We call this matroid the *trail matroid* of G. Theorem 1 is a special case of more general results obtained below.

Example 1. Let $G = (V, E)$ be the graph shown in Fig. 1. Since each vertex is odd, a minimum trail partition contains 3 trails. It is easy to check that the collection of minimum trail separated subsets of E consists of all subsets of cardinal at most 3, other than $\{a, d, f\}$, $\{b, d, e\}$, $\{c, e, f\}$.

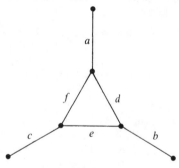

Figure 1

Example 2. When G is Eulerian its trail matroid has rank 1.

If $\mathscr{P}$ is a collection of trails in a graph (V, E), for $v \in V$ let $\mathscr{P}(v)$ denote the number of trails of $\mathscr{P}$ which end at v. As stated earlier we have

LEMMA. *Let $G = (V, E)$ be a graph with no Eulerian components. Then for any minimum trail partition $\mathscr{P}$ of G,*

$$\mathscr{P}(v) = \begin{cases} 1 \text{ if } \deg(v) \text{ is odd} \\ 0 \text{ if } \deg(v) \text{ is even.} \end{cases}$$

Now let f be a function: $V \to \mathbb{Z}$. We say that the collection of paths $\mathscr{P}$ is an *f-trail partition* on G if it is a trail partition of G and $\mathscr{P}(v) = f(v)$ for each $v \in V$.

Also we say that f is *compatible* with G if for $v \in V$, $f(v) \equiv \deg(v) \pmod 2$, and $f(v) \leqslant \deg(v)$, and $f \neq 0$ everywhere on a component of G.

LEMMA. *Let $G = (V, E)$ be a graph and f a function from V into $\mathbb{Z}$. Then there exists an f-trail partition of G if and only if f is compatible aith G.*

Proof. An easy extension of Euler's theorem by adding extra edges to G.

For any compatible function f of G define a dual function $f^*: V \to \mathbb{Z}$ by

$$f^*(v) = \tfrac{1}{2}(\deg(v) - f(v)) \qquad (v \in V).$$

We may now extend Theorem 1 in a very elegant way.

Let $G = (V, E)$ be a graph and let $f: V \to \mathbb{Z}$ be a compatible function. We say that $X \subseteq E$ is *f-trail separated* if $\exists$ an f-trail partition $\mathscr{P}$ of G such that no two elements of X are in the same trail of $\mathscr{P}$. We call the collection of f-trail separated subsets of E the *trail partition structure* on E of the pair G, f and denote it by $\mathscr{P}(G, f)$.

THEOREM 2. *Let $G = (V, E)$ be a graph and let f be a function compatible with G. Then $\mathscr{P}(G, f)$ is the collection of independent sets of a matroid on E. Moreover it is the dual matroid of the transversal matroid which has as its presentation the family of sets*

$$(A_{vj} : v \in V, 1 \leqslant j \leqslant f^*(v))$$

where A_{vj} is the set of edges adjacent with v.

We call a matroid M a *trail matroid* if there exists a graph G, and a compatible function f on G such that M is isomorphic to the matroid whose independent sets are $\mathscr{P}(G, f)$.

By Theorem 2 we know:

(1) Every trail matroid is a strict gammoid.

However not every strict gammoid is a trail matroid as McDiarmid also proves.

(2) The cycle matroid of a graph G is a trail matroid if and only if G has no subgraph homeomorphic from K_4 or $K_{2,3}$.

As we shall see in the next chapter this implies that the set of trail matroids is a proper subset of the set of strict gammoids.

Related to this theory of trail partitions is the work of Greene and Kleitman [74] about partially ordered sets. Consider for example the following theorem.

THEOREM 3. *If P is a partially ordered set let $\mathscr{T}(P)$ be the collection of subsets of P which are the tops of chains in some minimal partition of P into chains. Then $\mathscr{T}(P)$ is the set of bases of a matroid on P.*

EXERCISES 4

1. Prove that the smallest strict gammoid which is not a trail matroid is the cycle matroid of the graph G_0

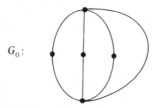

G_0:

2. Show that neither the restriction nor the contraction of a trail matroid need be trail.

3. Show that every uniform matroid with positive rank is a trail matroid.

NOTES ON CHAPTER 13

The origin of the word *gammoid* can be traced to a paper by Pym and Perfect [70] where it is used to denote a digraph with two specified vertex subsets—an obvious generalization of the term *deltoid* used by Mirsky and Perfect [67]. It seems that Mason [72] was the first to use gammoid in the sense it is used here. He also introduced strict gammoids, and his paper, with that of Ingleton and Piff [73] are the main sources of the material of sections 1 and 2.

Section 3 is based on the work of Brualdi [71a], [71c], McDiarmid [75] and Piff [72]. An extension to undirected graphs, edge linkings, etc., is given by Woodall [75].

Section 4 is a brief summary of the work by McDiarmid [75d], we use "trail" for what he calls "path".

For a review of recent Sperner-type theorems about partially ordered sets and the relevance in this theory of the matroid theorems of Greene and Kleitman [74] see the paper by Greene [74a].

More general linking results have been obtained recently by Schrijver [74, 75, 75a].

CHAPTER 14

Transversal Matroids and Related Topics

1. BASE ORDERABLE MATROIDS

In this chapter we gather together a collection of results (mostly very recent) on the structure of transversal matroids and other classes of matroids whose origin seems to have arisen through the study of transversal matroids.

The concept of a base orderable matroid originated in the work of Brualdi and Scrimger [68]. Since it seems difficult to find necessary and sufficient conditions for a matroid to be transversal it is useful to have a set of properties which are possessed by transversal matroids and can be easily checked. Base orderability satisfies the first criterion if not the second, but apart from this is an attractive property of intrinsic interest in itself.

A matroid M on S is *base orderable* if for any two bases B_1, B_2 of M there exists a bijection $\pi : B_1 \to B_2$ such that both $(B_1 \backslash x) \cup \pi(x)$ and $(B_2 \backslash \pi(x)) \cup x$ is a base of M for each $x \in B_1$. We call such a π an *exchange ordering* for B_1, B_2.

First notice that $M(K_4)$ is not base orderable for if labelled as in Fig. 1 it is easy to check that no such π exists for the bases $B_1 = \{1, 2, 3\}$ and $B_2 = \{4, 5, 6\}$.

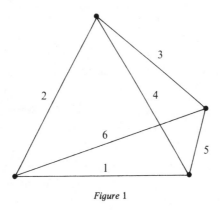

Figure 1

231

However, it turns out that the class of base orderable matroids is well-behaved in the sense that it is invariant under the standard matroid operations.

(1)   If M is base orderable on S and $T \subseteq S$ then $M \mid T$ is base orderable.

Proof. If A_1, A_2 are bases of $M \mid T$ let B_1, B_2 be bases of M containing A_1, A_2 respectively. Then if π is the exchange ordering for B_1, B_2, its restriction to A_1 is an exchange ordering for A_1, A_2.

(2)   If M is base orderable its dual M^* is base orderable.

Proof. Let B_1, B_2 be distinct bases of M^*. Then $S \backslash B_1$ and $S \backslash B_1$ are bases of M. There exists an exchange ordering $\pi: S \backslash B_1 \to S \backslash B_2$. Define $\pi^*: B_1 \to B_2$ by

$$\pi^*(x) = \begin{cases} \pi^{-1}(x) & x \in B_1 \backslash B_2, \\ x & x \in B_1 \cap B_2. \end{cases}$$

Then π^* is easily seen to be an exchange ordering for B_1 and B_2.

(3)   If M is base orderable, any minor of M is base orderable.

Proof. An immediate consequence of (1) and (2).

(4)   The union of base orderable matroids is base orderable.

Proof. It is clearly sufficient to prove that if M_1, M_2 are base orderable then so is $M_1 \vee M_2$.

We know that if $M = M_1 \vee M_2$ has no coloops any base of M is the union of disjoint bases of M_1, and M_2. Hence when M has no coloops consider any bases B_1, B_2 of M. We can write

$$B_1 = C_1 \cup D_1, \qquad B_2 = C_2 \cup D_2$$

where

$$C_1 \cap D_1 = \phi, \qquad C_2 \cap D_2 = \phi$$

and C_1, C_2 are bases of M_1, D_1, D_2 are bases of M_2. Let $\pi_1: C_1 \to C_2$ and $\pi_2: D_1 \to D_2$ be exchange orderings for C_1, C_2 and D_1, D_2 respectively. Then the map $\pi: C_1 \cup D_1 \to C_2 \cup D_2$ defined by

$$\pi(x) = \begin{cases} \pi_1(x) & x \in C_1 \\ \pi_2(x) & x \in D_1 \end{cases}$$

is an exchange ordering for B_1, B_2. Thus (4) holds when M has no coloops. When M has a coloop a simple modification of the above argument proves that (4) still holds.

Since every matroid of rank one is base orderable and a transversal matroid is the union of such matroids, we know that transversal matroids are base orderable. By duality so are strict gammoids and hence by (1) we have proved:

THEOREM 1. *Gammoids are base orderable; in particular transversal matroids are base orderable.*

Now the class of gammoids and the class of base orderable matroids possess similar properties in the sense that both are closed under the taking of minors and duals, and the cycle matroid of K_4 is the smallest matroid not belonging to each of them. However it is easy to see that the class of base orderable matroids is in fact much larger. We know that any gammoid is representable over any field of large enough cardinal, but there exist many base orderable matroids which are not representable over any field—the Non-Pappus configuration of Section 9.2 is base orderable.

The following property is closely related to the property of being base orderable. We say a matroid M is *strongly base orderable* or has the *full exchange property* if for any two bases B_1, B_2 there exists a bijection $\pi: B_1 \to B_2$ such that for all subsets $A \subseteq B_1$, $(B_1 \backslash A) \cup \pi A$ is a base of M. It is easy to show that for any such π, $(B_2 \backslash \pi A) \cup A$ is also a base of M.

Clearly if M is strongly base orderable it is base orderable and for some time it was not known whether there existed a matroid which was base orderable but not strongly base orderable. Such an example has recently been found.

Example (Ingleton [75]). Let $S = \{a_1, a_2, a_3, a_4, b_1, b_2, b_3, b_4\}$, and let M be a matroid on S which has as bases all 4-sets except

$$\{a_1, b_1, b_2, b_4\}, \qquad \{a_2, b_1, b_2, b_3\}, \qquad \{a_2, a_3, a_4, b_4\},$$
$$\{a_1, a_3, a_4, b_3\}, \qquad \{a_1, a_2, b_3, b_4\}.$$

The bases $B_1 = \{a_1, a_2, a_3, a_4\}$, $B_2 = \{b_1, b_2, b_3, b_4\}$ prevent M being strongly base orderable but it is easy to see that there are not enough circuits in M to prevent M being base orderable.

Thus the property of being strongly base orderable is significantly more restrictive than base orderability. It has however the same characteristics—the reader can modify the proofs of the base orderable case to prove:

(5) If M is strongly base orderable any minor of M is strongly base orderable, and so is the dual M^*.

(6) The union of strongly base orderable matroids is strongly base orderable.

THEOREM 2. *Any gammoid is strongly base orderable; in particular transversal matroids are strongly base orderable.*

For strongly base orderable matroids Davies and McDiarmid [75] have proved the following very useful and interesting result.

THEOREM 3. *Let M_1, M_2 be strongly base orderable matroids on a finite set S with rank functions ρ_1 and ρ_2 respectively and let k be a positive integer. Then M_1 and M_2 have d disjoint common independent sets of cardinality k if and only if for all subsets X and Y of S*

$$d(\rho_1 X + \rho_2 Y) + |S\backslash(X \cup Y)| \geqslant dk.$$

Since transversal matroids are strongly base orderable this gives as a corollary the result of Fulkerson and Davies that $\mathscr{A} = (A_i : i \in I)$ and $\mathscr{B} = (B_i : i \in I)$ have d disjoint CTS if and only if $|A(J) \cap B(K)| \geqslant d(|J| + |K| - |I|)$, for all subsets J, K of I.

EXERCISES 1

1. Prove that the Fano matroid is not base orderable.

2. Prove that if M on S is (strongly) base orderable and $f : S \to T$ then $f(M)$ is (strongly) base orderable.

3. Find the smallest base orderable matroid which is (a) not transversal (b) not a gammoid.

4. Prove that the truncation of a (strongly) base orderable matroid is (strongly) base orderable. (J. Davies).

5. Prove that the intersection of (strongly) base orderable matroids is (strongly) base orderable.

6. Prove that if G is a directed graph and M on $V(G)$ is (strongly) base orderable then the induced matroid $L(G, M)$ is (strongly) base orderable.

7. Find a matroid which is not base orderable and which does not contain $M(K_4)$ as a minor. (A. W. Ingleton)

○8. Does there exist an excluded minor condition for a matroid to be (strongly) base orderable?

2. SERIES PARALLEL NETWORKS AND EXTENSIONS OF MATROIDS

There is a simple type of electrical network termed a *series-parallel connection* which occurs frequently in both theoretical and applied electrical engineering. One reason for the importance of series-parallel connection stems from the fact that the joint resistance is easily evaluated by the laws:

Os: Resistance is additive for resistors in series.

Op: Reciprocal resistance is additive for resistors in parallel.

For example consider Fig. 1.

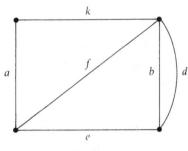

Figure 1

Let r_a, etc., denote the resistance of branch a. If R_a denotes the joint resistance of the network as measured by a battery inserted in branch a, repeated
application of Os and Op gives

$$R_a = r_a + r_k + \{r_f^{-1} + [r_e + (r_b^{-1} + r_d^{-1})^{-1}]^{-1}\}^{-1}.$$

A branch a is a *series parallel connection* if the joint resistance R_a can be
evaluated by Ohm's two rules Os and Op.

A network in which every branch is in series parallel connection shall be
termed a *series parallel network*. We shall generalize these concepts to
matroids, as suggested by Minty [66].

Let M be a matroid on S let $x \in S$ and suppose $y \notin S$. The *series extension*
of M at x by y is the matroid $sM(x, y)$ on $S \cup y$ which has as its bases the sets
of the form (1) or (2).

(1) $B \cup y$; B a base of M,
(2) $B \cup x$; B a base of M, $x \notin B$.

Similarly the *parallel extension* of M is the matroid $pM(x, y)$ on $S \cup y$ which
has as its bases the sets of the form (3) or (4).

(3) B a base of M
(4) $(B \backslash x) \cup y$; $x \in B$, B a base of M.

It is clear that when M is the cycle matroid of a graph G and that for a given
edge e we (a) replace it by a pair of edges e_1, e_2 in "series" in the electrical
sense or (b) replace it by a pair of edges e_1, e_2 in "parallel" in the electrical

sense, and if G_1, G_2 denote respectively these series and parallel extensions of G then

(5) $M(G_1) \simeq sM(G)(e, e_1)$

(6) $M(G_2) \simeq pM(G)(e, e_1)$.

A basic result which simplifies proofs of theorems about series parallel extensions is the following, easily checked, duality theorem.

(7)   If M is a matroid on S and $x \in S$ and $y \notin S$, then

$$(sM(x, y))^* = pM^*(x, y).$$

Proof. Routine check that their bases coincide.

We define a *series-parallel extension* of a matroid M to be a matroid. M_1 which can be obtained from M by successive series and parallel extensions.

From (7), for example, it is easy to prove the following lemmas which we need in subsequent sections.

(8)   A series parallel extension of a gammoid is a gammoid.

Proof. Let M be a gammoid on S and let $pM(x, y)$ be the parallel extension of M at x. Suppose M can be represented as a gammoid in the form $L(G, B)|S$ for some digraph G.

Replace G by a new digraph $G' = G[x, y]$ obtained from G by replacing the vertex x by two vertices x, y and letting the edges of G' be as in G except that each edge of the form (x, v) in G is deleted and a corresponding edge (y, v) inserted.

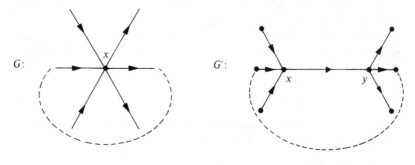

Figure 2

Also insert an additional edge (x, y). It is clear that $L(G', B)$ is a representation of $pM(x, y)$ as a gammoid. Thus the parallel extension of a gammoid is a gammoid.

Since the dual of a gammoid is a gammoid it now follows by duality and (7) that a series extension of a gammoid is a gammoid.

(9)   A series parallel extension of a base orderable matroid is base orderable.
(10) Let M be base orderable on S and let x, y be in series in M. Then $M.(S \backslash y)$, the series contraction of M at y is base orderable.

Both (9) and (10) are easy to prove and are left to the reader.

A *series–parallel matroid* was defined by Minty [66] as a matroid which can be expressed as a series parallel extension of a matroid on one element. As pointed out by Minty it is an easy exercise to prove:

(11) A series parallel matroid is the cycle matroid of a planar graph.

Hence a series parallel matroid is nothing more than the cycle matroid of a series parallel network. From now on we shall use *series parallel network* for both the graph forming the network and the cycle matroid of this graph.

A basic theorem of Duffin [65] (though as Duffin points out the result is also implicit in the work of Dirac [52], [57]) is the following characterization of series parallel networks.

THEOREM 1. *A block graph G is a series parallel network if and only if it contains no subgraph homeomorphic from K_4.*

For a proof we refer the reader to Duffin [65], with a warning that it has a slightly electrical flavour—for example K_4 is throughout referred to as a Wheatstone Bridge!

The following alternative characterization of series parallel networks follows easily.

THEOREM 2. *A connected matroid M is a series parallel network if and only if it is binary and has no minor isomorphic with $M(K_4)$.*

Proof. Note that by Theorem (6.2.1) that since each vertex of K_4 is joined to only three other vertices, the properties P_1, P_2 are equivalent.

P_1: G possesses a subgraph homeomorphic from K_4;
P_2: $M(G)$ possesses a minor isomorphic with $M(K_4)$.

Hence if M is a series parallel network, it is graphic by (11) and does not possess $M(K_4)$ as a minor by Theorem 1 and the above note.

Conversely if M is binary and does not have $M(K_4)$ as a minor *a fortiori* satisfies Tutte's excluded minor conditions (10.5.1) and must be graphic (indeed planar). By the equivalence of P_1, P_2, $M = M(G)$ where G has no subgraph homeomorphic from K_4. Hence by Duffin's result M is a series parallel network.∎

We finally state an important property of graphs whose cycle matroids do not have $M(K_4)$ as a minor. As we shall see it is of crucial importance to results of subsequent sections.

THEOREM 3. *A 2-connected simple graph in which the degree of every vertex is at least 3 has a subgraph homeomorphic from the complete graph K_4.*

For a proof see Dirac [52].

EXERCISES 2

1. Prove that if M is a graphic matroid any series parallel extension of M is graphic.

2. Prove that if M is representable over some field then any series–parallel extension of M is representable over some field.

3. Find the smallest matroid which is not a proper series-parallel extension of any matroid.

4. Prove that the Fano matroid is not a proper series–parallel extension of any matroid.

3. GRAPHIC TRANSVERSAL MATROIDS

Although the problem of finding a nice characterization of transversal matroids is still unsolved, Bondy [72a] and Las Vergnas [70] have given a very attractive characterization of those matroids which are both graphic and transversal.

Let C_k^2 be the graph obtained from a cycle of length k by replacing each edge by a pair of parallel edges (see Fig. 1). We know that $M(K_4)$ is not transversal and that $M(C_3^2)$ is not transversal (see Section 7.3). It is also easy to check that $M(C_k^2)$ is not transversal for $k > 2$. The following result shows that essentially these are the only graphic matroids which are not transversal.

THEOREM 1. *Let G be a graph. Then its cycle matroid $M(G)$ is transversal if and only if G contains no subgraph homeomorphic from K_4 or $C_k^2 (k > 2)$.*

We prove Theorem 1, by a series of lemmas.

LEMMA 1. *Let M be transversal and let $sM(x, y)$ be a series extension of M. Then $sM(x, y)$ is transversal.*

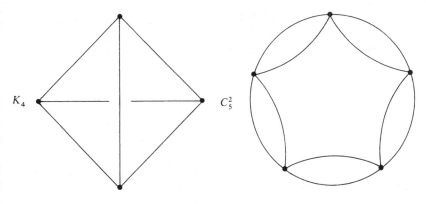

Figure 1

Proof. Let M have the presentation $\mathscr{D} = (D_1, \ldots, D_r)$. Put

$$D_{r+1} = \{x, y\}$$

Then $\mathscr{D}' = (D_1, \ldots, D_{r+1})$ is a presentation of $sM(x, y)$.

LEMMA 2. *Let the graph* G *be a block with more than 2 vertices and suppose it has no subgraph homeomorphic from* K_4 *or* C_k^2 ($k > 2$). *Then* G *contains a vertex adjacent to exactly two vertices, and joined by just one edge to one of these vertices.*

Proof. We use induction on $|V(G)|$. The lemma clearly holds if $|V(G)| = 3$. Suppose now $|V(G)| = N > 3$. By Dirac's Theorem 2.3 there exists a vertex u of G which is adjacent to exactly 2 vertices, v and w. Assume u is joined to v by edges $e_1, \ldots, e_m$ and to w by edges $f_1, \ldots, f_n$. If either m or n is 1, u is the required vertex. Otherwise m, n are at least 2.

Case (i) $(v, w) \notin E(G)$. Let $G' = G . (E \backslash \{f_1, \ldots, f_n\})$. Then G' is a block of $N - 1$ vertices and has no subgraph homeomorphic from K_4 or C_k^2, $(k > 2)$ since G does not. Hence by the induction hypothesis there is a vertex x ($\neq u$) in G' adjacent to exactly 2 vertices in G', and joined by just one edge to one of these vertices. The vertex x has the same property in G.

Case (ii) $(v, w) \in E(G)$. Since G contains no homeomorph of C_3^2 there is only one edge g joining v and w, and moreover, if H denotes the graph obtained by deleting the vertex u, its incident edges $e_1, \ldots, e_m, f_1, \ldots, f_n$ and the edge g, then H must be disconnected, with v and w in different components for the same reason.

Let x be some other vertex in G (x exists since $N > 3$), and let C be the component of G which contains x.

Then either $v \notin C$ and w is a cut vertex of G or $w \notin C$ and v is a cut-vertex of G. In either case the hypothesis that G is a block is contradicted.

This completes the proof of the lemma.

LEMMA 3. *Let M be a transversal matroid on S and let x, y be in series in M. If $T = S \backslash y$ the contraction of M to T is a transversal matroid.*

Bondy's original proof of Lemma 3 is quite intricate. However as Piff [72] noticed, because of the duality between transversal matroids and strict gammoids it is easy to prove the dual statement.

LEMMA 3*. *Let M be a strict gammoid on S and let x, y be in parallel in M. If $T = S \backslash y$, $M \,|\, T$ is a strict gammoid.*

Proof. Let $M = L(G, B)$ where G is a digraph and $B \subseteq V(G)$. If x, y are not members of B then since they are parallel in M there must exist a vertex u of G such that u separates $\{x, y\}$ from B. Let D be the set of vertices of G which are separated from B by u. Then the elements of D are all parallel to u (and also to x and y) in M. Hence an alternative representation of M as a strict gammoid is $M = L(G', B)$ where G' is the graph obtained from G by deleting all edges joining vertices of $D \backslash u$ and inserting a new edge (v, u) for each vertex $v \in D \backslash u$. Provided $y \neq u$, $M \,|\, T \simeq L(G' \backslash y, B)$. If $y = u$ then $x \neq u$ and $L(G' \backslash x, B)$ is a strict gammoid representation of $M \,|\, (S \backslash x)$. Since the roles of x and y in the above lemma are symmetric we have proved the assertion that $M \,|\, T$ is a strict gammoid. In the case where either x or y is a member of B (they cannot both be in B), a similar construction gives a strict gammoid presentation of $M \,|\, T$.

Lemma 3 follows from Lemma 3* by duality.

Proof of Theorem 1. Suppose that $M(G)$ is transversal but that G contains a subgraph H homeomorphic from K_4 or C_k^2 ($k > 2$).

Then $M(H)$ is not transversal since it is possible to obtain either K_4 or C_k^2 from H by a sequence of series contractions and we get from Lemma 3 that either $M(K_4)$ or $M(C_k^2)$ is transversal, which is a contradiction.

Conversely suppose that G contains no subgraph homeomorphic from K_4 or C_k^2. We will show by induction on $|E(G)|$ that $M(G)$ is transversal.

This is trivially so for graphs with 1 or 2 edges. Suppose it is true for graphs with less than N edges. By Lemma 2 there is a vertex u in G such that $\partial(u) = \{v, w\}$ and u is joined to v by one edge e and to w by edges $f_1, \ldots, f_n$.

Let G' be the graph obtained from G by deleting the edge e. Then G' can contain no subgraph homeomorphic from K_4 or C_k^2. Hence $M(G')$ is transversal. Now consider any presentation $(A_1, \ldots, A_r)$ of $M(G')$.

Since u is only adjacent to exactly one vertex w in G', it is easy to check that there must exist an A_i, say A_1 such that

$$A_1 = \{f_1, \ldots, f_n\}$$

and

$$f_i \notin A_j, \qquad 1 \leqslant i \leqslant n, \quad j \neq 1.$$

But now as in the proof of Lemma 1, $M(G)$ is a transversal matroid with presentation $M(G) = M[A_1, A_2', \ldots, A_r']$ where

$$A_1' = A_1 \cup e, \qquad A_2' = A_2, \ldots, A_r' = A_r.$$

This completes the proof of the theorem.■

A second theorem of Bondy [72a] characterizes graphic base orderable matroids.

THEOREM 2. *Let G be a graph. Then its cycle matroid $M(G)$ is base orderable if and only if G contains no subgraph homeomorphic from K_4.*

Proof. Suppose G contains a subgraph H homeomorphic from K_4. Then $M(H)$ is a series extension of $M(K_4)$ or $M(K_4)$ is a series contraction of $M(H)$.

But if $M(G)$ is base orderable, $M(H)$ is base orderable and by (1.3) $M(K_4)$ would be base orderable which is a contradiction. Conversely, suppose G contains no subgraph homeomorphic from K_4. Then by Theorem 2.1 each block B_i of G is a series parallel network and by (2.9) and (2.10) is base orderable which completes the proof of the theorem. ■

EXERCISES 3

1. Prove that for $n \geqslant 4$ the wheel W_n is not transversal but that whirls are transversal.

2. Let G be a graph. Prove that its cycle matroid $M(G)$ is a gammoid if and only if G contains no subgraph homeomorphic from K_4.

3. Characterize those graphs whose cycle matroids are strict gammoids.

4. AN EQUIVALENT CLASS OF BINARY STRUCTURES

The main theorem of this section will show that for a binary matroid the properties of being a gammoid, being base orderable, and being a series parallel network are all equivalent to the property of not possessing $M(K_4)$ as a minor.

The results of this section are based on de Sousa and Welsh [72] and Narayanan and Vartak [73].

(1)   If M is binary and base orderable then it is graphic.

Proof. If M is not graphic it contains as a minor either M (Fano) or its dual, or $M^*(K_5)$ or $M^*(K_{3,3})$. But each of these contains as a minor $M(K_4)$, and since this is not base orderable M is not base orderable. Contradiction.

(2)   If M is binary and transversal it is graphic.

Proof. Since M is transversal it is base orderable. Hence it is graphic by (1).

(3)   A binary matroid M is base orderable if and only if it does not contain $M(K_4)$ as a minor.

Proof. Clearly if M is base orderable it cannot contain $M(K_4)$ as a minor. Suppose M does not contain $M(K_4)$ as a minor but is not base orderable.

Since M is binary and does not contain $M(K_4)$ then M is graphic. Hence by Bondy's theorem $M = M(G)$ where G contains a subgraph homeomorphic from K_4. By Theorem 6.2.1 $M(G)$ contains $M(K_4)$ as a minor which is a contradiction.

We can now prove our main theorem.

THEOREM 1. *Let M be a binary matroid. Then the following statements about M are equivalent.*

(a)   *M is a gammoid.*
(b)   *M is base orderable.*
(c)   *M is a series parallel network.*
(d)   *M does not contain $M(K_4)$ as a minor.*
(e)   *M is the cycle matroid of a graph G which does not contain a subgraph homeomorphic from K_4.*

Proof. If M is a gammoid it is base orderable by Theorem 1.1. Property (b) is equivalent to (d) by (3), and (d) is equivalent to (e) since K_4 has only vertices of degree 3 and we can use Theorem 6.2.1. Properties (c) and (d) are equivalent by Theorem 2.1. Hence we have (a) $\Rightarrow$ (b) $\Leftrightarrow$ (c) $\Leftrightarrow$ (d) $\Leftrightarrow$ (e). To prove that properties (b)–(e) imply (a) we use induction on the cardinal of the ground set S. It is clearly true for $|S| = 1, 2$. We may assume that M is connected and is the cycle matroid of a block graph G.

By Theorem 2.3 G contains a vertex u adjacent to at most two other vertices v, w. Let $e_1, \ldots, e_k$ be the edges joining v, u and let $T = S \backslash \{e_1, \ldots, e_k\}$. Then $M \cdot T$ again satisfies (b)–(c) and by the induction hypothesis must be a gammoid. But M is just a series parallel extension of $M \cdot T$, and hence by (2.8) must also be a gammoid. This proves the theorem. ∎

EXERCISES 4

1. Prove that a transversal matroid is binary if and only if it has a presentation such that the associated bipartite graph is a forest. (J. Edmonds (see Bondy [72a])).

2. Show that every binary matroid which is base orderable is strongly base orderable. (Narayanan and Vartak [73]).

5. PROPERTIES OF TRANSVERSAL MATROIDS

Although transversal graphic and binary matroids have the simple characterization given in Sections 3 and 4, there is no known characterization of transversal matroids in general which is substantially easier to check than by referring back to the definitions. The results of this section are rather technical and will be stated without proof.

Probably the most useful theorem for proving that a given matroid is not transversal is the consequence of Theorem 9.6.3 namely:

(1) A necessary condition that a matroid M is transversal is that it is representable over fields of every characteristic.

The proof of (1) depends on the rather obvious remark:

(2) A matroid M is transversal if and only if it is the union of matroids of rank one.

A first set of conditions for M to be transversal was found by Mason [70]. Although at first sight unwieldy, they have been useful for proving that certain matroids are not transversal. They dualise Theorem 13.2.4.

A geometrical characterization of transversal matroids has been given by Ingleton [71a]. He defines a *quasi-simplex* of a matroid M to be a family of hyperplanes $(H_i : i \in I)$ with $|I| = \rho M$ satisfying

$$(3) \quad \rho \left(\bigcap_{i \in I \backslash J} H_i \right) \leqslant |J| \quad (J \subseteq I).$$

THEOREM 1. *A matroid M is transversal if and only if it contains a quasi-*

simplex $(H_i : i \in I)$ *which also satisfies*:

(a) *for each circuit C of M, there exists J, with* $|J| = |C| - 1$ *and*

$$C \subseteq \bigcap_{i \in I \setminus J} H_i;$$

(b) *for each* $x \in S$, $x \in \bigcap_{i \in I \setminus J} H_i$ *for some* $J \subseteq I$.

Closely related to this last theorem are theorems about presentations of transversal structures given by Bondy and Welsh [71], Bondy [72], and Las Vergnas [70].

Recall that if M is transversal, any family of sets $(A_i : i \in I)$ such that $M = M[(A_i : i \in I)]$ is a presentation of M. Presentations of transversal matroids possess a surprising amount of structure.

First note that a given transversal matroid can have several presentations.

Example. Let $S = \{x_1, \ldots, x_n\}$. The trivial matroid in which every subset is independent has presentations

$$M[(A_i : 1 \leqslant i \leqslant m)] \qquad A_i = S, \quad m \geqslant |S|,$$

$$M[(A_i : 1 \leqslant i \leqslant |S|)] \qquad A_i = \{x_i\}.$$

As a first, rather easy result, we leave the reader to prove:

(4) If M is a transversal matroid of rank r then it has a presentation $(A_1, \ldots, A_r)$.

Suppose that M is a transversal matroid on S of rank r with presentation $(A_1, A_2, \ldots, A_r)$.

Then consider any base B of M, it must be a transversal of the family $(A_i : 1 \leqslant i \leqslant r)$ and hence A_i must intersect B. Thus we have immediately that A_i must be dependent in M^*, and therefore contains a cocircuit C_i^* of M. The first important result on presentations is the following.

THEOREM 2. *If M is a transversal matroid of rank r with presentation* $(A_1, \ldots, A_r)$, *then for each i*, $1 \leqslant i \leqslant r$, *there exists a cocircuit* C_i^* *of M such that*

(a) $C_i^* \subseteq A_i$
(b) $M = M[C_1^*, \ldots, C_r^*]$.

Clearly, this cocircuit presentation obtained is minimal in the sense that $\forall x \in S$, and any i, $1 \leqslant i \leqslant r$,

$$M \neq M[C_1^*, \ldots, C_{i-1}^*, \quad C_i^* \setminus x, \quad C_{i+1}^*, \ldots, C_r^*].$$

For a proof see Bondy and Welsh [71].

Now we shall concern ourselves with *maximal presentations*. If M is a transversal matroid of rank M a presentation $(A_1, \ldots, A_r)$ is *maximal* if for any i, $1 \leqslant i \leqslant r$, and $x \notin A_i$,

$$M \not\models M[A_1, \ldots, A_{i-1}, A_i \cup x, A_{i+1}, \ldots, A_r]$$

THEOREM 3. *Let M be a transversal matroid of rank r on S. Then M has a unique maximal presentation.*

This result leads to the following, even more striking, property of presentations which shows that transversal matroids have far more structure than one might expect.

THEOREM 4. *Let M be a transversal matroid with maximal presentation $(M_1, \ldots, M_r)$. Suppose that $\mathscr{C} = (C_1, \ldots, C_r)$ and $\mathscr{D} = (D_1, \ldots, D_r)$ are cocircuit presentations of M such that $C_i \cup D_i \subseteq M_i$, $1 \leqslant i \leqslant r$. Then $|C_i| = |D_i|$, $1 \leqslant i \leqslant r$.*

In fact more can be said; in the terminology of Theorem 4, the common cardinality of C_i and D_i is given by

$$|C_i| = |M_i| - (r - 1) + \rho(S \backslash M_i).$$

Theorem 3 was proved by Mason [70]. Theorem 4 by Bondy [72]. Extensions of some of these theorems are proved by Brualdi and Dinolt [72] who also obtain an algorithm for determining whether or not a given matroid is transversal. When the given matroid is transversal the algorithm gives explicitly the unique maximal presentation.

EXERCISES 5

1. If G is a 4-connected graph and A_v is the set of edges of G incident with v prove that the transversal matroid $M[(A_v : v \in V(G))]$ has a unique presentation. (Piff [72]).

2. Prove that if M has rank r and there exist $n > r$ distinct circuits each of rank $r - 1$ such that $\rho(C_i \cup C_j) = r(i \neq j)$ then M is not transversal. (Mason [70])

3. If the transversal matroid M has a cocircuit representation in which each cocircuit is fundamental with respect to some cobase we say that it is a *fundamental transversal matroid*. A matroid M is the *base intersection* of matroids $M_1, \ldots, M_t$ if the bases of M are precisely the common bases of $M_1, \ldots, M_t$. Prove that:

   (a) Every matroid of rank r is the base intersection of fundamental transversal matroids of rank r. (Bondy and Welsh [71])

(b) The dual of a fundamental transversal matroid is also fundamental transversal. (Las Vergnas [70]).

(c) Uniform matroids are the only connected matroids which are fundamental transversal with respect to every base. (Brualdi [74]).

6. MATCHINGS IN GRAPHS

In their fundamental paper [65], Edmonds and Fulkerson besides discovering the class of transversal matroids proved that there was a natural matroid associated with matchings in arbitrary non-bipartite graphs.

Let $G = (V, E)$ be a graph with no loops or parallel edges. A *matching* in G is a set of edges with the property that no two members of this set are incident.

Note. Throughout this section we shall be speaking of an edge and a vertex, being *incident* if the vertex is an endpoint of the edge, and of two edges being *incident* if they have a common endpoint.

We shall also say that a set X of vertices *meets* a matching U if X is a subset of the set of endpoints of the matching.

The basic result of Edmonds and Fulkerson is the following:

THEOREM 1. *If G is a graph let* match (G) *be the collection of subsets of $V(G)$ defined by $X \in$ match (G) if X meets the edges of some matching of G. Then* match (G) *is the collection of independent sets of a matroid on $V(G)$.*

Proof. For any $A \subseteq V = V(G)$ let T_1 and T_2 be maximal subsets of A which meet matchings, say N_1 and N_2 respectively.

Consider the subgraph H of G which has edge set $N_1 \triangle N_2$ and vertex set the endpoints of $N_1 \triangle N_2$. The connected components of H are either paths or cycles because each vertex of H meets at most 2 edges.

The vertex set $T_1 \triangle T_2$ consists precisely of those vertices which are the ends of paths in H and which are in A.

Suppose $|T_2| > |T_1|$; then $|T_2 \backslash T_1| > |T_1 \backslash T_2|$.

Thus since $T_1 \triangle T_2 = (T_2 \backslash T_1) \cup (T_1 \backslash T_2)$ some component of H must be a path P with one end v in $T_2 \backslash T_1$ and the other end not in $T_1 \backslash T_2$. Considering the path P as an edge set, $N_1 \triangle P$ is a matching which meets T_1 in A and in addition it meets v in A. Thus we contradict the hypothesis that T_1 is a maximal subset of A which meets a matching. Hence T_1 and T_2 have the same cardinality and thus we have proved that match (G) is the collection of independent sets of a matroid on $V(G)$. ∎

We will use match (G) to denote the matroid and also its collection of independent sets.

Example 1. Let G be the star graph of Fig. 1. Then match (G) has as its bases the sets $\{1, x\}$ $(2 \leqslant x \leqslant 6)$.

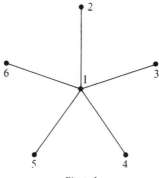

Figure 1

Example 2. Match (K_n) has as its bases $V(K_n)$ or all $(n-1)$-subsets of $V(K_n)$ depending on whether n is even or odd.

Clearly for $A \subseteq V(G)$ if we define match $(G; A)$ to be the collection of subsets X of A which meet some matching in G, then match $(G; A)$ is the collection of independent sets of a matroid on A, which we shall also call match $(G; A)$. In fact

$$\text{match}(G; A) = \text{match}(G \,|\, A).$$

Note that in general

$$\text{match}(G; A) \neq \text{match}(G) \,|\, A.$$

We call a matroid a *matching matroid* if there exists some graph G and a subset A of $V(G)$ such that M is isomorphic to match $(G; A)$.

Every transversal matroid is a matching matroid. For consider the transversal matroid $M[A_1, \ldots, A_n]$ on S. Let Δ denote the associated bipartite graph with vertex set $\{1, 2, \ldots, n\} \cup S$.

Then it is clear that

$$M[A_1, \ldots, A_n] \simeq \text{match}(\Delta; S).$$

A surprising result, proved by Edmonds and Fulkerson [65] (though the proof is almost entirely dependent on a difficult theorem of Edmonds [65b]) is the following.

THEOREM 2. *A matroid is a matching matroid if and only if it is transversal.*

We know of no short proof of this theorem, nor of a simple routine for constructing a "bipartite representation" of the matching matroid of a non-bipartite graph.

Given an arbitrary graph, a natural question to ask is "when does it have a matching which is incident with every vertex—called in the literature a *perfect matching*?". Clearly it must have an even number of vertices. Another obvious necessary condition is that for any subset $T \subseteq V(G)$ the number of odd components of $G|(V \backslash T)$ must be not greater than $|T|$. In what is surely one of the most beautiful theorems of graph theory, Tutte [47a] proved that this was in fact also a sufficient condition.

THEOREM 3. *A graph $G = (V, E)$ has a perfect matching if and only if for all $T \subseteq V$, the number of odd components of $G|(V \backslash T)$ is not greater than $|T|$.*

Note. An *odd component* of a graph is a connected component with an odd number of vertices.

If G is a graph with vertex set V we define a function $O:V \times V \to \mathbb{Z}$ by $O(A, T)$ is the number of odd component of the subgraph of G on $V \backslash T$ which are contained in the subset A. Thus if $A \subseteq T$, $O(A, T) = 0$.

We now state a generalization of Tutte's theorem which was proved by Berge [58].

THEOREM 4. *If G is a graph with vertex set V, the number of vertices not incident with a maximum matching in G is given by*
$$\max_{T \subseteq V} (O(V, T) - |T|).$$

We shall obtain Theorems 3 and 4 as special cases of a more general rank formula proved by Brualdi [71].

THEOREM 5. *If G is a graph the rank function ρ of the matching matroid match(G) on $V = V(G)$ is given for $A \subseteq V$ by*
$$\rho A = \min_{T \subseteq V} (|A| + |T| - O(A, T)).$$

Before proving Brualdi's theorem we shall show how it includes the previous theorems of Tutte and Berge.

Proof of Theorems 3 and 4. The number of vertices of G not incident with a maximum matching is clearly.
$$|V| - \rho(\text{match } G) = |V| - \rho(V)$$

By Brualdi's theorem

$$|V| - \rho(V) = |V| - \min_{T \subseteq V} (|V| + |T| - O(V, T))$$

$$= \max_{T \subseteq V} (O(V, T) - |T|),$$

which gives Berge's theorem. Tutte's theorem follows since a perfect matching exists if and only if $\rho V = |V|$. ∎

The proof of Brualdi's theorem which we give is due to Milner [74].

Proof. For $T \subseteq V$, let $\mathscr{C}(T)$ be the set of connected components of $G|(V \backslash T)$. For $A, T \subseteq V$ let $\theta(A, T)$ be the set of odd components of $G|(A \backslash T)$ and let

$$m(A, T) = |A| + |T| - O(A, T)$$

$$m(A) = \inf_{T \subseteq V} m(A, T).$$

LEMMA 1. $\rho A \leqslant m(A, T)$.

Proof. Let W be any matching. For each component $C \in \theta(A, T)$ there is a vertex v such that either v is not incident with W or there is $t \in T$ such that the edge $(v, t) \in W$. There are at most $|T|$ such points v and hence the number of vertices incident with W and not in A is at least $O(A, T) - |T|$.

LEMMA 2. $\rho A \geqslant m(A, T)$.

Proof. If $A = \varnothing$ the result is obvious. We assume $A \neq \varnothing$ and use induction on $|A|$.

Suppose $\rho A < m = m(A, T)$. Then there is a matching $W \subseteq E$ such that every edge of W is incident with A and W is maximal. Let T_0 be the vertex set of W, then $|T_0| < 2m$ and $m(A, T_0) \leqslant 2|T_0|$ since every vertex of $A \backslash T_0$ is an isolated vertex of $G|(V \backslash T_0)$.

Now consider any subset T of V such that

(α) $m(A, T) = m(A)$.

For $t \in T$ let $F_t = \{C \in \theta(A; T) : t \text{ is joined to } C\}$. We claim that

(β) $|F(K)| \geqslant |K|$ $K \subseteq T$

where $F(K) = \bigcup_{t \in K} F_t$.

For suppose that (β) is false for some $K \subseteq T$. Let $T_1 = T \backslash K$. Each component in $\mathscr{C}(T) \backslash F(K)$ is also a component of $G|(V \backslash T_1)$. Therefore

$$m(A) \leqslant m(A, T_1) \leqslant m(A, T) + |F(K)| - |K| < m(A).$$

This contradiction proves (β). Hence by Hall's theorem, for each $t \in T$ there exists $C_t \in \theta(A, T)$ such that t is joined to $x_t \in C_t$ and the C_t are pairwise disjoint. Now the edges (x_t, t) $(t \in T)$, form a matching and so $m > \rho A \geq |T|$. Hence we can assume that T is chosen so that (α) holds and also so that:

(γ) $|T|$ is maximal subject to T satisfying (α).

For each $C \in \mathcal{C}(T)$ define $D = D(C)$ as follows. If $C = C_t$ for some $t \in T$ put $D = C_t \backslash x_t$. If $C \in \theta(A, T) \backslash \{C_t : t \in T\}$ put $D = C \backslash x_c$ where c is any point of C. If $C \in \mathcal{C}(T) \backslash \theta(A, T)$ put $D = C$. It will be enough to prove that

(δ) $A \cap D$ is independent in match $(G | D)$ $(D = D(C), C \in \mathcal{C}(T))$.

For if (δ) holds, then there is a matching of G which is incident with every vertex of

$$A' = A \backslash \{x_c : C \in \theta(A, T) \backslash \{C_t : t \in T\}\},$$

and hence $\rho A \geq |A'| = m(A, T) \geq m$—a contradiction. We now prove (δ).

Case 1 $A \cap D \neq A$. Then $|A \cap D| < |A|$ and so by the induction hypothesis (applied to this graph $G' = G | D$) there is $T' \subseteq D$ such that

$$\rho_{G'}(A \cap D) = |A \cap D| + |T'| - q$$

where q is the number of odd components of G' in $A \cap D$.

Case 1(a) $C \in \theta(A, T)$. Then $D = D \cap A$ is even and if (δ) is false then

$$\rho'(A \cap D) \leq |A \cap D| - 2.$$

That is $q - 2 - |T'| \geq 0$. Put $T_1 = T \cup T' \cup (C \backslash D)$. Then

$$\mathcal{C}(T_1) = (\mathcal{C}(T) \backslash \{C\}) \cup \mathcal{C}_{G'}(T')$$

and so

$$m(A, T) - m(A, T_1) = |C| - 1 - (|C| - q) - (|T'| + |C \backslash D|)$$
$$= q - 2 - |T'| \geq 0.$$

Hence $m(A, T_1) = m(A, T) = m(A)$. But this contradicts ($\gamma$).

Case 1(b) $C \in \mathcal{C}(T) - \theta(A, T)$. If ($\delta$) is false in this case, then

$$\rho_{G'}(A \cap D) \leq |A \cap D| - 1$$

and hence if $T_1 = T \cup T'$ we have

$$m(A, T) - m(A, T_1) = |D \cap A| - (|D \cap A| - q) - |T'| \geq 1.$$

Hence $m(A, T_1) < m(A, T) = m(A)$, a contradiction.

Case 2 $A \cap D = A$. In this case $D = C$ and so

$$T = \varnothing, \qquad \theta(A, \varnothing) = \varnothing.$$

Case 2(a)   A is a component of G. Then $|A|$ is even since $\theta(A, \varnothing) = \varnothing$. But then for every vertex $x \in A$,

$$|\theta(A, \{x\})| \geqslant 1$$

and so

$$m(A, \{x\}) \leqslant |A| = m(A, \varnothing) = m(A)$$

and this contradicts the assumed maximality of $|T| (= 0)$.

Case 2(b)   There exists $x \in A$ and $y \in V \backslash A$ such that (x, y) is an edge of G.

Consider the graph $G_1 = G|(V \backslash y)$ and the set $A_1 = A \backslash x$. By the induction hypothesis there is T_1 such that

$$\rho_1(A_1) = |A_1| - |\theta_1(A, T_1)| + |T_1|$$

(where index 1 refers to graph G_1).

Let $T = T_1 \cup y$. Then

$$|\theta_1(A_1, T_1)| \leqslant |\theta(A, T)| \leqslant |\theta_1(A_1, T_1)| + 1.$$

By (γ)

$$|\theta(A, T)| < |T| = |T_1| + 1.$$

Hence $|\theta_1(A_1, T_1)| \leqslant |T_1|$.

This proves that $\rho_1 A_1 = |A_1|$.

Hence there is a matching W_1 in G_1 which is incident with all vertices of A_1. If x is a vertex incident with W_1, then $\rho A = |A|$. If not, then $W_1 \cup (x, y)$ is a matching incident with all vertices of A. Hence $\rho A = |A|$ in either case and in case 2(b) $m(A) = |A| = \rho A$ contrary to our original assumption. ∎

EXERCISES 6

1. If X is the set of edges of a matching in a graph G show that in general X can not be augmented to a matching of maximum cardinality.

2. What is the matching matroid of the Petersen graph?

3. Is the cycle matroid $M(K_4)$ isomorphic to the matching matroid of any graph?

4. Show that the Fano matroid is not a matching matroid.

5. Prove that a matching matroid is algebraic over every field.

○6. Find a direct construction which exhibits the linear representability of matching matroids over say the rationals.

○7. Characterize those transversal matroids M which can be regarded as *strict* matching matroids in the sense that there exists a graph G such that $M \simeq$ match (G).

NOTES ON CHAPTER 14

Brualdi and Scrimger [68], Brualdi [69] and Mason [70] were the first to study (strongly) base orderable matroids. Some interesting examples are given by Ingleton [75]. Minty [66] suggested extending Duffin's work on series parallel networks to matroids—for a detailed study see Brylawski [71], Narayanan and Vartak [73]. They also study the *series-parallel connection* of matroids on sets with a distinguished point (called *pointed matroids*). This is closely connected with the operation of welding introduced by Smith [72], [74]. For a discussion of the interrelationship between these operations see Bixby [75].

Sections 3 and 4 are based on Las Vergnas [70], Bondy [72], de Souza and Welsh [72] and Narayanan and Vartak [73].

The necessary condition 5.1 for transversality follows from the theorem of Piff and Welsh [70] that transversal matroids are representable over every sufficiently large field. Theorems 5.1 and 5.2 can be dualized to give conditions of a similar form for a matroid to be a strict gammoid (see Mason [70a], [72] or Ingleton and Piff [73]).

Another criterion similar to that of Ingleton's Theorem 5.3 has been discovered by Brylawski [73].

Apart from the papers listed other criteria (of a more technical nature) for matroids to be transversal have been given by Brown [74], Brualdi [74a] and Dowling and Kelly [74].

Section 5 on matchings is based on the papers of Edmonds and Fulkerson [65], Brualdi [71] and unpublished work of E. C. Milner. Alternative proofs of Brualdi's theorem are given by McCarthy [75], and Las Vergnas [73]. This last paper also contains many extensions of the ideas of Section 5 to matroids determined by degree-constrained subgraphs of a graph.

CHAPTER 15

Polynomials, Colouring Problems, Codes and Packings

1. THE CHROMATIC POLYNOMIAL OF A GRAPH

Probably the most famous unsolved problem in graph theory, if not in the whole of combinatorics is that of settling the four colour conjecture. In this chapter we attempt to put this problem in a more geometrical context. The basic idea was known to Veblen (1912) but until recently seems to have been ignored.

A *k-colouring* of a graph G is the assignment to each vertex of G one of a specified set of k colours. A *proper k-colouring* of G is a colouring in which no two adjacent vertices are the same colour. The *chromatic polynomial* $P(G; \lambda)$ is the function, defined on $\mathbb{Z}^+$ associated with G, which when λ is a positive integer equals the number of different ways of properly colouring G in λ colours.

Example. For the complete graph K_n

$$P(K_n; \lambda) = \lambda^{(n)} = \lambda(\lambda - 1) \ldots (\lambda - n + 1).$$

The *chromatic number* of a graph G, $\chi(G)$ is the smallest integer k for which there exists a proper k-colouring of G. Thus the four colour conjecture, that every planar graph has a 4-colouring, can be restated as:

4-colour conjecture. If G is a planar graph $P(G; 4) > 0$.

The origin of the idea of attacking colouring problems through the chromatic polynomial seems to be the paper by Whitney [32b], though as early as 1912 G. D. Birkhoff [12] was implicitly employing the same technique.

The most basic property of chromatic polynomials is contained in the following theorem. It also gives a routine (though tedious) method of calculating the chromatic polynomial of a given graph.

THEOREM 1. *If G is a graph without loops or parallel edges and e is an edge of G then*

(1) $P(G; \lambda) = P(G'_e; \lambda) - P(G''_e; \lambda).$

where G'_e and G''_e denote respectively the graph obtained by the deletion and contraction of the edge e from G.

Proof. Very easy, Divide the proper λ-colourings of G'_e into these which assign the endpoints of e he same colour and those which assign the endpoints of e different colours. ∎

An immediate consequence of (1) is

THEOREM 2. *The function $P(G; \lambda)$ is a polynomial in λ of maximum degree $|V(G)|$.*

Proof. Use (1) and induction. ∎

Even more obvious are the following propositions:

(2)   If G has connected components $G_1, G_2, \ldots, G_k$, then

$$P(G; \lambda) = P(G_1; \lambda)P(G_2; \lambda). \ldots . P(G_k; \lambda).$$

(3) $P(G, \lambda)$ has no constant term, and the coefficient of $\lambda^{|V(G)|}$ is 1.
(4) If G is connected then $P(G; \lambda) \leq \lambda(\lambda - 1)^{|V(G)|-1}$ for λ a positive integer.

The following properties of the coefficients of the chromatic polynomial are proved in Biggs [74]. If

$$P_G(\lambda) = \sum_{i=1}^{|V(G)|} a_i \lambda^i;$$

(5) the a_i alternate in sign;
(6) if G is connected then

$$|a_i| \geq \binom{|V(G)| - 1}{i - 1};$$

(7) the coefficient a_p is given by

$$a_p = \sum_{j=0}^{|E(G)|} (-1)^j m_{pj},$$

where m_{pj} is the number of spanning subgraphs of G with p connected components and j edges.

Theorem 1 can also be used (see Read [68]) to obtain the following result:

THEOREM 3. *The chromatic polynomial $P(G; \lambda)$ has a representation*

$$P(G; \lambda) = \sum_{i=\chi(G)}^{|V(G)|} b_i \lambda^{(i)}$$

where b_i is the number of ways of properly colouring G in exactly i colours with colour indifference.

Note: To colour G with *colour indifference* is to regard two colourings the same if they are isomorphic; for example a triangle has 6 proper 3-colourings but only 1 3-colouring with colour indifference.

For further properties of chromatic polynomials of graphs we refer to Biggs [74] or Read [68].

EXERCISES 1

1. Prove that G is a tree with n vertices if and only if
$$P(G; \lambda) = \lambda(\lambda - 1)^{n-1}.$$
(Read [68])

2. Prove that if G is a cycle with n vertices then
$$P(G; \lambda) = (\lambda - 1)^n + (-1)^n (\lambda - 1).$$

3. Prove that if G is the wheel $\mathscr{W}_n$ then
$$P(G; \lambda) = \lambda(\lambda - 2)^n + (-1)^n \lambda(\lambda - 2).$$

4. Find a graph G with a chromatic polynomial $P(G; \lambda)$ such that $P(G; \alpha) < 0$ for some α strictly greater than the chromatic number of G. (P. D. Seymour)

5. If G is a connected graph with n vertices and chromatic polynomial
$$P(G; \lambda) = \lambda^n - c_1 \lambda^{n-1} + \ldots + (-1)^{n-1} c_{n-1} \lambda$$
then for all $i \leqslant (n - 1)/2$,
$$c_{i-1} \leqslant c_i.$$
(Heron [72])

2. THE MÖBIUS FUNCTION OF A PARTIALLY ORDERED SET

The material of this section is based on the fundamental paper of Rota [64] on Möbius inversion. The first section has nothing to do with matroids and is a general technique of great importance in combinatorics, particularly in the theory of combinatorial identities.

Let P be a finite partially ordered set. The *Möbius function* μ of P is a map $\mu: P \times P \to \mathbb{Z}$ defined recursively by

$$\mu(x, x) = 1 \qquad (x \in P)$$
$$\mu(x, y) = 0 \qquad (x \not\leqslant y)$$
$$\mu(x, y) = - \sum_{x \leqslant z < y} \mu(x, z). \qquad (x \leqslant y).$$

Example 1. If P is the lattice of Fig. 1 then μ is given by

$$\mu(x, y) = (-1)^{h(y) - h(x)} \qquad x \leqslant y$$
$$= 0 \qquad x \not\leqslant y.$$

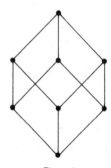

Figure 1

Example 2. The classical Möbius function $\mu(n)$ is defined to be $(-1)^k$ if n is as integer and the product of k distinct primes and zero otherwise. The classical inversion formula devised by Möbius in 1832 is: if

$$g(m) = \sum_{n: n \mid m} f(n)$$

then

$$f(m) = \sum_{n: n \mid m} \mu\left(\frac{m}{n}\right) g(n).$$

Suppose now we take P to be the partial order of positive integers $\leqslant N$ ordered by

$$x \leqslant y \Leftrightarrow x \mid y.$$

Then P has Möbius function $\hat{\mu}$ given by:

$$\hat{\mu}(n, m) = \mu\left(\frac{m}{n}\right) \qquad (n \leqslant m \leqslant N),$$

$$= 0 \qquad \text{otherwise.}$$

THEOREM 1. *Let P be a finite partially ordered set, and let $f: P \to \mathbb{R}$ be a function. Then if g is defined by:*

$$g(x) = \sum_{y \leqslant x} f(y),$$

we have

$$f(x) = \sum_{y \leqslant x} g(y)\mu(y, x).$$

Before proving this we note the following alternative definition of μ.
Let $\zeta: P \times P \to \mathbb{Z}$ be the *zeta function* defined by

$$\zeta(x, y) = \begin{cases} 1 & x \leqslant y \\ 0 & \text{otherwise.} \end{cases}$$

Then in the *incidence algebra* of P defined to be the set of all real functions $f: P \times P \to \mathbb{R}$ such that $f(x, y) = 0$ if $x \nleqslant y$, define

$$(f + g)(x, y) = f(x, y) + g(x, y),$$

$$(af)(x, y) = af(x, y) \qquad a \in \mathbb{R}$$

$$(f . g)(x, y) = \sum_{x \leqslant z \leqslant y} f(x, z)g(z, y).$$

Then the identity function of this algebra is δ defined by

$$\delta(x, y) = \begin{cases} 1 & x = y \\ 0 & \text{otherwise,} \end{cases}$$

and then

$$(\zeta . \mu)(x, y) = \sum_{x \leqslant z \leqslant y} \zeta(x, z)\, \mu(z, y)$$

$$= \sum_{x \leqslant z \leqslant y} \mu(z, y)$$

$$= \delta(x, y).$$

Thus we have proved:

(1) The Möbius function μ is the inverse of ζ in the incidence algebra on $P \times P$.

Proof of Theorem 1. For $x \in P$

$$g(x) = \sum_{y \leqslant x} f(y).$$

Consider the expression

$$\sum_{y \leqslant x} g(y)\, \mu(y, x) = \sum_{y \leqslant x} \mu(y, x) \sum_{z \leqslant y} f(z)$$

$$= \sum_{y \leqslant x} \sum_{z} f(z)\, \zeta(z, y)\, \mu(y, x)$$

where the second summation is over all $z \in P$.

Interchanging the order of summation, the right-hand side becomes

$$\sum_{z} f(z) \sum_{y \leqslant x} \zeta(z, y)\, \mu(y, x) = \sum_{z} f(z)\, \delta(z, x) = f(x)$$

as required, and the proof is complete. ∎

What can be regarded as the result dual to the above basic theorem is the following.

THEOREM 2. *Let* $f : P \to \mathbb{R}$ *and suppose that for all* $x \in P$

$$g(x) = \sum_{y \geqslant x} f(y).$$

Then

$$f(x) = \sum_{y \geqslant x} g(y)\, \mu(x, y).$$

Proof. Completely analogous to the proof of Theorem 1. ∎

Using the above theorems, or directly, it is easy to prove the following properties of μ.

(2) The Möbius function of an interval $[x, y]$ equals the restriction of μ to the interval.
(3) Let P^* be the dual poset obtained by inversion of P. Let μ^* be its Möbius function. Then

$$\mu^*(x, y) = \mu(y, x).$$

(4) If P, Q are two partially ordered sets with Möbius functions μ_P, μ_Q respectively then the Möbius function $\mu_{P \times Q}$ of their product is given by

$$\mu_{P \times Q}((x, y), (u, v)) = \mu_P(x, u)\, \mu_Q(y, v).$$

A more important result is Weisner's theorem [35].

THEOREM 3. *If $\mathscr{L}$ is a finite lattice and $b > o \in \mathscr{L}$ then for any $a \in \mathscr{L}$ we have*

(5)
$$\sum_{x:\, x \vee b = a} \mu(o, x) = 0.$$

If $b < I$ then

(6)
$$\sum_{x:\, x \wedge b = a} \mu(x, I) = 0.$$

Proof. If $a = b$

$$\sum_{x:\, x \vee b = b} \mu(o, x) = \sum_{o \,\leqslant\, x \,\leqslant\, b} \mu(o, x) = 0.$$

Let a be a minimal element of the lattice $\mathscr{L}$ for which the result is not true. Then

$$\sum_{x:\, x \vee b \,\leqslant\, a} \mu(o, x) = \sum_{x \vee b = a} \mu(o, x) + \sum_{x\cdot v\, b\, <\, a} \mu(o, x)$$

But by the minimality of a this last term is zero while the left hand side is

$$\sum_{o \,\leqslant\, x \,\leqslant\, a} \mu(o, x) = 0.$$

The result (5) follows and (6) is proved by a dual argument. ∎

Despite the power in principle of the inversion theorems, they suffer in practice from the drawback that the Möbius function is usually very tedious to compute and is only known in closed form for a few special well-behaved lattices.

Examples
(7) If $\mathscr{L}$ is the lattice of subsets of a set S

$$\mu(A, B) = (-1)^{|B|-|A|} \qquad (A \subseteq B).$$

(8) If $\mathscr{L}$ is the lattice of flats of $PG(n, q)$, for any pair of flats A, B, $A \subseteq B$, with $\dim B - \dim A = k$,

$$\mu(A, B) = (-1)^k q^{\binom{k+1}{2}}$$

Notice that for fixed A, μ alternates in sign. This is a special case of a more general result which seems to be one of the few known properties of the Möbius function of a geometric lattice.

THEOREM 4. *Let μ be the Möbius function of a finite geometric lattice $\mathscr{L}$. Then*

(i) *if $x \leqslant y$, $\mu(x, y) \neq 0$;*
(ii) *if y covers z and $x \leqslant z$ then $\mu(x, y)$ and $\mu(x, z)$ have opposite signs.*

Proof. Since every interval of a geometric lattice is geometric it suffices to prove the theorem for $x = o$, $y = I$, and z a copoint.

The result is true for lattices of rank $1(\mu(o, I) = -1)$. We use induction on the height of $\mathscr{L}$ assuming it to be true for geometric lattices of height $r - 1$.

Take $a = I$ in Weisner's theorem and let b be an atom of $\mathscr{L}$. Then

$$\mu(o, I) = - \sum_{\substack{x: x \vee b = I \\ x \neq I}} \mu(o, x).$$

But if $x \vee b = I$, with b an atom the submodularity of the rank (height) function shows that $\rho(x) \geqslant r - 1$. Hence the sum on the right hand side above is over a non empty set of copoints of $\mathscr{L}$ and by the induction hypothesis all the terms $\mu(o, x)$ are non-zero and of the same sign. Thus $\mu(o, I)$ is non-zero and of opposite sign as required. ■

As a nice application of Möbius inversion we obtain the following alternative form of the chromatic polynomial of a graph. Let G be a simple graph and let $\mathscr{L}$ be the geometric lattice determined by its cycle matroid $M(G)$.

For any colouring ω of the vertices of G let $X(\omega)$ be the set of edges of G with the property that $e \in X(\omega)$ if and only if both endpoints of e have the same colour under ω. Call $X(\omega)$ the *bond* of ω.

LEMMA. *For any ω, $X(\omega)$ is a flat in $M(G)$.*

Proof. Suppose not, so that $f \in \sigma(X(\omega)) \backslash X(\omega)$. Then there is a cycle $\{f, e_1, \ldots, e_k\}$ of G with

$$\{e_1, \ldots, e_k\} \subseteq X(\omega)$$

$$f \notin X(\omega).$$

Thus the two endpoints of f have the same colour under ω, and $f \in X(\omega)$.

Now for any integer λ and any flat X of $M(G)$ let $Q(\lambda; X)$ be the number of λ-colourings of G with bond equal to X.

For any flat X let $F(\lambda; X)$ be the number of λ-colourings of G with bond containing X. Thus

$$F(\lambda; X) = \sum_{T \geqslant X} Q(\lambda; T)$$

where the right-hand summation and ordering refers to the geometric lattice $\mathscr{L}(M(G))$. Then by the Möbius inversion formula

$$Q(\lambda; \varnothing) = \sum_{\varnothing \leqslant Y} F(\lambda; Y) \mu(\varnothing, Y).$$

But $Q(\lambda; \varnothing)$, the number of λ-colourings with bond $\varnothing$, is the number of proper λ-colourings of G, namely $P(G; \lambda)$. On the other hand for any flat $Y \in \mathscr{L}$

$$F(\lambda; Y) = \lambda^{k(Y)}\lambda^{|V(G)| - |V(Y)|}$$

where $k(Y)$ is the number of connected components in the subgraph determined by Y.

That is

$$F(\lambda; Y) = \lambda^{|V(G)| - \rho Y}.$$

where ρ is the rank function of $M(G)$, that is $\rho(Y) = |V(Y)| - k(Y)$.

Thus since this is true for all integers λ, we have:

THEOREM 5. *For any graph G*

$$P(G; \lambda) = \lambda^{|V(G)|} \sum_{Y \leqslant E(G)} \lambda^{-\rho Y}\mu(\varnothing, Y).$$

where μ is the Möbius function of $\mathscr{L}(M(G))$, and the sum on the right is over all flats Y of the cycle matroid of G.

EXERCISES 2

1. If $\mathscr{L}$ is the lattice of flats of $M(K_n)$ prove that $\mu(o, I)$ in $\mathscr{L}$ is $(-1)^n (n - 1)!$

2. Prove properties (2)–(4) of the Möbius function of a partially ordered set.

3. Let P be a finite partially ordered set. Let $c_k(x, y)$ be the number of chains in P between x and y which have length k. Prove that for $x, y \in P$,

$$\mu(x, y) = \sum_{k \geqslant 1} (-1)^k c_k(x, y).$$

4. Prove that if $\mathscr{L}$ is a finite lattice and $x \in \mathscr{L}$ then

$$\mu(o, I) = \sum_{y, z} \mu(o, y)\, \zeta(y, z)\, \mu(z, I)$$

where the sum on the right hand side is taken over all complements y, z of x. (Crapo [68])

5. Prove that if $\mathscr{L}$ is a finite lattice in which I is not the join of atoms then $\mu(o, I) = 0$.

6. A *crosscut* of a finite lattice $\mathscr{L}$ is a subset C such that

   (i) C does not contain o or I,
   (ii)   no two elements of C are comparable,
   (iii) any maximal chain between o and I intersects C.

   A *spanning set* of $\mathscr{L}$ is a set X of elements such that
   $$\vee X = I, \qquad \wedge I = 0.$$

   Prove that if C is a crosscut of $\mathscr{L}$ and q_k is the number of spanning subsets of C containing k distinct elements then
   $$\mu(o, I) = q_2 - q_3 + q_4 - q_5 + \ldots. \text{ (Rota [64])}$$

3. THE CHROMATIC OR CHARACTERISTIC POLYNOMIAL OF A MATROID

Rota [64] defines the *characteristic polynomial* of a finite partially ordered set P satisfying the Jordan–Dedekind chain condition (and having o by)

$$P(\lambda) = \sum_{x \in P} \mu(o, x)\, \lambda^{n - h(x)}$$

where h is the height function of P, and $n + 1$ is the length of any maximal chain in P. We will see that when P is the geometric lattice of flats of a graphic matroid, the characteristic polynomial of P is essentially the chromatic polynomial of the underlying graph.

From (1.7) we see that for a simple graph G each subset A of $E = E(G)$ contributes a term

$$(-1)^{|A|} \lambda^{k(G) + \rho(E) - \rho(A)}$$

to the chromatic polynomial $P(G; \lambda)$ where $k(G)$ denotes the number of connected components of G and ρ denotes the rank function of the cycle matroid $M(G)$.

Accordingly for any matroid M on a set S we define the *chromatic polynomial* of M, $P(M; \lambda)$ by

(1) $$P(M; \lambda) = \sum_{A \subseteq S} (-1)^{|A|} \lambda^{\rho(S) - \rho(A)}$$

where again ρ is the rank function of M.

Thus for a graph G with no loops or parallel edges

(2) $$\lambda^{k(G)} P(M(G); \lambda) = P(G; \lambda).$$

It follows that, since for positive λ, $P(G; \lambda) > 0$ if and only if $P(M(G); \lambda) > 0$, the chromatic number of G, $\chi(G)$ is given by

$$\chi(G) = \inf_{n \in Z^+} \{n : P(G; n) > 0\} = \inf_{n \in Z^+} \{n : P(M(G); n) > 0\}.$$

Hence we define the *chromatic number* $\chi(M)$ of a matroid M by

$$\chi(M) = \inf_{n \in Z^+} \{n : P(M; n) > 0\}.$$

Notice that an immediate consequence of (2) is:

(3)   if G_1, G_2 are cycle-isomorphic graphs (that is $M(G_1)$ is isomorphic to $M(G)$) then $\chi(G_1) = \chi(G_2)$.

Notice also that the definition (1) does not exclude those matroids with loops or parallel elements. In fact we can now state some basic properties of

$P(M; \lambda)$ which show that it behaves roughly like the chromatic polynomial of a graph.

(4) If e is neither a loop nor a coloop of M on S, and $T = S \backslash e$, then

$$P(M; \lambda) = P(M | T; \lambda) - P(M . T; \lambda).$$

(5) If M has a loop then $P(M; \lambda) = 0$.
(There is no way of properly colouring the vertices of a graph which contain a loop!)

(6) If e is a coloop of M on S and $T = S \backslash e$, then

$$P(M; \lambda) = (\lambda - 1)P(M | T; \lambda).$$

(The corresponding result for $P(G; \lambda)$ when e is an isthmus of the graph G is obvious).

We see also that Theorem 2.5 is a special case of the following identity between the chromatic polynomial of M and the characteristic polynomial of the associated geometric lattice $\mathcal{L}(M)$. For a proof see Rota [64].

THEOREM 1. *If M is a simple matroid on S and $\mathcal{L}$ is its lattice of flats*

$$P(M; \lambda) = \sum_{F \in \mathcal{L}} \mu(\varnothing, F) \lambda^{\rho(S) - \rho(F)}$$

where μ is the Möbius function of $\mathcal{L}$, and ρ is the rank function of M.

Thus if G is a simple graph its chromatic polynomial is essentially the characteristic polynomial of the geometric lattice $\mathcal{L}(M(G))$.

Examples. We now list some known chromatic polynomials and chromatic numbers of matroids.

Recall $U_{k,n}$ denotes the uniform matroid of rank k on n elements. Let $P_{k,n}(\lambda)$ be its chromatic polynomial. Then

$$P_{2,n}(\lambda) = \lambda^2 - n\lambda + (n - 1) \qquad n \geqslant 2$$

and it is easy to work out $P_{r,n}$ in general from its definition. We display in the following table the chromatic number of $U_{k,n}$ for all uniform matroids on $\leqslant 8$ elements.

$\|S\| = n$ \ rank	1	2	3	4	5	6	7	8
2								
3		3	2					
4		4	2	2				
5		5	2	3	2			
6		6	2	3	2	2		
7		7	2	4	2	3	2	
8		8	2	4	2	3	2	2

If M_F is the Fano matroid

$$P(M_F; \lambda) = \lambda^3 - 7\lambda^2 + 14\lambda - 8$$

and it has chromatic number 5.

The matroid of the Non-Pappus configuration has chromatic polynomial $\lambda^3 - 9\lambda^2 + 28\lambda - 20$ and it has chromatic number 2.

Its dual has chromatic polynomial $\lambda^6 - 9\lambda^5 + 36\lambda^4 - 84\lambda^3 + 126\lambda^2 - 118\lambda + 48$ and chromatic number 3.

The reader can use the above list of polynomials to show that when M is not graphic the chromatic polynomial does *not* have certain "nice" properties possessed by the chromatic polynomial of a graph.

Examples.
1. If M is not graphic and N is the restriction of M to any subset then $\chi(N)$ need not be less than $\chi(M)$.
2. If M is not graphic and α is an integer greater than the chromatic number then $P(M; \alpha)$ is not necessarily > 0.
3. The chromatic polynomial of a graphic matroid is a positive integral combination of the Stirling polynomials $\lambda^{(i)}$. Again this is not true for general non-graphic matroids.

The only "colouring" type theorem which does seem to carry over from graph theory is the following.

THEOREM 2. *If a matroid M is binary its chromatic number is $\leqslant 2$ if and only if all its circuits have even length, that is if it is bipartite.*

For graphs this is well known and very easy to prove. For binary non-graphic matroids the proof is harder—see Brylawski [72].

Much easier to prove is the following result.

THEOREM 3. *If M is a disconnected matroid with components* $\{M_i : i \in I\}$ *then*

$$P(M; \lambda) = \prod_{i \in I} P(M_i; \lambda).$$

Proof. A consequence of a more general result for Tutte polynomials proved in Section 4. ∎

As with chromatic polynomials of graphs little is known about the coefficients of the chromatic polynomials of a matroid. We do have the following result which we state without proof (see Rota [64]).

THEOREM 4. *For any matroid M the coefficients of* $P(M; \lambda)$ *alternate in sign.*

The corresponding result (1.5) for $P(G; \lambda)$ is a special case of this.

Very little is known about the chromatic number of a matroid. Obviously the chromatic number of a graph is $\leqslant |V(G)|$. Since (provided G is connected), $|V(G)| = \rho(M(G)) + 1$ we might expect a similar sort of result for matroids. However yet again this fails. The only result of note in this direction is the following due to Heron [72a].

(7)   The chromatic number of a simple matroid of rank ρ on n elements is not greater than $n - \rho + 2$.

The coefficient w_k of $\lambda^{\rho(M)-k}$ in $P(M; \lambda)$ is known as the kth *Whitney number of the first kind*. Dowling and Wilson [74] have proved:

THEOREM 5. *Let* $\mathscr{L}$ *be a finite geometric lattice of rank r with n atoms. Then*

$$|w_k| \geqslant \binom{r-1}{k-1}(n-r) + \binom{r}{k}, \qquad (0 \leqslant k \leqslant r),$$

and equality holds for some k, $2 \leqslant k \leqslant r$ *if and only if* $\mathscr{L}$ *is isomorphic to the direct product of a line and a free matroid.*

EXERCISES 3

1. Prove that it is impossible to deduce the number of bases of a matroid from a knowledge of its chromatic polynomial.

2. If M is a matroid on S with chromatic polynomial

$$P(M:\lambda) = \sum_{k=0}^{\rho(M)} (-1)^k a_k \lambda^{\rho(S)-k}$$

prove that if $|S| = n$,

$$a_k \leqslant \binom{n}{k}.$$
(Heron [72])

3. Find a non-graphic matroid which has a chromatic polynomial which can be expressed as a positive integral combination of the Stirling polynomials $\lambda^{(n)}$.

4. An element x of a geometric lattice $\mathscr{L}$ is a *modular element* if it forms a modular pair with every $y \in \mathscr{L}$, that is

$$\rho(x) + \rho(y) = \rho(x \vee y) + \rho(x \wedge y) \qquad \forall y \in \mathscr{L}.$$

Prove that:

(a) every atom of a geometric lattice is a modular element,
(b) if $P_x(\lambda)$ denotes the chromatic polynomial of the interval $[o, x]$,

$$P(M;\lambda) = P_x(\lambda)\left(\sum_{b:\, x \wedge b = o} \mu(o, b)\lambda^{\rho M - \rho x - \rho b} \right)$$

where M is the simple matroid determined by $\mathscr{L}$ and μ, ρ are respectively its Möbius function and rank function,
(c) an element $x \in \mathscr{L}$ is modular if and only if no two complements of x in $\mathscr{L}$ are comparable. (Stanley [71])

$^\circ$5. It has been conjectured by Read [68] that the sequence of coefficients $(a_i : 1 \leqslant i \leqslant |V(G)|)$ of the chromatic polynomial of a graph G is unimodal. We make the stronger conjecture that the sequence of coefficients $(a_i : 0 \leqslant i \leqslant \rho(M))$ of the chromatic polynomial of a matroid M is log-concave, that is

$$a_k^2 \geqslant a_{k-1} a_{k+1} \qquad (1 \leqslant k \leqslant \rho(M) - 1).$$

4. THE TUTTE POLYNOMIAL AND WHITNEY RANK GENERATING FUNCTION

In 1932 and 1947 H. Whitney and W. T. Tutte respectively introduced two polynomials in the independent variables x, y associated with a given graph. Both polynomials were closely related to the chromatic polynomial but contained much more information about the graph in question. It is possible to mimic the approaches of Tutte and Whitney and introduce analogous polynomials for matroids (in fact Crapo [69] does this). However, with the benefit of hindsight it seems less painful to proceed as follows.

The (Whitney) *rank generating function* $R(M; x, y)$ of a matroid M on the set S is defined by

(1) $\quad R(M; x, y = \sum_{A \subseteq S} x^{\rho S - \rho A} y^{\rho^* S - \rho^*(S \setminus A)}$,

where ρ, ρ^* as usual are the rank functions of M, M^* respectively. Using the relation between ρ, ρ^* we can rewrite (1) in the form

(2) $\quad R(M; x, y) = \sum_{A \subseteq S} x^{\rho S - \rho A} y^{|A| - \rho A}$.

Now we define the *Tutte polynomial* of M, $T(M; x, y)$ by

(3) $\quad T(M; x, y) = R(M; x - 1, y - 1)$.

From their definitions we immediately get

(4) $\quad \begin{aligned} R(M; x, y) &= R(M^*; y, x), \\ T(M; x, y) &= T(M^*; y, x). \end{aligned}$

The crucial properties of R and T are contained in the following theorem.

THEOREM 1. *If M is a matroid on S, $e \in S$, and $T = S \setminus e$ then*

$$\begin{aligned} R(M; x, y) &= (1 + y) R(M|T; x, y) & \text{(e a loop of M)} \\ &= (1 + x) R(M . T; x, y) & \text{(e a coloop of M)} \\ &= R(M|T; x, y) + (R(M . T; x, y) & \text{(otherwise).} \\ T(M; x, y) &= y T(M|T; x, y) & \text{(e a loop of M)} \\ &= x T(M . T; x, y) & \text{(e a coloop of M)} \\ &= T(M|T; x, y) + T(M . T; x, y) & \text{(otherwise).} \end{aligned}$$

Proof. Straightforward substitutions. ∎

Now to relate T and R with the chromatic polynomial.

THEOREM 2. *For any M on S the chromatic polynomial is given by*

$$P(M; \lambda) = (-1)^{\rho S} R(M; -\lambda, -1) = (-1)^{\rho S} T(M; 1 - \lambda, 0).$$

Proof. Immediate from the definitions. ∎

Routine verification gives

THEOREM 3. *If a matroid M on S has connected components $(S_i : i \in I)$ then*

$$R(M; x, y) = \prod_{i \in I} R(M|S_i; x, y)$$

$$T(M; x, y) = \prod_{i \in I} T(M|S_i; x, y).$$

The corresponding result (Theorem 3.3) for chromatic polynomials now follows as a special case.

Example. Let M be the uniform matroid of rank 2 on 3 elements. Then

$$R(M; x, y) = x^2 + 3x + y + 3$$
$$P(M; \lambda) \quad = \lambda^2 - 3\lambda + 2.$$

A natural question to ask is—how much information about the matroid is contained in its Tutte polynomial? First we notice:

(5) $T(M; 2, 2) = R(M; 1, 1) = 2^{|S|}$,

(6) $T(M; 1, 1) = R(M; 0, 0) = $ number of bases of M,

(7) $T(M; 2, 1) = R(M; 1, 0) = $ number of independent sets of M,

(8) $T(M; 1, 2) = R(M; 0,1) = $ number of spanning sets of M,

(9) $T(M; 0, 0) = R(M; -1, -1) = 0$,

(10) $T(M; 1, 0) = R(M; 0, -1) = (-1)^{\rho S} \mu(S)$ where μ is the Möbius function of the geometric lattice of M and $\mu(S) = \mu(\varnothing, S)$.

The proofs of (5) to (9) are immediate from inspection of the form of R. To prove (10) note that by Theorem 2

$$T(M; 1, 0) = (-1)^{\rho S} P(M; 0)$$

and then use Theorem 3.1.

However, as one would expect, two non-isomorphic matroids can have the same Tutte polynomial.

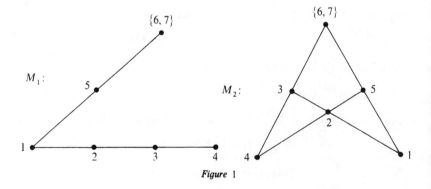

Figure 1

Example (Brylawski [72]). Let $S = \{1, 2, \ldots, 7\}$. Let M_1 be of rank 3 on S and have as its non-trivial hyperplanes

$$\{1, 2, 3, 4\} \qquad \{1, 5, 6, 7\} \qquad \{2, 6, 7\} \qquad \{3, 6, 7\} \quad \text{and} \quad \{4, 6, 7\}.$$

Let M_2 be of rank 3 on S with 6 parallel to 7 and with underlying simple matroid the cycle matroid of K_4.

Geometrically we have the Euclidean representations of Figure 1.

Then both M_1 and M_2 have the same Tutte polynomial

$$T(x, y) = x^2 + 3x^2 + x^2y + 2x + 5xy + 2xy^2 + 2y + 4y^2 + 3y^3 + y^4.$$

Also the following differences exist between M_1, M_2:

M_1 is not binary,	M_2 is a planar matroid;
M_1 is transversal,	M_2 is not even a gammoid;

moreover M_1 and M_2 differ in the number of circuits, flats, hyperplanes.

Closely connected with the Tutte polynomial, though it clearly contains far less information is the invariant $\beta(M)$, defined for any matroid M on S by

$$\beta(M) = (-1)^{\rho S} \sum_{A \subseteq S} (-1)^{|A|} \rho A.$$

This quantity β, studied by Crapo [67b] is alternatively given by

$$(11) \quad \beta(M) = (-1)^{\rho S + 1} \frac{dP(M; \lambda)}{d\lambda}\bigg]_{\lambda = 1}$$

$$= \frac{\partial T(M; x, y)}{\partial x}\bigg]_{x = y = 0}.$$

From (4) we get

$$(12) \quad \beta(M) = \beta(M^*).$$

Another consequence of (11) is that provided e is not a coloop or loop of M, and $T' = S \backslash e$,

$$(13) \quad \beta(M) = \beta(M \,|\, T) + \beta(M \,.\, T).$$

From this an induction argument shows

$$(14) \quad \beta(M) \geqslant 0,$$

and it is not difficult to prove the stronger result (see Crapo [67b]),

THEOREM 4. *A matroid M is connected if and only if $\beta(M) > 0$, (apart from the trivial case when M is a loop).*

Proof. Use the fact that if M is connected and $T = S \backslash e$ for some element e of S then not both $M | T$ and $M . T$ can be disconnected, (see Exercise 5.2.7).

The following characterization theorem is proved by Crapo [67b].

THEOREM 5. *A matroid M is regular if and only if $\beta(M_1) \leqslant 1$ for all four element minors M_1 of M and $\beta(M_2) \leqslant 2$ for all seven element minors M_2 of M.*

Proof. Since β is a non-negative function let M_1 be any four element minor, $M_1 \not\cong U_{2,4}$ since it is easy to check $\beta(U_{2,4}) = 2$. Hence M is binary by Tutte's criterion Theorem 10.2.1.

Similarly if M_2 is any seven element minor, $M_2 \not\cong M(\text{Fano})$ since $\beta(M(\text{Fano})) = 3$. Hence since $\beta(M_2) = \beta(M_2^*)$, M is regular. The converse is obvious. ∎

Brylawski [72] proves the more interesting result:

THEOREM 6. *A connected binary matroid M has no minor isomorphic to $M(K_4)$ if and only if $\beta(M) = 1$.*

In [72] Brylawski defines an *invariant* as a function f, defined on the class of all matroids such that $f(M_1) = f(M_2)$ if M_1 is isomorphic to M_2. A *Tutte–Grothendieck* (T–G) invariant is an invariant f taking values in a commutative ring and such that if M is a matroid on S,

(15) $f(M) = f(M | (S \backslash e)) + f(M . (S \backslash e))$,

for any element e of S which is neither a loop nor a coloop of M, and also if M has a connected component M_1 on $T \subseteq S$,

(16) $f(M) = f(M_1) f(M | S \backslash T)$.

An example of a Tutte–Grothendieck invariant is the Tutte polynomial. In fact it is the unique two-variable polynomial function defined on all matroids which is a (T–G) invariant and has the property that $T(M; x, y) = x$ if M is a coloop, and $T(M; x, y) = y$ if M is a loop.

In other words, he proves the following result.

THEOREM 7. *The only Tutte–Grothendieck invariants are evaluations of the Tutte polynomial.*

Proof. (Sketch) Every evaluation of a Tutte polynomial is a T–G invariant. Conversely, if f is any T–G invariant, and we know the values, say a and b, that f takes for loops and coloops then by applying (15) and (16) with (16) taking precedence we shall obtain $f(M)$ as a polynomial P in the two variables a and b. Since the Tutte polynomial of M also satisfies (15) and (16) P will be just the Tutte polynomial of M, evaluated at a and b. ∎

EXERCISES 4

1. Prove that the rank generating function $R(M; x, y)$ satisfies

$$R(M; t^{-1}, t) = t^{-\rho(M)}(t + 1)^{|S|}$$

2. Let Ω be any total ordering of the elements of S and let M be a matroid on S. An element $x \in B$ is *internally active* with respect to a base B of M if $\Omega(x)$ is greater than $\Omega(y)$ for all y in the fundamental cocircuit contained in $(S \backslash B) \cup x$. Similarly $y \in S \backslash B$ is *externally active relative to B* (and Ω) if $\Omega(y) > \Omega(z)$ for any element z of the fundamental circuit contained in $B \cup y$. The *internal (external) activity* of B is then the number of points internally (externally) active relative to B (and Ω). Prove that t_{ij}, the coefficient of $x^i y^j$ in $T(M; x, y)$ is the number of bases of M with internal activity i and external activity j. (Crapo [69])

   (Note that this implies the remarkable result that the above invariants are independent of the ordering Ω.)

3. Prove that the coefficients of the Tutte polynomial of any matroid are non-negative.

4. Show that in general the sum of two Tutte polynomials is not a Tutte polynomial.

5. Let G be a connected graph and let ω denote any orientation of G, so that ωG is the resulting digraph.
   Let D represent the vertex-edge incidence matrix of ωG in which
   $$d_{ij} = \begin{cases} +1 & \text{if } v_i \text{ is the positive end of } e_j, \\ -1 & \text{if } v_i \text{ is the negative end of } e_j, \\ 0 & \text{otherwise.} \end{cases}$$

   A *flow mod u* on G is a vector x with components in the ring $\mathbb{Z}_u$ of residue classes of integers modulo u in which $Dx' = 0$. A theorem of Tutte (see Biggs [74]) shows that the number of flows mod u which are non-zero on every edge of G is given by

   $$(-1)^{1-|V(G)|+|E(G)|} T(M(G); 0, 1-u).$$

   Show that an analogous result holds for regular matroids. (Crapo [69])

6. Prove that if M_1 is a series parallel extension of M, $\beta(M_1) = \beta(M)$. Deduce that if G_1, G_2 are homeomorphic graphs $\beta(M(G_1)) = \beta(M(G_2))$.

7. Prove that if M is a connected matroid and M_1 is a minor of M and
   $$T(M; x, y) = \sum t_{ij} x^i y^j, \quad T(M_1; x, y) = \sum t'_{ij} x^i y^j$$
   then
   $$t'_{ij} \leq t_{ij}$$
   In particular if M is connected and M_1 is a minor of M then $\beta(M_1) \leq \beta(M)$. (Brylawski [72])

8. Let u be a matroid invariant defined by

$$u(M) = (-1)^{\rho M} \quad \text{if all the circuits of } M \text{ are even,}$$
$$= 0 \quad \text{if } M \text{ has an odd circuit.}$$

Prove that if M is on S and $e \in S$, $T = S \backslash e$, then

$$u(M) = u(M \mid T) + u(M \, . \, T).$$

(Brylawski [72])

9. Prove that for the graph K_n, the wheel $\mathscr{W}_n$ and the whirl $\mathscr{W}^n$

$$\beta(M(K_n)) = (n-2)!, \; \beta(\mathscr{W}_k) = k-1, \; \beta(\mathscr{W}^k) = k.$$

(Crapo [67b])

10. Prove that if M is a matroid on S,

$$\beta(M) \leqslant \binom{|S|-2}{\rho S - 1}$$

with equality if and only if M is uniform. (Brylawski [72])

11. Show that the two graphs G, H of Fig. 2 have the property that

$$T(M(G); x, y) = T(M(H); x, y), \text{ and } M(G) \neq M(H).$$

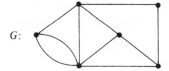
G:

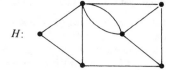
H:

Figure 2

This example was discovered by Dr Marion C. Gray (see Tutte [73]). Other examples, containing no parallel edges and with vertex connectivity as high as 5 have been constructed by Tutte [73].

12. Prove that the Tutte polynomial of a paving matroid can be obtained from a knowledge of the number of blocks of cardinal t, for all t. (Brylawski [72]).

°13. Find an (interesting) property P such that if M_1, M_2 have property P and $T(M_1; x, y) = T(M_2; x, y)$ then $M_1 \simeq M_2$. (Brylawski [72])

14. If H is the region of $\mathbb{R}^2_+$ outside the hyperbola $xy = 1$ prove that for all $(x, y) \in H$ and any matroid M

$$R(M; x, y) \frac{\partial^2 R}{\partial x \, \partial y} \geqslant \cdot \frac{\partial R}{\partial x} \frac{\partial R}{\partial y}. \quad \text{(Seymour and Welsh [75])}$$

5. THE CRITICAL PROBLEM

If $V(r, q)$ is the vector space $GF(q)^r$ and $A \subseteq V(r, q)$, a k-tuple $(f_1, \ldots, f_k)$ of linear functionals on $V(r, q)$ is said to *distinguish A* if for every $e \in A$, $f_i(e) \neq 0$ for some i, $1 \leq i \leq k$.

Now if M is a matroid of rank r on S, which is representable in $V(r, q)$ by the map ϕ let $N(M, q, k, \phi)$ be the number of k-tuples of linear functionals of $V(r, q)$ which distinguish ϕS.

The *critical number* $c(M; q, \phi)$ is the minimum number k such that $N(M, q, k, \phi) > 0$. When no such k exists, for example if $\phi(S) \ni 0$ then we say that $c(M; q, \phi) = \infty$.

Equivalently, since the kernel of any non-trivial linear functional is a hyperplane of the vector space we see that $c(M, q, \phi)$ is the minimum number k such that there exists a k-tuple $(H_1, \ldots, H_k)$ of hyperplanes of $V(r, q)$ with the property that

$$H_1 \cap \ldots \cap H_k \cap \phi(S) = \varnothing.$$

The main result of this section and probably of this whole chapter is the following remarkable theorem of Crapo and Rota [70].

THEOREM 1. *$N(M, q, k, \phi)$ is independent of the map ϕ and is given by $P(M; q^k)$.*

To prove this we need the following lemmas.

LEMMA 1. *If M has a loop $N(M, q, k, \phi) = 0$*

Proof. Obvious since every linear functional maps the zero vector to zero.

LEMMA 2. *If e is not a loop nor a coloop and $T = S \backslash e$ then*

$$N(M, q, k, \phi) = N(M \,|\, T, q, k, \phi \,|\, T) - N(M \,.\, T, q, k, \phi')$$

where ϕ' is the natural map of $M \,.\, T \to V(r - 1, q)$.

Proof. Let E be the subspace of $V = V(r, q)$ generated by $\phi(e)$ and let $g: V \to V/E$ be the canonical map. $M \,.\, T$ is represented in V/E by $g\phi = \phi'$. Now $(f_1, \ldots, f_k)$ distinguishes ϕS if and only if $(f_1, \ldots, f_k)$ distinguishes $\phi(T)$ and $f_i\phi(e) \neq 0$ for some i, $1 \leq i \leq k$. Now if $f_i\phi(e) = 0$ and if f_i' on V/E is defined by $f_i'(y + E) = f_i(y)$ then f_i' is a linear functional on V/E and every such functional can be got in this way.

Hence the sequence $(f_1, \ldots, f_k)$ distinguishes $\phi(T)$ and $f_i\phi(e) = 0$, $1 \leq i \leq k$ if and only if $g\phi T$ is distinguished by $(f_1', \ldots, f_k')$ in V/E. The result follows.

LEMMA 3. *If e is a coloop of M and T = S\e then*

$$N(M, q, k, \phi) = (q^k - 1) N(M \,|\, T, q, k, \phi).$$

Proof. Since e is a coloop, ϕT is contained in a hyperplane $H \subseteq V = V(r, q)$. Let E be the subspace generated by $\phi(e)$. Then $V = H \oplus E$. But f is a linear functional of V if and only if $f(x, y) = g(x) + h(y)$ where g, h are linear functionals on H, E respectively. Hence the number of k-tuples of linear functions which distinguish ϕS is the product of $N(M \,|\, T; q, k, \phi \,|\, T)$ and the number of k-tuples of linear functions of E which distinguish $\phi(e)$. But there are q different linear functionals on E only one of which maps e to zero. Thus there are $q^k - 1$ k-tuples of linear functions on E which distinguish $\phi(e)$, and this completes the proof.

Proof of Theorem 2. We use induction on $|S|$. Let the result be true for $|S| < n$, it is trivially true for $|S| = 1$. Let $N = N(M, q, k, \phi)$. If e is a loop $N = 0$ and $P(M; q^k) = 0$. If e is a coloop and $T = S\backslash e$, by induction

$$N = (q^k - 1) N(M \,|\, T; q, k, \phi) = P(M \,|\, T; q^k)(q^k - 1)$$

and the result holds by Lemma 3.

If e is neither a loop nor a coloop using Lemma 2 and induction we get with $T = S\backslash e$

$$N = P(M \,|\, T; q^k) - P(M . T; q^k)$$

and the result follows by (3.4). ∎

Henceforth, therefore, we can write $N(M; q, k)$ for $N(M; q, k, \phi)$. A consequence of Theorem 2 is:

THEOREM 3. *The critical number $c(M; q, \phi)$ is independent of the map ϕ.*

We henceforth denote $c(M; q, \phi)$ by $c(M; q)$.

Thus an alternative characterization of the critical number is given by

(4) $c(M; q) > k \Leftrightarrow P(M; q^k) = 0,$

   $c(M; q) \leqslant k \Leftrightarrow P(M; q^k) > 0.$

Thus for example if M is the cycle matroid of a simple graph G since G is 4-colourable if and only if $P(M(G); 4) > 0$, we may restate the 4-colour conjecture in the following form:

4-*Colour Conjecture:* If M is the cycle matroid of a simple planar graph $c(M; 2) \leqslant 2$ and $c(M; 4) \leqslant 1$.

Crapo and Rota [70] describe the critical problem as "the central problem of 'extremal' combinatorial theory". However very little in the way of general results seems to be known. We list here the properties of c known to us:

(5) Provided M on S is representable over $GF(q)$, for any $T \subseteq S$,

$$c(M \,|\, T \,; q) \leqslant c(M \,; q).$$

(6) If M can be embedded in $PG(n, q)$ then $c(M, q) = 1$ if and only if M can be embedded in $AG(n - 1, q)$.

Proofs of these results are straightforward, see Crapo and Rota [70].

We illustrate the above ideas in the following example.

Example. Let M on S be the Fano matroid $PG(2, 4)$. Then easy calculation gives

$$P(M \,; \lambda) = (\lambda - 1)(\lambda - 2)(\lambda - 4).$$

Thus

$$c(M \,; 2) = 3$$
$$N(M \,; 2, 1) = N(M \,; 2, 2) = 0,$$
$$N(M \,; 2, 3) = P(M \,; 2^3) = 168.$$

In other words no single linear functional L_i, nor pair of functionals L_i, L_j can distinguish S but there are 168 ordered triples (L_i, L_j, L_k) of linear functionals which distinguish S. This can be seen more directly since there are 7 ways of selecting L_i, by symmetry 6 ways of selecting $L_j \neq L_i$ and now only 4 ways of selecting L_k so as to avoid the point of intersection of L_i and L_j. Thus there are $7 . 6 . 4 = 168$ ways of selecting our triple and this agrees with $P(M \,; 2^3)$.

Now over $GF(2)$ there is an obvious $1 - 1$ correspondence between linear functionals and hyperplanes and it is instructive to see how the "distinguishing hyperplane" interpretation of the chromatic polynomial ties up with the colouring interpretation when M is the cycle matroid of a simple graph G.

Recall from Theorem 9.5.2 that if G has vertex set $\{v_1, \ldots, v_m\}$ and edge set $\{e_1, \ldots, e_n\}$ then it has a representation over $V(m, GF(2))$ in which each e_i is represented by the ith column of the $m \times n$ incidence matrix A where

$$A_{ij} = \begin{cases} 1 & v_i \text{ incident with } e_j \\ 0 & \text{otherwise.} \end{cases}$$

Suppose there are k hyperplanes $H_1, \ldots, H_k$ of V which distinguish $E(G)$, or more precisely the set of vectors in V corresponding to $E(G)$. Let H_j have equation

$$h_{j1}x_1 + \ldots + h_{jm}x_m = 0 \qquad (1 \leqslant j \leqslant k).$$

Then assign to the vertex v_i of G the k-tuple $(h_{1i}, \ldots, h_{ki})$. Since the k-tuple

$(H_1, \ldots, H_k)$ distinguishes $E(G)$ it is clear that if v_i, v_j are adjacent vertices of G then the corresponding k-tuples $(h_{1i}, \ldots, h_{ki})$ and $(h_{1j}, \ldots, h_{kj})$ are different. In other words corresponding to any k-tuple of distinguishing hyperplanes we have a proper 2^k-colouring of the vertices of G. A similar argument shows that the converse holds. This algebraic formulation of the colouring problem for graphs can be traced back at least to Veblen's paper in 1912. Tutte [66a] has shown that other well known graph-colouring problems can be given similar algebraic formulations but the geometrical problems involved seem no easier in general.

The reader will notice that so far $c(M; q)$ has been defined only for matroids M which are representable over the field $GF(q)$. We can widen the definition to matroids which are not representable over $GF(q)$ by defining

$$c(M; q) = \inf k : P(M; q^k) > 0$$

for any prime power q. By (4) this definition coincides with our previous definition whenever M is representable over $GF(q)$. We know of no combinatorial interpretation of $c(M; q)$ in this more general case, however it does supply us with the following interesting representability criterion.

THEOREM 4. *A necessary condition that a simple matroid M of rank r be representable over $GF(q)$ is that*

$$P(M; q^r) \geqslant \prod_{i=0}^{r-1} (q^r - q^i)$$

Proof. (Sketch). The left-hand side is $N(M; q, r)$ which is clearly not less than the number of r-tuples of linear functions which distinguish $V(r, q)\backslash 0$. But the latter is $N(M^1; q, r)$ where M^1 is the matroid $PG(r - 1, q)$. Evaluating the chromatic polynomial of M_1 (a straightforward exercise, see below) gives us the required result. ∎

Example. Suppose M is the uniform matroid of rank 3 on a 5-set. Then we can evaluate

$$P(M; 8) = 44 < (8 - 1)(8 - 2)(8 - 4) = 168$$

and hence $U_{3,5}$ is not representable over the field $GF(2)$.

EXERCISES 5

1. Prove that if M has a loop $c(M; q) = \infty$ for any prime power q.

2. Prove that if M is the projective geometry $PG(n, q)$ then
$$P(M; \lambda) = \prod_{i=0}^{n} (\lambda - q^i)$$

and hence find $c(M, p')$ for any prime p' coprime with q.

3. Let M be a matroid on S and let $T \subseteq S$, is $c(M; q) \geqslant c(M \mid T; q)$ when M is not representable over $GF(q)$?

6. CODES, PACKINGS, AND THE CRITICAL PROBLEM

This section is based on a very interesting paper of the same title by Dowling [71]. It shows how a fundamental problem of coding theory can be considered as a particular case of the critical problem for matroids.

For t a positive integer $\geqslant 2$, the t-packing problem for a matroid M on S can be defined as follows:

Call $X \subseteq S$, *t-independent* if:

(i) $|X| \geqslant t$,
(ii) each t-element subset of X is independent in M.

The *t-packing problem* for M is to find the maximum cardinal of a t-independent set in M. We denote this number by $I(M; t)$ and call a t-independent set of cardinal $I(M; t)$ a *t-packing* of M.

A t-independent set is clearly $(t - 1)$-independent so

(1) $I(M; t) \leqslant I(M; t - 1)$.

Example 1. When M is the cycle matroid of a graph G, $I(M; t)$ is the maximum number of edges of G which contain no cycle of cardinal $\leqslant t$. In a classic paper of extremal graph theory Turan [41] proved that for the complete graph K_n the maximum number of edges containing no triangle is $[n^2/4]$. Thus

$$I(M(K_n); 3) = [n^2/4].$$

For $t \geqslant 4$ this problem even is, as far as we know, still unsolved.

Example 2. Let M be the projective geometry $PG(n - 1, q)$ of rank n over the field $GF(q)$.

Since every two element subset of M is independent

$$I(PG(n - 1, q); 2) = I(n; q; 2) = (q^n - 1)/(q - 1).$$

Also given a t-packing of any hyperplane H of M, any singleton not in H can be added to yield a t-independent set of M. Since a hyperplane H is isomorphic to $PG(n - 2, q)$,

$$I(PG(n-1; q); t) = I(n, q; t) \geqslant I(n-1, q, t) + 1.$$

If T is a t-packing $(t \geqslant 3)$ in $PG(n-1, q)$ and $a \in T$ then the matroid obtained by contracting a out of M is $PG(n - 2, q)$ and has $T \backslash a$ as a $(t - 1)$-independent subset. Hence

$$I(n, q; t) \leqslant I(n - 1, q; t - 1) + 1.$$

When $q = 2$ and t is odd it is easy to see that this becomes the equality

$$I(n, 2; 2s + 1) = I(n - 1, 2; 2s) + 1.$$

Further known values of $I(r, q; t)$ are given by

(2)
$$
\begin{aligned}
I(3, q; 3) &= \begin{cases} q + 1 & q \text{ odd} \\ q + 2 & q \text{ even} \end{cases} \\
I(4, q; 3) &= q^2 + 1 & q > 2, \\
I(n, q; 3) &= 2^{n-1} & q = 2, \\
I(r, q; r) &= r + 1 & q = 2.
\end{aligned}
$$

The coding problem

Let V_n denote the vector space $V(n, q)$ and let $w(a)$, $a = (a_1, \ldots, a_n)$ be the number of non-zero coordinates a_i of a. Then w is a norm on V_n, called *Hamming weight* inducing the metric d (*Hamming distance*) on V_n defined by $d(a, b) = w(a - b)$, that is the number of indices i such that $a_i \neq b_i$.

An (n, k)-(*linear*) code of distance $t + 1$ is a k-dimensional subspace C of V_n that $d(a, b) \geqslant t + 1 \ \forall \ a, b \in C$, $a \neq b$.

The *redundancy* of an (n, k)-code is $n - k$.

An (n, k)-code of distance $t + 1$ is *optimal* if for given n, t, k is a maximum. Let $R(n, t)$ denote the redundancy of an optimal code. One of the major problems of algebraic coding theory is to determine $R(n, t)$.

THEOREM 2. $R(n, t) \leqslant r \Leftrightarrow I(r, q; t) \geqslant n$.

Proof. Suppose C is a subspace of V_n such that $d(a, b) \geqslant t + 1$ for all $a, b \in C$, $a \neq b$. Then for $a \neq 0$,

$$w(a) = d(a, 0) \geqslant t + 1.$$

Thus C contains no elements of the *t-ball*

$$S_{n, t} = \{e : 1 \leqslant w(e) \leqslant t\}.$$

Conversely, if C is a subspace such that $C \cap S_{n, t} = \varnothing$ then $d(a, b) \geqslant t + 1$ for $a \neq b$ in C, for $d(a, b) = w(a - b)$. Thus an (n, k)-linear code of distance $t + 1$ is equivalently a k-dimensional subspace of V_n containing no elements of the *t*-ball $S_{n, t}$.

Let C be an $(n, n - r)$ code of distance $t + 1$. Let A be an $r \times n$ matrix with ker $A = C$, then $C \cap S_{n, t} = \varnothing$ implies $Ax' \neq 0$ for all $x \in S_{n, t}$. By the definition of $S_{n,t}$ this implies that any subset of t or fewer columns of A is linearly independent over $GF(q)$, (call such a matrix *t-independent*). Conversely, given any $r \times n$ t-independent matrix A, its kernel, of dimension at least $n - r$, can contain no element of $S_{n,t}$. Thus there exists an $(n, n - r)$ code of distance $t + 1$ if and only if there exists in V_r an n-point t-independent set. ∎

Now to relate this coding problem with the critical problem.

The coding problem is to find the maximum rank of a subspace of V_n which has null intersection with the t-ball $S_{n,t}$.

THEOREM 3. *The critical number $c(S_{n,t}; q)$ equals the redundancy $R(n, t; q)$.*

Proof. Let $H_1, \ldots, H_c$ be a minimum set of hyperplanes of V_n such that
$$H_1 \cap \ldots \cap H_c \cap S_{n,t} = \varnothing.$$
Then $X = H_1 \cap \ldots \cap H_c$ is a subspace of V_n of rank $n - c$ and hence
$$c(S_{n,t}; q) \geqslant R(n, t; q).$$

Conversely, let Z be a subspace of V_n with rank equal to k and such that $Z \cap S_{n,t} = \varnothing$.

Since Z is a subspace of rank k we can find hyperplanes $H_1, H_2, \ldots, H_u$ of $V_u (u = n - k)$, such that
$$Z = H_1 \cap \ldots \cap H_u.$$
Thus $c(S_{n,t}; q) \leqslant u$ and the theorem follows. ∎

We illustrate the ideas of this section in the following example

Example. Consider the vector space $V = V(5, 2)$ of dimension 5 over $GF(2)$. From (2) we know that $I(5, 2; 3) = 2^{5-1} = 16$ and thus there exists a set of 16 vectors no three of which are dependent. Hence $R(16, 3) \leqslant 5$ or in other words, in the vector space $V(16, 2)$ there exists a subspace of dimension at least 11 in which the distance between any pair of words (vectors) is at least 4.

Although interesting, we cannot pretend that Theorem 3 simplifies the determination of $R(n, t; q)$ for as we observed earlier, obtaining $c(S_{n,t}; q)$ entails obtaining the chromatic polynomial of the matroid $S_{n,t}$, or the characteristic polynomial of its geometric lattice.

EXERCISES 6

1. Prove that
$$I(n, q; t) \leqslant \frac{q^{n-t+2} - 1}{q - 1} + t - 2.$$

2. Prove that $R(n, t) \geqslant R(n - 1, t - 1) + 1$ and that if t is odd and $q = 2$,
$$R(n, 2s + 1) = R(n, 2s) + 1.$$

3. Prove that
$$I(r, q; t) = \max n : c(S_{n, t}, q) = r.$$

4. Prove that $I(3, 2, 2) = 7$. Deduce that there exists a $(7, 4)$ code over $GF(2)$ of distance at least 3.

7. WEIGHT ENUMERATION OF CODES: THE MACWILLIAMS' IDENTITY

Suppose U is an (n, k) linear code over $GF(q)$. The *weight enumerator* $A_U(z) = \sum_{u \in U} z^{w(u)} = \sum A_i z^i$ where w is the (Hamming) weight and A_i is the number of vectors in U with weight exactly i.

The weight enumerator can be used to find the probabilities of decoding error and decoding failure for decoding algorithms which correct all patterns of up to a specified number of errors. For a discussion of this and other applications of the weight enumerator see Berlekamp [68] or Van Lint [71].

Because of their importance a lot of research has gone into the study of weight enumerators; in particular to methods of finding weight enumerators. One particularly useful and elegant method is to determine the weight enumerator of U from the weight enumerator of its dual code $U^\perp$ when the dual code is smaller.

The *dual code* $U^\perp$ is the $(n, n - k)$ linear code consisting of all vectors $v \in V(n, q)$ such that $vu' = 0$ for all $u \in U$.

The MacWilliams [63] identity relating the weight enumerators of U and $U^\perp$ is given in the next theorem.

THEOREM 1. *If U is an (n, k) code over $GF(q)$ and $U^\perp$ is its dual code then*

$$A_{U^\perp}(z) = q^{-k}(1 + (q - 1)z)^n A_U\left(\frac{1 - z}{1 + (q - 1)z}\right).$$

We shall prove this theorem as a corollary of our results relating the weight enumerator of a code with the Tutte polynomial of a related matroid on the columns of a code. Although this is not the most direct proof (see for example Van Lint [71]) it is useful in that it gives insight into the relationship between codes and matroids. It will also show that the problem of determining the weight polynomial of a code is a special case of the problem of finding the Tutte polynomial of a matroid.

Our treatment is that of Greene [74a].

With each $k \times n$ matrix N of rank k over $GF(q)$ associate the following two objects.

(1)   The row space of N, denoted by $U(N)$.
(2)   The column matroid of N denoted by $G(N)$.

The subspace $U(N)$ is an (n, k) linear code over $GF(q)$ while $G(N)$ is a matroid of rank k on a set of n elements.

Now if U is any (n, k) linear code over $GF(q)$, $U = U(N)$ for any $k \times n$ matrix N whose row vectors form a basis of U. Such a matrix is a *generating matrix* for U.

By straightforward linear algebra we have:

(3)   If N_1, N_2 are two generating matrices for U, then $G(N_1)$ and $G(N_2)$ are isomorphic matroids.

If we now define two linear representations of a matroid M on S over $GF(q)$ to be *equivalent* if they are related by a non-singular linear transformation of $(GF(q))^k$, we see from (3) that there is a 1–1 correspondence between (n, k) linear codes over $GF(q)$ and equivalence classes of representations of simple matroids of rank k with n points. It is routine to check, and is implicit in our representation theory of Chapter 9, that:

(4)   If an (n, k) code is generated by a matrix N and its dual code is generated by a matrix $N^\perp$ then the matroids $G(N)$, $G(N^\perp)$ are dual matroids.

Now given a $k \times n$ matrix N whose columns are indexed by S, we call a *reduction* of N with respect to $e \in S$ any matrix $\bar{N}$ which is row equivalent to N and has its eth column $(1, 0, \ldots, 0)'$, and is such that the removal of this column from $\bar{N}$ does not reduce its rank. (Such a reduction may not exist; give examples!).

For any such reduction $\bar{N}$ we write $\bar{N}\backslash e$ to denote the matric got by removing the eth column and write $\bar{N}.e$ for the matrix got by removing the eth column and also the first row from N.

If $M(N)$ is the matroid on S induced by the columns of N, from the representation theory of Chapter 9 we have

(5) $M(N)|S\backslash e \simeq M(\bar{N}\backslash e)$
(6) $M(N).S\backslash e \simeq M(\bar{N}.e)$

LEMMA 1. *Let N be a matrix over a field F with columns indexed by S and for some $e \in S$ let the corresponding column be non-zero. If $\bar{N}$ is a reduction of N with respect to e then*

$$A_{U(N)} = z A_{U(\bar{N}\backslash e)} + (1 - z) A_{U(\bar{N}.e)}$$

Proof. The code words in $U(N)$ fall into two classes: those which have a zero in position e and those which do not. The former have a weight distribution identical with $U(\bar{N} . e)$, and the latter correspond to vectors in $U(\bar{N}\backslash e)$ but not in $U(\bar{N} . e)$ with a single non-zero coordinate added.

Hence

$$A_{U(N)} = A_{U(\bar{N} . e)} + z[A_{U(\bar{N}\backslash e)} - A_{U(\bar{N} . e)}]$$

and the lemma is proved.

Lemma 1 suggests strongly that some function of A is a Tutte–Grothendieck invariant and this we see in the following main theorem.

THEOREM 2. *If U is an (n, k) code over $GF(q)$ and N is any generating matrix for U and M is the matroid induced on the columns of N by linear independence,*

$$A_U(z) = (1 - z)^k z^{n-k} T\left(M; \frac{1 + (q - 1)z}{1 - z}, \frac{1}{z}\right).$$

Proof. Write

(7) $$g(M) = (1 - z)^{-k} z^{-n+k} A_U(z).$$

Then by Lemma 1 and (5) and (6) we see that $g(M)$ satisfies

$$g(M) = g(M|(S\backslash e)) + g(M . (S\backslash e))$$

where $e \in S$, the set of columns of N, and e is not a loop or coloop of $M(N)$.

It is easy to check that for matroids M_1, M_2 which are representable over $GF(q)$ $g(M_1 + M_2) = g(M_1) g(M_2)$ and hence g is a T–G invariant over this class. Since such an invariant is uniquely determined by its values on a loop and coloop we will have proved the theorem if we show that:

$$g(\text{loop}) = T\left(\text{loop}; \frac{1 + (q - 1)z}{1 - z}, \frac{1}{z}\right)$$

$$g(\text{coloop}) = T\left(\text{coloop}; \frac{1 + (q - 1)z}{1 - z}, \frac{1}{z}\right)$$

and both of these are trivial. ∎

As an application we prove the MacWilliams' identity.

Proof of Theorem 1

By Theorem 2

$$A_U\left(\frac{1-z}{1+(q-1)z}\right) = \left(\frac{qz}{1+(q-1)z}\right)^k \left(\frac{1-z}{1+(q-1)z}\right)^{n-k}$$

$$T\left(M;\frac{1}{z},\frac{1+(q-1)z}{1-z}\right)$$

$$= \frac{q^k}{(1+(q-1)z)^n}(1-z)^{n-k}z^k\,T\left(M^*;\frac{1+(q-1)z}{1-z},\frac{1}{z.}\right)$$

$$= \frac{q^k}{(1+(q-1)z)^n}\,A_{U^{\perp}}(z)$$

which completes the proof. ∎

We illustrate these ideas in the following simple example.

Example. Let U be the code with generating matrix over $GF(2)$ given by

$$N = \begin{pmatrix} 1 & 0 & 1 \\ 0 & 1 & 1 \end{pmatrix}.$$

Then the column matroid is the uniform matroid of rank 2 on 3 elements. Hence its Tutte polynomial is

$$T(x, y) = x^2 + x + y.$$

Thus by Theorem 1 the code has weight polynomial

$$A(z) = (1-z)^2\,zT\left(\frac{1+z}{1-z},\frac{1}{z}\right)$$

$$= (1-z)^2\,z\left[\frac{(1+z)^2}{(1-z)^2} + \frac{(1+z)}{(1-z)} + \frac{1}{z}\right]$$

$$= z(1+z)^2 + z(1-z^2) + (1-z)^2$$

$$= 1 + 3z^2$$

which agrees with the fact that the code generated by N has 3 codewords of weight 2 and 1 of weight zero.

EXERCISES 7

1. Let U be the $(7, 3)$ code over $GF(2)$ with generating matrix given by

$$\begin{vmatrix} 1 & 0 & 0 & 1 & 1 & 0 & 1 \\ 0 & 1 & 0 & 1 & 0 & 1 & 1 \\ 0 & 0 & 1 & 0 & 1 & 1 & 1 \end{vmatrix}$$

prove that

$$A_U(z) = 1 + 7z^4.$$

2. Let U be an $(n. k)$ code over $GF(2)$. Prove that

   (i) U has 2^k code words,
   (ii) there exist $(2^k - 1)(2^k - 2)(2^k - 2^2)\ldots(2^k - 2^{k-1})$ different generator matrices for U,
   (iii) if N_1, N_2 are generator matrices for U then $N_1 = DN_2$ where D is a non-singular $k \times k$ matrix over $GF(2)$,
   (iv) if $GL(k, 2)$ is the general linear group of dimension k over $GF(2)$ then for any generator N of U, as D runs through $GL(k, 2)$ the product DN gives all the distinct generator matrices of U.

   (Slepian [60])

3. Prove that a necessary condition for the existence of an (n, k) code containing words of weight $w_i, i = 1, 2, \ldots, s$ only is that there exists a set of integers $\alpha_i, i = 1, 2, \ldots, s$, such that the expression

$$(1 + \gamma z)^n + \gamma \sum_{j=1} \alpha_i (1 + \gamma z)^{n-w_i}(1 - z)^{w_i}$$

   takes the form

$$q^k + \gamma q^k \sum_{i=1}^{n} \beta_i z^i$$

   when expanded in powers of z, where $\gamma = q - 1$. Show that this condition is not sufficient. (MacWilliams [63])

4. Prove that any linear (n, k) code has a generator matrix of the form (I_k, P) where P is a k by $n - k$ matrix and I_k is the $k \times k$ identity matrix.

5. A generator matrix of the dual code $U^\perp$ is called a *parity check matrix* of the code U. Prove that if U has generator (I_k, P) then a parity check matrix of U is given by $(-P', I_{n-k})$.

NOTES ON CHAPTER 15

The material of Section 1 is well-known graph theory, see for example the review paper of Read [68] or the book of Biggs [74]. Section 2 is based on the paper of Rota [64]. A nice application of this theory is found in the proof by Dilworth [54] that the number of elements of a modular lattice covering k elements is the same as the number covered by k elements. Rota [71] gives an interesting account of the relationship between Möbius functions and Euler characteristics; see also Crapo [66] and [68].

The characteristic polynomial has sometimes been called the *Poincaré polynomial*. For properties of this polynomial see Brylawski [72], or Heron [72], [72a]. These papers also contain much of Section 4 on the Tutte and Whitney polynomials of a matroid. Crapo [69] was the first to study the Tutte polynomial of a matroid. Crapo's invariant β reduces for graphic matroids to the invariant θ which Biggs [74] calls the *chromatic invariant* of a graph. An interesting application of the β-function to enumerating the partitions of Euclidean space by arrangements of hyperplanes is given by Zaslavsky [75]. See also Cardy [73] for an alternative interpretation of β.

Smith [72] gives an interpretation of the Tutte polynomial of matroids in terms of resistance and conductance polynomials of electrical networks.

The interpretation of colouring problems as a particular case of the critical problem is due to Crapo and Rota [70] though the (algebraic) geometrical approach to colouring had been pointed out by Veblen [12] and Tutte [66a]. The treatment given here is more longwinded than that of Crapo and Rota. We have gone to some, perhaps pedantic lengths, to emphasise that until proved otherwise c is a function of the embedding ϕ.

Sections 6 and 7 are based respectively on the papers of Dowling [71] and Greene [74]. Greene's result linking the Tutte polynomial with the Mac-Williams' identity for codes is particularly appealing.

Crapo and Rota's interpretation of the chromatic polynomial is a remarkable result. It would be extremely interesting to know whether $P(M, \lambda)$ counts anything (for suitable λ) when M is not representable over a field.

The Whitney rank generating function is essentially the *Whitney polynomial* in statistical mechanics or percolation theory. Many classic unsolved problems in these fields would be solved if one could evaluate $R(M(G_n); x, y)$ where G_n denotes the $n \times n$ portion of the plane square lattice, or more especially evaluate

$$\lim_{n \to \infty} \left(R(M(G_n); x, y)\right)^{1/n}.$$

CHAPTER 16

Extremal Problems

1. INTRODUCTION

In this chapter we shall study various quantitative problems about matroids and geometric lattices. The early sections will be concerned mainly with such problems as determining the maximum and minimum number of hyperplanes, circuits and bases in a matroid of given rank on a given set of elements. The last section deals with enumerative problems about matroids.

If M is a matroid of rank r on the finite set S of cardinality n and $0 \leqslant k \leqslant r$ we let

(1) $b(M)$ denote the number of bases of M,
(2) $i_k(M)$ denote the number of independent sets of cardinal k,
(3) $i(M)$ denote the number of independent sets of M,
(4) $c(M)$ denote the number of circuits of M,
(5) $h(M)$ denote the number of hyperplanes of M,
(6) $sp(M)$ denote the number of spanning sets of M.

There are obvious relations among some of these quantities. For example

(7) $i(M) = \sum_{k=0}^{r} i_k(M),$

(8) $c(M) = h(M^*),$
(9) $i(M) = sp(M^*).$

Trivial, but in one sense best possible bounds for the various quantities are given by

(10) $1 \leqslant b(M) \leqslant \binom{n}{r},$

(11) $1 \leqslant i_k(M) \leqslant \binom{n}{k}.$

Slightly less obvious is the following result.

$$(12) \quad c(M) \leqslant \binom{n}{r + 1}.$$

Proof. Let M be a matroid on S, $|S| = n$ and let $e \in S$ not be a loop of M. If no such e exists (12) holds. Let $T = S \backslash e$. Then consider any circuit C of M.

If C does not contain e then C is a circuit of $M | T$, if C contains e then $C \cap T$ contains a circuit of $M . T$. Hence

$$(13) \quad c(M) \leqslant c(M | T) + c(M . T).$$

But $M | T$ has rank r, $M . T$ has rank $r - 1$. Now using induction on $|S|$, we know

$$c(M | T) \leqslant \binom{n - 1}{r}, \qquad c(M . T) \leqslant \binom{n - 1}{r - 1}$$

and (12) follows. It is clearly best possible, consider the uniform matroid $U_{r,n}$.

Comparison of (13) with the basic equality

$$(14) \quad b(M) = b(M | T) + b(M . T)$$

proved in Chapter 15 suggests that on average a matroid has more bases than circuits. Indeed W. Quirk and P. D. Seymour (private communication) have shown that if M is the cycle matroid of a simple graph

$$(15) \quad b(M) \geqslant c(M)$$

However (15) clearly does not hold for arbitrary matroids nor even for non-simple graphs.

EXERCISES 1

1. A matroid M is *identically self dual* (ISD) if $M = M^*$.
   Prove that:
   (a) The direct sum of ISD matroids is also ISD.
   (b) The only connected graphic ISD matroid is $U_{1,2}$.
   (c) If M is ISD on $2n$ elements it has at least 2^n bases.
   (d) There exists a connected binary ISD matroid on a set of $2n$ elements provided there exists a symmetric (n, k, λ) design with k odd and λ even. (Bondy and Welsh [71]).

2. Prove that there exist connected binary ISD matroids of rank d for all d except 2, 3, and 5. (Graver [73]).

3. If M is a matroid of rank $r \geqslant 1$ on a set of n elements prove that

$$h(M) \leqslant (_{r-1}^{n})$$

and characterize those matroids for which equality holds.

4. Prove that $PG(n, q)$ has $b(n, q)$ bases where

$$b(n, q) = \frac{1}{(n+1)!} \frac{(q^{n+1} - 1)(q^{n+1} - q) \ldots (q^{n+1} - q^{n})}{(q - 1)^{n+1}}.$$

○5. Let $b(r, n)$ $c(r, n)$ be respectively the average number of bases and circuits of a rank r matroid on n elements. For what values of r and n is it true that

$$b(r, n) \geqslant c(r, n)?$$

6. For integers $n, k, 1 \leqslant k \leqslant n/2$ let

$$p_n(k) = 1 + \binom{n}{1} + \ldots + \binom{n}{k}.$$

Prove that if M is such that it has rank r on n elements and t is such that

$$\binom{n}{t} \leqslant b(M) \leqslant \binom{n}{t+1}$$

then

$$p_n(t) \leqslant i(M) \leqslant p_n(t + 1).$$

7. Let M be a matroid on S, prove that

$$2^{|S|} b(M) \leqslant i(M) sp(M).$$

(Seymour and Welsh [75]).

2. THE WHITNEY NUMBERS AND THE UNIMODAL CONJECTURE

If $\mathscr{L}$ is a finite geometric lattice of rank r we denote the number of lattice elements of rank i by $W_i = W_i(\mathscr{L})$. $W_0, W_1, \ldots, W_r$ are called the *Whitney numbers* of the lattice.

There has been a great deal of interest in the Whitney numbers in recent years, and there are several interesting conjectures about them. Foremost among these is the unimodality conjecture of G. C. Rota:

Conjecture 1. For $i \leqslant j \leqslant k$, $W_j \geqslant \min\{W_i, W_k\}$.

Many of the known unimodal sequences possess the stronger property of being log convex/concave, and examination of special well known examples (such as projective or affine geometries, perfect matroid designs, paving matroids) suggests

Conjecture 2. The Whitney numbers are log concave, that is

$$W_k^2 \geqslant W_{k-1}W_{k+1} \qquad (2 \leqslant k \leqslant r - 1).$$

Mason [72a] makes the plausible conjecture that the ratio $W_k^2/W_{k+1}W_{k-1}$ is minimized over all matroids on n points by the free matroid, and makes the stronger conjectures:

Conjecture 3. $W_k^2 \geqslant \dfrac{k+1}{k} W_{k-1}W_{k+1}$;

Conjecture 4. $W_k^2 \geqslant \dfrac{(k+1)}{k} \dfrac{(n-k+1)}{n-k} W_{k-1}W_{k+1}$ where $n = W_1$.

A special case of the unimodal conjecture was formulated recently, but even this seems very difficult to prove.

The Points-Lines-Planes Conjecture. Given p points in Euclidean 3-space let l lines and π planes be determined by these points (2 distinct points determine a line, 3 non-collinear points determine a plane). Then

$$(1) \qquad\qquad l^2 \geqslant 3p\pi/2.$$

(The role of the constant 3/2 is that we suspect that the ratio $l^2/p\pi$ is minimized by p points in general position).

From the point of view of conjecture 1 it would be very interesting to have a proof of the weaker result

$$(2) \qquad\qquad l^2 \geqslant cp\pi.$$

for some constant $c > 0$.

Turning now from conjectures to known results, the first theorem of interest in this area was proved by de Bruijn and Erdös [48].

This was the following result about configurations of points and lines.

THEOREM 1. *Suppose* $\{F_1, \ldots, F_d\}$ *is a collection of subsets of a set* S *satisfying the conditions*

(i) *each pair of points of* S *is contained in exactly one* F_i

(ii) *no* F_i *contains all the points of* S

(iii) *each* F_i *contains at least two points of* S.

Then $d \geqslant |S|$ *and if* $d = |S|$ *the configuration is either a projective plane or one of the degenerate structures represented by*

$$F_1 = \{2, 3, \ldots, m\},\, F_2 = \{1, 2\},\, F_3 = \{1, 3\}, \ldots, F_m = \{1, m\}.$$

Translated into a theorem about matroids, the de Bruijn–Erdös theorem can be restated as follows.

THEOREM 1. *Let M be a simple matroid of rank 3 on $S = \{1, 2, \ldots, m\}$. Then M has at least as many hyperplanes as points and if $h(M) = |S|$ then either M is a projective plane or is isomorphic to a matroid whose hyperplanes are of the form $\{2, 3, \ldots, m\},\, \{1, 2\},\, \{1, 3\}, \ldots, \{1, m\}$.*

Thus for $m = 6$, for example, since we know there is no projective plane with 6 points the only simple rank 3 matroid with 6 points and 6 hyperplanes has Euclidean representation of a form shown in Fig. 1.

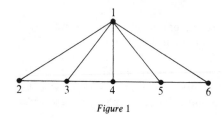

Figure 1

An extension of the theorem of de Bruijn and Erdös was the following result proved by Motzkin [51].

THEOREM 2. *Any n distinct points in Euclidean d-space, not all on one hyperplane determine at least n hyperplanes.*

Motzkin's theorem is a special case of the more general result:

THEOREM 3. *A geometric lattice has at least as many hyperplanes (copoints) as atoms.*

This result has been proved independently by several authors Basterfield and Kelly [68], Greene [70], Heron [73]. It should also be mentioned that Mason [72a] has pointed out that the combinatorial lemma used by Motzkin to prove his theorem needs only slight modification to give a proof of the more general Theorem 3.

Now if we consider matroids of rank >3 with the same number of hyperplanes as points (that is equality holds in Theorem 3), we notice that trivial examples of such structures are (a) Boolean algebras and (b) projective geometries.

Moreover if M_1, M_2 are simple matroids on the disjoint sets, S_1, S_2 and

$$h(M_1) = |S_1|, \qquad h(M_2) = |S_2|$$

then it is clear that the direct sum $M_1 + M_2$ on $S_1 \cup S_2$ satisfies

$$h(M_1 + M_2) = |S_1| + |S_2|.$$

Since Boolean algebras, projective spaces and "de Bruijn–Erdös configurations" are matroids whose associated geometric lattice is modular, and the direct product of modular lattices is modular this suggests the following result first proved by Greene [70].

THEOREM 4. *A simple matroid has the same number of hyperplanes as points if and only if its associated geometric lattice is modular*

It turns out that Theorems 1–4 are special cases of more general results about geometric lattices proved by Dowling and Wilson [73]. Their methods are extremely elegant and interesting and we devote the next section to proving their main theorems.

EXERCISES 2

1. Prove that for each of the following classes of matroid (a) projective geometries, (b) affine geometries (c) paving matroids, the Whitney numbers W_k satisfy

$$W_k^2 \geqslant W_{k+1} W_{k-1}.$$

2. Construct a semimodular lattice such that if W_k is the number of elements of height k the sequence W_k is not a unimodal sequence.

3. Call a polynomial $A(x) = \sum_{i=0}^{n} a_i x^i$ *log concave* if $a_i \geqslant 0$ and $a_i^2 \geqslant a_{i+1} a_{i-1}$. Prove that the product of two log concave polynomials is log concave. (Mason [72a]).

4. Define the polynomial $W(M; x)$ of a matroid M by

$$W(M; x) = \sum W_k x^k$$

where W_k is the number of k-flats of M. Prove that if M, N are matroids on disjoint sets

$$W(M; x) W(N; x) = W(M + N : x).$$

5. Let M be a simple matroid on S with the property that for any two distinct hyper-planes H_i, H_j, $|H_i \cap H_j| = \lambda \geq 1$. Prove that M is isomorphic to a matroid of one of the following types:

(a) a projective space
(b) the direct sum of a single independent element and a simple matroid of rank 2
(c) the free matroid.

(Heron [73])

3. THE DOWLING–WILSON THEOREMS

Let $\mathscr{L}$ be a finite lattice, and let $\mathbb{Q}$ denote the rationals. Define $V(\mathscr{L})$ to be the free vector space over $\mathbb{Q}$ generated by the lattice elements. That is $V(\mathscr{L})$ is the set of all mappings $f : \mathscr{L} \to \mathbb{Q}$ with the usual addition and scalar multiplication.

Let I_x $(x \in \mathscr{L})$ be the *indicator* function of x, that is

$$I_x(y) = \begin{cases} 1 & y = x \\ & \\ 0 & y \neq x \end{cases} \qquad y \in \mathscr{L}.$$

Clearly the vectors $\{I_x : x \in \mathscr{L}\}$ form a basis for $V(\mathscr{L})$.

For $x \in \mathscr{L}$ let J_x be the indicator function of the set $\{y : y \vee x = I\}$. That is for $x \in \mathscr{L}$,

$$J_x = \sum_{y : y \vee x = I} I_y$$

Similarly let K_x be the indicator function of the set $\{y : y \in \mathscr{L}, y \leq x\}$. That is

$$K_x = \sum_{y : y \leq x} I_y.$$

Now if μ is the Möbius function of $\mathscr{L}$ (see Section 15.2) we have the following linear relations between the three sets of vectors $\{I_x\}$, $\{J_x\}$, $\{K_x\}$.

LEMMA. *Let $\mathscr{L}$ be a finite lattice. Then for each $x \in \mathscr{L}$ the following equations hold in $V(\mathscr{L})$.*

(1) $\displaystyle I_x = \sum_{y : y \leq x} \mu(y, x) K_y.$

(2) $\displaystyle J_x = \sum_{y : y \geq x} \mu(y, I) K_y.$

(3) $\displaystyle \mu(x, I) K_x = \sum_{y : y \geq x} \mu(x, y) J_y.$

(4)   *If $\mu(\alpha, I) \neq 0$ for all $\alpha \in \mathscr{L}$ then*

$$I_x = \sum_y \lambda(x, y) J_y$$

where

$$\lambda(x, y) = \sum_{\alpha: \alpha \leqslant x \wedge y} \frac{\mu(\alpha, x)\mu(\alpha, y)}{\mu(\alpha, I)}.$$

Proofs. For (1) apply Möbius inversion to the definition of K_x. For (2) observe that

$$\sum_{y: y \geqslant x} \mu(y, I) K_y = \sum_{y: y \geqslant x} \mu(y, I) \sum_{z: z \leqslant y} I_z$$

$$= \sum_z \left(\sum_{y: y \geqslant x \vee z} \mu(y, I) \right) I_z = J_x,$$

since

$$\sum_{y: y \geqslant x \vee z} \mu(y, I) = \begin{cases} 1 & \text{if } x \vee z = I \\ 0 & \text{otherwise.} \end{cases}$$

Applying Möbius inversion to (2) we get (3).

Now assuming $\mu(\alpha, I)$ is never zero, (1) and (3) give

$$I_x = \sum_{\alpha: \alpha \leqslant x} \mu(\alpha, x) K_\alpha$$

$$= \sum_{\alpha: \alpha \leqslant x} \frac{\mu(\alpha, x)}{\mu(\alpha, I)} \sum_{y: y \geqslant \alpha} \mu(\alpha, y) J_y$$

$$= \sum_y \lambda(x, y) J_y.$$

Note. With $x = o$ in (3) and $o \neq \alpha \in \mathscr{L}$ we get

$$\mu(o, I) K_0(\alpha) = \sum_y \mu(o, y) J_y(\alpha)$$

$$0 = \sum_{y: y \vee \alpha = I} \mu(o, y)$$

which we recognize as one form of Weisner's Theorem 15.2.3.

We now use the above lemma to prove:

THEOREM 1. *For any finite geometric lattice $\mathscr{L}$, the Whitney numbers satisfy*

(5) $W_1 + W_2 + \dots + W_k \leqslant W_{r-k} + \dots + W_{r-2} + W_{r-1}$

where r is the rank of $\mathscr{L}$, and $1 \leqslant k \leqslant r - 1$.

    If equality holds in (5) for some k then $\mathscr{L}$ is a modular lattice.

Proof.

Let $\mathscr{L}$ be geometric with rank (height) function ρ, and let U_k be the subspace of $V(\mathscr{L})$ spanned by the vectors $\{I_x : \rho(x) \leqslant k\}$.

Let $\Pi : V(\mathscr{L}) \to U_k$ be the projection map associating to each $f \in V(\mathscr{L})$ its restriction to the subset $\{x \in \mathscr{L} : \rho(x) \leqslant k\}$. That is Π is the linear map defined by

$$\Pi(I_x) = \begin{cases} I_x & \rho(x) \leqslant k \\ 0 & \rho(x) > k. \end{cases}$$

If $\rho(y) < r - k$, then

$$\Pi(J_y) = \sum_{x : x \vee y = I} \Pi(I_x) = 0$$

since if $x \vee y = I$, the semimodular inequality implies

$$\rho(x) \geqslant \rho(I) + \rho(x \wedge y) - \rho(y) > k.$$

Also if $\rho(x) \leqslant k$, by (4) we have since $\mu(a, I) \neq 0$ in a geometric lattice,

$$I_x = \Pi(I_x) = \sum_y \lambda(x, y) \Pi(J_y)$$

$$= \sum_{y : \rho(y) \geqslant r - k} \lambda(x, y) \Pi(J_y).$$

Hence the $W_{r-k} + \ldots + W_{r-1} + W_r$ vectors $\{\Pi(J_y) : \rho(y) \geqslant r - k\}$ span the subspace U_k. But the subspace U_k has dimension $W_0 + W_1 + \ldots + W_k$. Since $W_0 = W_r = 1$ the inequality (5) is proved. The case of characterizing when equality occurs in (5) is harder. We refer to Dowling and Wilson [73]. ∎

This technique gives a succession of very interesting properties of geometric lattices.

THEOREM 2. *Let $\mathscr{L}$ be a finite lattice such that $\mu(a, 1) \neq 0$ for all $a \in \mathscr{L}$. Then there exists a permutation $f : \mathscr{L} \to \mathscr{L}$ such that*

$$x \vee f(x) = I,$$

for all $x \in \mathscr{L}$.

Proof. By (4) $\{J_x : x \in \mathscr{L}\}$ is a basis for $V(\mathscr{L})$. Hence the matrix whose rows and columns are indexed by $\mathscr{L}$ and such that the entry in row x and column y is

$$J_x(y) = \begin{cases} 1 & \text{if } x \vee y = I \\ 0 & \text{otherwise} \end{cases}$$

is nonsingular. Thus some term in the determinant expansion is non-zero. In other words for some permutation f of $\mathscr{L}$, $J_x(f(x)) = 1$ for all $x \in \mathscr{L}$. ∎

Note. Inequality (5) is an immediate consequence of Theorem 2 since by the submodular law of geometric lattices f must map the set of elements of rank $\leqslant k$ injectively into the set of elements of rank $\geqslant r - k$.

COROLLARY. *Let $\mathscr{L}$ be a finite lattice such that $\mu(o, a) \neq 0$ for all $a \in \mathscr{L}$. Then there exists a permutation $f^*\mathscr{L} \to \mathscr{L}$ such that*

$$x \wedge f^*(x) = o \qquad (x \in \mathscr{L}).$$

Proof. Apply Theorem 2 to the dual lattice. ∎

THEOREM 3. *Let $\mathscr{L}$ be a finite geometric lattice of rank r and $0 \leqslant k \leqslant r$. Then there exists an injection*

$$g: \{x \in \mathscr{L}: \rho(x) \leqslant k\} \to \{y \in \mathscr{L}: \rho(y) \geqslant r - k\}$$

such that $x \leqslant g(x)$ for all x in the domain of g.

Proof. With the notation as in the proof of Theorem 1, we have seen that $\{\Pi(J_y): \rho(y) \geqslant r - k\}$ spans the subspace U_k.
By (2), (4)

$$\Pi(J_y) = \sum_{z: z \geqslant y} \mu(z, I)\Pi(K_z)$$

so that $\{\Pi(K_z): \rho(z) \geqslant r - k\}$ also spans U_k.
Hence the matrix whose rows are indexed by $\{x \in \mathscr{L}: \rho(x) \leqslant k\}$ and columns by $\{y \in \mathscr{L}: \rho(y) \geqslant r - k\}$, the entry in row x and column y being

$$K_y(x) = \begin{cases} 1 & \text{if } x \leqslant y \\ 0 & \text{otherwise,} \end{cases}$$

has rank equal to the number of rows. The existence of the injection g now follows since some maximal square submatrix is non-singular. ∎

Let $\text{Top}_k(\mathscr{L})$ and $\text{Bot}_k(\mathscr{L})$ denote respectively, the sum of the top and bottom $k + 1$ Whitney numbers of $\mathscr{L}$, that is

$$\text{Bot}_k(\mathscr{L}) = W_0 + W_1 + \ldots + W_k$$

$$\text{Top}_k(\mathscr{L}) = W_r + W_{r-1} + \ldots + W_{r-k} \qquad (r = \rho(\mathscr{L}))$$

THEOREM 4. *If $\mathscr{L}$ is a finite geometric lattice of rank r and a is an element of $\mathscr{L}$ with $\rho(a) \leqslant r - k$ then*

$$\text{Top}_k(\mathscr{L}) \geqslant \text{Top}_k(\mathscr{L}) - \text{Top}_k[a, I] + \text{Bot}_k[a, I] \geqslant \text{Bot}_k(\mathscr{L}).$$

COROLLARY. *In a finite geometric lattice of rank $\geqslant 2$ the number of lines (2-flats) containing a given atom plus the number of hyperplanes not containing that atom cannot be less than the total number of atoms.*

Proof. Take a to be an atom and $k = 1$ in Theorem 4. ∎

Proof of Theorem 4

Apply Theorem 1 to the geometric lattice $[a, I]$ and we get $\text{Top}_k[a, I] \geqslant \text{Bot}_k[a, I]$ and the first inequality follows.

For $z \in \mathscr{L}$ define a function $A_z \in V(\mathscr{L})$ by

$$A_z = \sum_{x:\, x \vee a = z} I_x.$$

Then whenever $y \geqslant a$ in $\mathscr{L}$,

(6) $$K_y = \sum_{z:\, a \leqslant z \leqslant y} A_z.$$

Now to prove the second inequality it suffices to show that the set of vectors $S = S_1 \cup S_2$ where

$$S_1 = \{\Pi(K_y): \rho(y) \geqslant r - k,\ y \not\geqslant a\}$$
$$S_2 = \{\Pi(A_z): \rho(z) \leqslant \rho(a) + k,\ z \geqslant a\}$$

spans the subspace U_k of $\mathscr{L}$. From the proof of Theorem 3 $\{\Pi(K_y): \rho(y) \geqslant r - k\}$ spans U_k so it remains only to show that $\Pi(K_y)$ is in the span of S for $y \geqslant a$. But this follows from (6) and the observation that the semimodular law implies $\Pi(A_z) = 0$ whenever $\rho(z) > \rho(a) + k$. ∎

Another result of Dowling and Wilson [73a] which we state without proof is:

THEOREM 5. *Let $\mathscr{L}$ be a finite geometric lattice of rank r with n atoms. Then*

(7) $$W_k \geqslant \binom{r-2}{k-1}(n-r) + \binom{r}{k}   \qquad 0 \leqslant k \leqslant r.$$

When $r \geqslant 4$, equality holds in (7) for some k, $2 \leqslant k \leqslant r - 2$ if and only if $\mathscr{L}$ is isomorphic to the direct product of a modular plane and a free geometry.

Note. A *free geometry* with j atoms is the geometric lattice of all subsets of a j-set, that is the geometric lattice of a free matroid.

A *modular plane* is a geometric modular lattice of rank 3.

EXERCISES 3

1. Let x be a copoint of the geometric lattice $\mathscr{L}$. If $\eta(x)$ is the number of atoms $\notin x$ prove that

$$|\mu(o, 1)| \leqslant |\mu(o, x)| \eta(x).$$

(Greene [71])

2. Let x be an element of the geometric lattice $\mathscr{L}$ and write $y \perp x$ if

$$x \vee y = I, \quad x \wedge y = o, \quad \rho(x) + \rho(y) = \rho(I).$$

Prove that

$$|\mu(o, x)| \sum_{y \perp x} |\mu(o, y)| \leqslant |\mu(o, 1)|.$$

(Greene [71])

3. Let $\mathscr{L}$ be a finite modular lattice. If $\mathscr{V}_k$ denotes the set of elements of $\mathscr{L}$ covered by precisely k elements and $\mathscr{W}_k$ denotes the set of elements of $\mathscr{L}$ covering exactly k elements, then $|\mathscr{V}_k| = |\mathscr{W}_k|$. (Dilworth [54]).

4. Deduce that in a finite modular lattice the number of join irreducibles equals the number of meet irreducibles.

5. Let $\mathscr{L}$ be a geometric lattice of rank r and $1 \leqslant k \leqslant r - 2$. Prove that if $x \wedge y > o$ whenever $\rho(x) = k + 1$ and $\rho(y) = r - k$ then the lattice $\mathscr{L}$ is modular. (Dowling and Wilson [73])

6. Prove that in Theorem 3 with $k < r$, the injection g can be chosen so that $g(o) = I$.

7. Let x, y be elements of a geometric lattice such that x covers y. Prove that the number of elements covering y is at least one more than the number of elements covering x.

8. Let M be a simple matroid and let A be a k-flat of M, and B be a $(k + 1)$-flat of M with $A \subseteq B$. Prove

   (a) B is contained in fewer $(k + 2)$-flats than A is contained in $(k + 1)$-flats.
   (b) A contains fewer $(k - 1)$ flats than B contains k-flats.

9. Prove that in a geometric lattice with n atoms there exist n disjoint maximal chains linking the atoms of the lattice into the set of copoints. (Mason [73])

10. A theorem of J. J. Sylvester can be stated as follows. If a set of points in n-dimensional real projective space, not all points on a line, is such that on the line joining any two points of the set there is always a third point of the set then the number of points cannot be finite.

    Call a matroid a *Sylvester matroid* if any two points are contained in a circuit of cardinality 3. Prove if M is a Sylvester matroid on S has rank $r \geqslant 2$, then

$$|S| \geqslant 2^r - 1.$$

(Murty [69, 70a])

4. BASES AND INDEPENDENT SETS

In this section we study the analogous extremal problems for bases and independent sets.

The following conjecture Welsh [71b] was made in 1969.

Conjecture 1. $i_k \geqslant \min(i_{k-1}, i_{k+1}) \, 2 \leqslant k \leqslant r - 1$.

Mason [72a] has more recently made the successively stronger conjectures,

Conjecture 2. $i_k^2 \geqslant i_{k-1} i_{k+1}$,

Conjecture 3. $i_k^2 \geqslant \dfrac{k+1}{k} i_{k-1} i_{k+1}$,

Conjecture 4. $i_k^2 \geqslant \dfrac{k+1}{k} \cdot \dfrac{n-k+1}{n-k} i_{k-1} i_{k+1}$.

for $2 \leqslant k \leqslant r - 1$, where again $n = |S|$ and r is the rank of the matroid.

We first give a more general result which lends some support to the truth of these conjectures.

THEOREM 1. *Let $\mathscr{C}$ be an incomparable collection of subsets of a finite set S each of whose elements has cardinal r. Let d_k be the number of subsets of S of cardinal k which are contained in some member of $\mathscr{C}$. Then for all t such that $k \leqslant t \leqslant r - k$,*

$$\binom{t}{k} d_t \geqslant \binom{r}{k} d_k.$$

Proof. (See Mason [72a]). ∎

Applying this to the case where $\mathscr{C}$ is the set of bases of a matroid it is easy to prove.

COROLLARY 1. *The sequence $\{i_k\}$ of independence numbers of a matroid M can fail to be unimodal only for $k > \rho(M)/2$.*

COROLLARY 2. *If M is a matroid of rank r*

$$i_k(M) \leqslant i_{r-k}(M) \qquad k \leqslant r/2.$$

The following natural extremal problem about the bases of a matroid has been settled by Dinolt [71] and Murty [71]. As pointed out in Section 1 there are trivial upper and lower bounds for $b(M)$ which can be achieved by trivial matroids.

THEOREM 2. *Let M be a connected matroid of rank r on a set of n elements. Then*

$$b(M) \geq r(n - r) + 1$$

Moreover for $n > r$ this minimum is achieved only when M is isomorphic to the cycle matroid of the graph $G(n, r)$ consisting of a cycle C_{r+1} with $n - r - 1$ other edges parallel to one edge.

$G(8, 4)$:

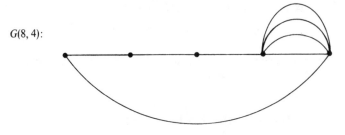

Figure 1

Dinolt [71] goes on to characterize the two "second minimal" connected matroids.

Let $H_1(n, r)$ be the matroid on $\{1, 2, \ldots, n\}$ with circuits given by

$$\{\{1, 2, \ldots, r + 1\}, \{1, 2, \ldots, r - 1, j\}, \{r, r + 1, j\}, \{i, j\}_{i \neq j} : i, j = r + 2, \ldots, n\}.$$

Let $H_2(n, n - r)$ be the matroid which satisfies

$$H_2(n, n - r) \simeq (H_1(n, r))^*.$$

It has circuits given by

$$\{\{r, r + 1\}, \{i, j\}_{i \neq j}, \{i, k, r + 2, \ldots, n\} : i, j = 1, 2, \ldots, r - 1, \quad k = r, r + 1\}$$

$G_1(9, 4)$:

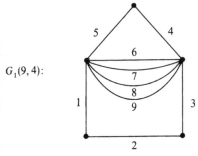

$G_2(9, 5)$

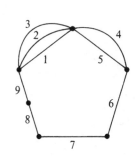

Figure 2

The reader will recognize that $H_1(n, r)$ is the cycle matroid of the graph $G_1(n, r)$, defined in the obvious way ($G_1(9, 4)$ is exhibited in Fig. 2).

$H_2(n, r)$ is also a graphic matroid but does not have a unique graphic representation. One graphic representation of $H_2(9, 5)$ is the graph $G_2(9, 5)$ of Fig. 2.

By considering their graphs, it is easy to check that

$$b(H_1(n, r)) = 2(n - r - 1)(r - 1) + r + 1$$
$$b(H_2(n, r)) = (2r - 1)(n - r - 1) + 2.$$

THEOREM 3. *Let M be a connected matroid of rank r on S, ($|S| = n$). Then there are three mutually exclusive possibilities:*

(a) $b(M) = r(n - r) + 1$

(b) $b(M) \geqslant 2(r - 1)(n - r - 1) + r + 1 \qquad n \geqslant 2r$

(c) $b(M) \geqslant (2r - 1)(n - r - 1) + 2 \qquad n < 2r$

and the right-hand sides are best possible.

Murty [71] also proves

THEOREM 4. *If M is a connected simple matroid on a set of $n \geqslant 4$ elements then either M is a circuit or it has at least $3n - 7$ bases. Moreover if $n \geqslant 5$ and M is not a circuit then $b(M) = 3n - 7$ if and only if M is the cycle matroid of the graph on n edges shown in Fig. 3.*

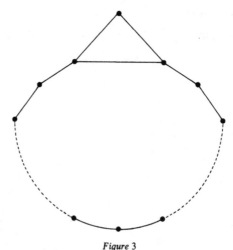

Figure 3

EXERCISES 4

1. Find an incomparable set $\mathscr{C}$ of subsets of S of cardinality m such that if d_k is the number of k-subsets of S which are contained in some member of $\mathscr{C}$ the sequence (d_k) is not unimodal. (Mason [72a])

2. Determine the sequence $i_k(M)$ when M is (a) a projective geometry, (b) an affine geometry.

3. Prove that a matroid M of rank r on a set of n elements has a sequence $\{i_k\}$ which satisfies

$$i_k = \binom{n}{k}, \quad 0 \leqslant k \leqslant r - 1,$$

if and only if M is a paving matroid.

4. Prove that if M is a matroid on S with no loops or coloops then $b(M) \geqslant |S|$. (Murty [71])

5. Let M be a matroid on S and for $A \subseteq S$ let $b(A)$ denote the number of bases of M which contain A. A graph theorem of Tutte shows that if M is graphic

$$b(A) \, b(B) \leqslant b(A \cup B) \, b(A \cap B).$$

Find a binary matroid for which this inequality does not hold. (Seymour and Welsh [75])

5. SPERNER'S LEMMA AND RAMSEY'S THEOREM

Since the simplest geometric lattice is the Boolean lattice of all subsets of a set it is natural to ask whether well known properties of Boolean lattices carry over to modular geometric lattices or even to general geometric lattices.

Two of the most celebrated extremal theorems of the lattice of subsets of a set are Sperner's lemma [28] and Ramsey's theorem [30]. In this section we show how analogous questions can be posed for geometric lattices.

Sperner's Lemma

Recall that distinct subsets of S are *incomparable* if one does not contain the other. Sperner's lemma is the following basic result.

THEOREM 1. *The maximum number of incomparable subsets of an n-set is given by*

$$\max_{0 \leqslant k \leqslant n} \binom{n}{k}.$$

An exceptionally simple proof of Sperner's lemma has been given by Lubell [66] who proves the slightly stronger result.

THEOREM 2. *Let $\mathscr{F}$ be a collection of mutually incomparable subsets of an n-set. Then*

$$\sum_{A \in \mathscr{F}} 1 \bigg/ \binom{n}{|A|} \leqslant 1.$$

Proof. Let $\mathscr{L}$ be the Boolean lattice of subsets of the n-set S. For each set $A \subseteq S$ there are exactly $|A|!\,(n - |A|)!$ maximal chains of $\mathscr{L}$ which contain A. Since none of the $n!$ maximal chains of $\mathscr{L}$ meets $\mathscr{F}$ more than once, we have

$$\sum_{A \in \mathscr{F}} |A|!\,(n - |A|)! \leqslant n!$$

which proves the theorem. ∎

If $\mathscr{L}$ is an arbitrary geometric lattice we say that $\mathscr{L}$ has the *Sperner property* if the maximum number of incomparable elements of $\mathscr{L}$ is given by

$$\max_{0 \leqslant k \leqslant \rho(M)} W_k(M)$$

where M is the simple matroid associated with $\mathscr{L}$.

Thus Sperner's lemma says that Boolean algebras have the Sperner property, and as we shall see the lattice of flats of any projective geometry has the Sperner property.

However Dilworth and Greene [71] disproved a conjecture of G. C. Rota that every geometric lattice had the Sperner property, by constructing the following infinite class of geometric lattices which fail to have the Sperner property.

Let G_n be the graph with vertex set $a, b, 1, 2, \ldots, n$ shown in Fig. 1.

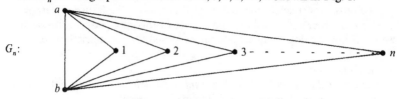

Figure 1

A simple counting argument shows that W_k, the number of k-flats in the geometric lattice $\mathscr{L}_n$ of the cycle matroid $M(G_n)$, is given by

$$W_k = 2k\binom{n}{k} + \binom{n}{k - 1}.$$

The sequence W_k reaches its maximum at approximately $k = 2n/3$.

Let $\mathcal{T}_k$ be the set of k-flats containing $e = (a, b)$ and let $\mathcal{U}_k$ be the set of k-flats not containing e. Then for $\mathcal{L}_{10}$ it is not difficult to show that $\mathcal{T}_6 \cup \mathcal{U}_7$ is an incomparable set of elements of $\mathcal{L}_{10}$ and

$$|\mathcal{T}_6 \cup \mathcal{U}_7| > \max_k W(k).$$

Thus $\mathcal{L}_{10}$ is a counterexample with 60,073 elements.

An extension of this idea shows that the lattices $\mathcal{L}_n$ do not have the Sperner property for all large n.

Baker [69] proves the following theorem.

THEOREM 3. *Let P be a finite partially ordered set with o and I which satisfies the Jordan–Dedekind chain condition and has height function h. Suppose also that for each k, all elements of height k*
(i)   *are covered by the same number a_k of elements of height $k + 1$, and*
(ii)   *cover the same number b_k of elements of height $k - 1$.*
Then any set of pairwise incomparable elements of P has at most W elements where $W = \max W_k$ and W_k is the number of elements of height k.

Proof. For each k let $E(k)$ be the set of elements of P of height k. For $x \in P$ let $p(x)$ be the probability that a randomly chosen maximal chain of P will contain x. That is $p(x) = s(x)/t$, where t is the total number of maximal chains of P and $s(x)$ is the number of those which contain x.

Now if $h(x) = k$, $s(x)$ is the product of the number $b_k \ldots b_1$ of maximal chains of the interval $[o, x]$ and the number $a_k \ldots a_{n-1}$ of maximal chains of the interval $[x, I]$ (where $n = h(I)$). Thus for all x in $E(k)$

$$p(x) = b_k \ldots b_1 a_k \ldots a_{n-1}/t.$$

Also for a given k, each maximal chain of P contains exactly one element of $E(k)$. Since

$$1 = \sum_{x \in E(k)} p(x)$$

we must have

$$p(x) = 1/|E(k)| = W(k)^{-1} \geqslant W^{-1}.$$

Finally if S is a set of incomparable elements of P, again no maximal chain of P can contain more than one element of S so

$$1 \geqslant \sum_{x \in S} p(x) \geqslant \sum_{x \in S} W^{-1} = |S|/|W|$$

and the theorem is proved. ∎

An immediate consequence of Baker's result is:

COROLLARY. *The lattice of flats of any perfect matroid design has the Sperner property, in particular the lattices of flats of projective and affine geometries have the Sperner property.*

Ramsey's theorem for matroids
  Ramsey's theorem for finite sets can be stated as follows.

THEOREM 4. *Let S_n be a set of n elements and suppose that the collection $\mathcal{T}_r$ of all r-element subsets of S_n is divided into t disjoint parts. $\alpha_1, \ldots, \alpha_r$. Let $k \geqslant r$, $r > 1$. Then there exists an integer $N = N(k, r, t)$ such that if $n \geqslant N$ there exists a subset A of k elements all of whose r-element subsets belong to α_i for some i.*

For a proof see the books of Ryser [63] or Hall [67].

A very attractive analogue of Ramsey's theorem has been formulated by G. C. Rota.
  Let $\mathcal{G} = \{\mathcal{L}_i : i = 0, 1, 2, \ldots\}$ be a collection of geometric lattices. For integers $k \geqslant 0$, $r \geqslant 0$, $t \geqslant 0$ consider the statement:

$L(k, r, t)$: There exists an integer $N = N(k, r, t)$, depending only on k, r, t such that if $n \geqslant N$ and if the elements of $\mathcal{L}_n$ of rank r are coloured with t colours, then there is an element x of rank k in $\mathcal{L}_n$ such that all elements y of $\mathcal{L}_n$ of rank r with $y \leqslant x$ have the same colour.

If we let $\mathcal{L}_i$ be the Boolean lattice of subsets of a set of i elements then this statement (which we denote in this case by $S(k, r, t)$) becomes Ramsey's theorem for k, r, t. Thus Ramsey's theorem can be restated as:

(1)   The statement $S(k, r, t)$ is true for all $k, r, t \geqslant 0$.

Rota conjectured that if one chooses $\mathcal{L}_i$ to be the geometric lattice of flats of $PG(i, q)$ then the corresponding statement (denoted in this case by $P_q(k, r, t)$) is true. This was partially proved by Graham and Rothschild [71].

THEOREM 5. *The statement $P_q(k, 1, t)$ is true for any k, t, and prime power q.*

Suppose also we denote by $A_q(k, r, t)$ the statement $L(k, r, t)$ with $\mathcal{L}_i$ the lattice of flats of the affine geometry $AG(i, q)$. In [71] Rothschild and Graham also show:

(2) $P_q(k, r, t) \Rightarrow A_q(k, r, t)$;

(3) $A_q(k + 1, r + 1, t) \Rightarrow P_q(k, r, t)$;

(4) $\forall k \; A_q(k, r, t) \Rightarrow \forall k \; P_q(k, r, t)$.

Clearly there are many unsolved problems arising from this work. Foremost is the extension of Theorem 5 to arbitrary r.

Another (possibly easier) problem is to find a new class $\mathscr{G}$ of lattices for which the statement $L(k, r, t)$ is true.

EXERCISES 5

°1. Do paving matroids have the Sperner property?

2. Prove that uniform matroids have the Sperner property.

6. ENUMERATION

In this section we shall obtain asymptotic estimates of $f(n)$, the number of nonisomorphic matroids on an n-set; $g(n)$ the number of nonisomorphic simple matroids on an n-set and $t(n)$ the number of nonisomorphic transversal matroids on an n-set.

The first results were obtained by Crapo [65] who proved that

(1) $f(n) \geqslant 2^n$

and Welsh [69] who showed

(2) $t(n) \geqslant 2^n$

Bollobas [69] improved these results by showing that for sufficiently large n

(3) $f(n) \geqslant 2^{n^2/12}.$

Since a matroid is defined by specifying a set of subsets of an n-set it is obvious that

(4) $f(n) \leqslant 2^{2^n}.$

In other words

(5) $\log_2\log_2 f(n) \leqslant n.$

Blackburn, Crapo and Higgs [73] catalogued nonisomorphic simple matroids on n-elements for $1 \leqslant n \leqslant 8$. We list their results.

	8 pts.	7 pts.	6 pts.	5 pts.	4 pts.	3 pts.	2 pts.	1 pts.
rank 1								1
2	1	1	1	1	1	1	1	
3	68	23	9	4	2	1		
4	617	49	11	3	1			
5	217	22	4	1				
6	40	5	1					
7	6	1						
8	1							
total	950	101	26	9	4	2	1	1

Piff [73] obtained the first non-trivial upper bound on $f(n)$ by proving

THEOREM 1. *There exists $k > 0$ such that for $2 \leqslant n < \infty$*

$$f(n) \leqslant n^{k 2^n n^{-1}}$$

which implies

(6) $$\log_2 \log_2 f(n) \leqslant n - \log_2 n + O(\log \log n)$$

Recently Knuth [74] has improved the lower bound (3) by proving

THEOREM 2. *For $1 \leqslant n < \infty$*

(7) $$n! g(n) \geqslant 2^{\left\{ \binom{n}{\lceil n/2 \rceil} / 2n \right\}}$$

from which we obtain

$$\log_2 \log_2 g(n) \geqslant n - (3/2) \log_2 n + O(\log \log n)$$

Proof. Note first that the factor $n!$ in (7) accounts for any isomorphisms between the matroids we shall construct so we can ignore isomorphisms in what follows.

Let $\mathscr{F}$ be a family of subsets of $\{1, 2, \ldots, n\}$, such that each subset contains exactly $[n/2]$ elements and where no two different subsets have more than $[n/2]$-2 elements in common.

This set $\mathscr{F}$, together with the set of all $[n/2]$-1 element subsets which are not contained in any element of $\mathscr{F}$ constitutes a set of blocks such that every subset of size $[n/2]$-1 is contained in a unique block; therefore it defines a paving matroid.

If $\mathscr{F}$ contains m members, each of the 2^m subfamilies of $\mathscr{F}$ will define a paving matroid in the same way. Hence (7) will follow if we can find such a family $\mathscr{F}$ of subsets, containing at least

$$m \geqslant \binom{n}{[n/2]} \Big/ 2n$$

members.

The problem is solved by realizing that it is the same as finding m binary code words (see Section 15.7) of length n, each containing exactly $[n/2]$-1 bits and which are single-error correcting. This characterization suggests the following "Hamming code" construction. Let $k = [\log_2 n] + 1$ and construct the $n \times k$ matrix H of 0's and 1's whose rows are the numbers from 1 to n expressed in binary notation. For $0 \leqslant j < 2^k$ consider the set U_j of all row vectors x of 0's and 1's such that x contains exactly $[n/2] - 1$ bits and the vector xH mod 2 is the binary representation of j. Note that if x and y are distinct elements of U_j they cannot differ in just two places; otherwise we would have $(x + y)H$ mod 2 = 0, contradicting the fact that no two rows of H are equal. Therefore U_j defines a family of subsets of $\{1, 2, \ldots, n\}$ having the desired property.

Furthermore, at least one of these 2^k families U_j will contain

$$\binom{n}{[n/2]} \Big/ 2^k$$

or more elements, since they are disjoint and they exhaust all of the $\binom{n}{[n/2]}$ possible $[n/2]$-element subsets. This completes the proof since $2^k \leqslant 2n.\blacksquare$

Despite these fairly precise estimates of $f(n)$ there are some intriguing open questions. For example the following conjectures were made in Welsh [71b] and as far as we know they are still unsettled.

Conjecture 1. For any integers $m, n, f(m + n) \geqslant f(m)f(n)$.

Conjecture 2. If $f_r(n)$ denotes the number of non-isomorphic matroids of rank r on an n-set then for fixed n, the sequence $(f_r(n): 1 \leqslant r \leqslant n)$ is unimodal.

EXERCISES 6

1. Prove that the number $R_d(n)$ of non-isomorphic matroids on an n-set which are representable over a finite field F with $|F| = d$, satisfies

$$k_1 2^{n^{2/4}}/n! < R_d(n) < k_2 d^{n^{2/4}}$$

for some $k_1, k_2 > 0$. (Piff [72])

2. Let $u(n)$ be the number of nonisomorphic matroids on an n-set which are not representable over any field. Prove that $u(n) \geqslant f(n - 8)$.

○3. Prove that the number $g(n)$ of simple matroids on a set of n elements satisfies

$$\lim_{n \to \infty} \frac{\log_2 \log_2 g(n) - n}{\log_2 n} = c$$

for some constant c.

4. Prove that the number $t(n)$ of nonisomorphic transversal matroids satisfies for $n \geqslant 17$, and suitable k

$$2^{3n^{2/4}}/(n - 1)! \geqslant t(n) \geqslant k2^{n^{2/4}}/n!$$

(Piff [72], Heron [72a])

NOTES ON CHAPTER 16

An interesting recent paper of Knuth [74a] gives some data such as the numbers of bases and circuits etc. from generating matroids "at random".

Motzkin [51] contains a wealth of results about extremal geometrical configurations. For some recent theorems about log-concave sequences see Kurtz [72] or Hoggar [74]. Stonesifer [75] proves the unimodal conjecture for the group partition lattices discussed in Section 11.4. Alternative proofs can be deduced from the more general theorems of Kurtz [72]. The Dowling–Wilson theorems have been slightly extended by Dowling [75]. For a discussion of the unimodal conjectures and Motzkin's theorem see Mason [72a].

In Section 5 we have selected two of the more famous results of extremal set theory and used them to illustrate the way in such results may have extensions to more general results for geometric lattices. Section 6 is an account of the history of the enumeration problem for matroids, see also Recski [74].

For more recent work on Sperner's theorem for general partially ordered sets see Harper [74].

Baclawski [75] has recently proved a conjecture of G. C. Rota by developing a homology theory for the category of partially ordered sets such that the Betti numbers of a geometric lattice are its Whitney numbers.

CHAPTER 17

Maps Between Matroids and Geometric Lattices

1. STRONG MAPS BETWEEN GEOMETRIC LATTICES

It is natural to seek mappings between matroids such that if ϕ is such a map and M_1 is the image of the matroid M under ϕ then M and M_1 share some of the same structure. Unfortunately for the class of matroids there is no immediate natural set of mappings analogous to say homomorphism between vector spaces. However in recent years the consensus of agreement seems to be converging towards the idea that the strong maps between geometric lattices introduced by Crapo [67] and Higgs [68] are most natural. Certainly most of the useful theory which has been developed is in terms of these maps. Accordingly the bulk of this chapter will be devoted to the study of these maps, and the various embedding and erecting problems arising in a natural way from them.

Throughout this section $\mathscr{L}_1$, $\mathscr{L}_2$ will denote two finite geometric lattices and M_1, M_2 will denote their associated simple matroids on the ground sets S_1, S_2 of atoms of $\mathscr{L}_1$, $\mathscr{L}_2$ respectively.

A function $f: \mathscr{L}_1 \to \mathscr{L}_2$ is a *strong map* if:

(a) for all $a, b \in \mathscr{L}_1$, $f(a \vee b) = f(a) \vee f(b)$,
(b) if x is an atom of $\mathscr{L}_1$, $f(x)$ is either an atom or the zero element of $\mathscr{L}_2$,
(c) $f(o) = o$.

A strong map is an *injection* if it is a one-to-one map and is a *surjection* if it is onto.

An obvious consequence of the definitions is that the composition of strong maps is a strong map. Hence the collection of geometric lattices and strong maps forms a category.

We now list some basic properties of a strong map f between $\mathscr{L}_1$ and $\mathscr{L}_2$.
(1)   If y covers x in $\mathscr{L}_1$ then $f(y)$ either covers or is equal to $f(x)$.

Proof: Follows from (a) and the property that every element of a geometric lattice is the join of atoms.

309

(2)   If ρ_1, ρ_2 are the rank functions of $\mathscr{L}_1$ and $\mathscr{L}_2$ respectively then $\rho_2 f(y) \leqslant \rho_1 y$ for all $y \in \mathscr{L}_1$.

Proof. Consider maximal chains from o to y in $\mathscr{L}_1$ and use (1).

(3)   The map f is a lattice isomorphism $\Leftrightarrow$ f is onto, strong and the ranks of $\mathscr{L}_1$ and $\mathscr{L}_2$ are equal.

Proof. Obviously any lattice isomorphism $f: \mathscr{L}_1 \to \mathscr{L}_2$ is onto, strong and $\rho(\mathscr{L}_1) = \rho(\mathscr{L}_2)$. Suppose firstly that f is $1 - 1$. Since f is $1 - 1$ and strong it has an order preserving inverse, because $x \not\leqslant y$ and $f(x) \leqslant f(y)$ implies that y is strictly less than $x \vee y$, and $f(y) = f(x) \vee f(y) = f(x \vee y)$, which contradicts the statement f is 1-1. Thus an order preserving inverse exists and any strong map which is both 1-1 and onto is an isomorphism.

Assume now $f: \mathscr{L}_1 \to \mathscr{L}_2$ is strong, and that $\rho(\mathscr{L}_1) = \rho(\mathscr{L}_2)$. Since f is onto

$$f(o) = o. \qquad f(I) = I.$$

If $f(x) = f(y)$ for $x, y \in \mathscr{L}_1$, $x \neq y$,

$$f(x \vee y) = f(x) \vee f(y) = f(x)$$

Choose a maximal chain C from o to I in $\mathscr{L}_1$ which passes through x, $x \vee y$. The image of this chain, $f(C)$ is a maximal chain from o to I in $\mathscr{L}_2$ since a strong map is cover preserving.

But $\mathscr{L}_1$, $\mathscr{L}_2$ have equal rank so f is 1-1 on C and hence $x = x \vee y$, contradiction.

Hence f is 1-1 on $\mathscr{L}_1$ and by the previous argument must be an isomorphism.

(4)   If $f: \mathscr{L}_1 \to \mathscr{L}_2$ is strong and $x \in \mathscr{L}_2$ there exists $y \in \mathscr{L}_1$ such that $f(y) = x$ and

$$\rho_2(x) = \rho_1(y).$$

That is every image element has a preimage of equal rank.

Proof. Left as an exercise for the reader.

Examples of strong maps

(1)   Let $\mathscr{L}_1$ be any geometric lattice. Let $\mathscr{L}_2$ be the trivial lattice of Fig. 1

$$\mathscr{L}_2: \quad \begin{array}{c} \bullet\, I \\ \big| \\ \bullet\, o \end{array}$$

Figure 1.

Let $\phi(x) = I \; \forall \; x \in \mathscr{L}_1$, $x \neq o$, $\phi(o) = o$. Then ϕ is a strong map.

(2) Let $\mathscr{L}_2$ be any geometric lattice with set of atoms S. Let $\mathscr{L}_1$ be the Boolean algebra of subsets of S ordered by inclusion. Let $\phi: \mathscr{L}_1 \to \mathscr{L}_2$ be defined by $\phi(x) = x(x \in S)$ and then extend ϕ in the obvious way. Clearly ϕ is strong. We call ϕ the *canonical map* determined by the geometric lattice $\mathscr{L}_2$.

(3) Let $\mathscr{L}_1$ be any geometric lattice of rank r. Let $\mathscr{L}_2$ be the truncation of $\mathscr{L}_1$ at level k. In other words the simple matroid M_2 of $\mathscr{L}_2$ is the truncation at k of the simple matroid M_1. Define a map $\phi: \mathscr{L}_1 \to \mathscr{L}_2$ by

$$\phi(x) = \begin{cases} x & \rho(x) \leqslant k - 1, \\ I & \rho(x) \geqslant k. \end{cases}$$

Then ϕ is a strong map: $\mathscr{L}_1 \to \mathscr{L}_2$, called the *truncation strong map*.

(4) Let M be the simple matroid associated with $\mathscr{L}$, let A be any flat of M. Consider $M|A$. This is simple and its associated geometric lattice is the interval $[o, A]$ of $\mathscr{L}$. Take $\mathscr{L}_1$ to be this interval and $\mathscr{L}_2 = \mathscr{L}$, and define $\phi: \mathscr{L}_1 \to \mathscr{L}_2$ by

$$\phi(x) = \begin{cases} x & x \text{ atom of } \mathscr{L}_1, x \in A, \\ o & x \text{ atom of } \mathscr{L}_1, x \notin A, \end{cases}$$

and extend ϕ to $\mathscr{L}_1$. Then ϕ is a strong map, called the *injection* or *restriction map* defined by A.

(5) Let M_1 be the simple matroid of $\mathscr{L}_1$, let A be any flat of M_1 and let $\mathscr{L}_2$ be the geometric lattice associated with the contraction minor $M.(S_1 \backslash A)$. Then $\mathscr{L}_2$ is isomorphic to the interval $[A, I]$ of the lattice $\mathscr{L}_1$.

Now define $\phi: \mathscr{L}_1 \to \mathscr{L}_2$ by

$$\phi X = X \vee A \qquad (X \in \mathscr{L}_1).$$

Then ϕ is a strong map, called the *contraction map* defined by A.

A possibly disturbing feature of strong maps is that if $\mathscr{L}(M_2)$ is the image of $\mathscr{L}(M_1)$ under a strong map very few properties of M_1 are shared by M_2. The reason for this is that *every* simple matroid on S is the image of the free matroid under the canonical map of example (2).

For example, we have:

(6) There exist simple matroids M_1 and M_2 with M_2 the image of M_1 under a strong map and such that M_1 has P but M_2 does not have P for each of the following properties P: being transversal, a gammoid, base orderable and representable over a field F.

EXERCISES 1

1. Find two geometric lattices $\mathscr{L}_1$, $\mathscr{L}_2$ such that there is no strong map $\phi: \mathscr{L}_1 \to \mathscr{L}_2$ nor is there a strong map $\psi: \mathscr{L}_2 \to \mathscr{L}_1$.

2. Give an example of two lattices $\mathscr{L}_1$, $\mathscr{L}_2$ and a map $\phi: \mathscr{L}_1 \to \mathscr{L}_2$ such that ϕ is a lattice homomorphism but $\phi(\mathscr{L}_1)$ is not a sublattice of $\mathscr{L}_2$.

3. If $\mathscr{L}_1$ is the geometric lattice of a graphic matroid and $\phi: \mathscr{L}_1 \to \mathscr{L}_2$ is a strong map, is $\phi(\mathscr{L}_1)$ the lattice of a graphic matroid?

4. Let $\mathscr{L}_1$, $\mathscr{L}_2$ be geometric lattices with M_1, M_2 their corresponding simple matroids. If S_1, S_2 are the sets of atoms of $\mathscr{L}_1$, $\mathscr{L}_2$ a surjection $\phi: S_1 \to S_2$ is a *geometric map* if it is a strong map which maps closed sets into closed sets. Prove:

    (a) a surjection $\phi: S_1 \to S_2$ is geometric if and only if $\sigma_2(\phi(X)) = \phi(\sigma_1 X)$ for all $X \subseteq S_1$.

    (b) if ϕ is a geometric map then $\mathscr{L}_1$ contains an interval $[o, t]$ which is isomorphic to $\mathscr{L}_2$ under the map ϕ. (Sachs [71]).

5. Find a lattice homomorphism $\psi: \mathscr{L}_1 \to \mathscr{L}_2$ such that ψ is not a strong map. Find an example of a strong map which is not a lattice homomorphism.

2. FACTORIZATION THEOREMS FOR STRONG MAPS

It is useful sometimes to consider strong maps as matroid maps.

To do this we need to adjoin a *zero element* o to all sets on which we define a matroid and let o be a loop of the extended matroid.

Thus if M is a matroid on S, and M' is a matroid on T we let $S' = S \cup o$, $T' = T \cup o$ and let M', N' be matroids on S', T' respectively in which o is a loop and $M' | S = M$, $M | T = N$.

The map $f: S' \to T'$ is said to define a *strong map* between M and N if for all $X, Y \subseteq S$, with $Y \subseteq X$,

(1) $\rho_2 f(X) - \rho_2 f(Y) \leqslant \rho_1 X - \rho_1 Y$

     where $\rho_1 \rho_2$ are the rank functions of M, N respectively.

We leave it to the reader to verify:

(2) f is a strong map if and only if for all flats F of N', $(f)^{-1}F$ is a flat of M'.

(3) f is a strong map if and only if for all subsets $X \subseteq S'$,

$$f(\sigma_1 X) \subseteq \sigma_2 fX$$

     where σ_1, σ_2 are the closure functions of M, N respectively.

Note. We use f to denote both the function $S' \to T'$ and its restriction to S. Also when talking about a map $g: S \to T$ we will without explicitly saying so mean a map

$$g = S \cup o \to T \cup o$$

in which $g(o) = o$.

The above propositions show that f is a strong map between two matroids if and only if the obvious extension of f to the flats of M is a strong map between the geometric lattices $\mathscr{L}(M)$, $\mathscr{L}(N)$. In other words we are essentially talking about the same sort of map here as in the preceding section.

In particular the special maps discussed previously are easily recognized in the present context. For example:

(a)   *Injection.* Let $S \subseteq T$ and let $M = N \,|\, T$. Then the inclusion map $i: S \to T$ defines a strong map—the injection map.

(b)   *Contraction.* Let $T \subseteq S$ and let $N = M.T$. Defining $c: S \to T$ by

$$c(x) = \begin{cases} x & x \in T \\ o & x \notin T, \end{cases}$$

we get the contraction map defined earlier.

As we shall see these two maps have a very important role in the theory of strong maps, since the main theorem of this section will show that every strong map is an injection followed by a contraction.

To prove this we need the following propositions.

THEOREM 1. *Let M_1, M_2 be matroids on disjoint sets S_1, S_2 and suppose there exist strong maps $f_i: M_i \to M(i = 1, 2)$. Then there is a unique map $f: M_1 + M_2 \to M$ such that the diagram below commutes where i_1, and i_2 are the inclusion maps*

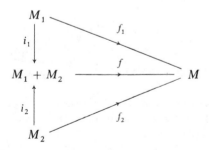

Diagram 1.

Proof. Let $T = S_1 \cup S_2$ and define, $f\colon S_1 \cup S_2 \to T$, by

$$f(x) = \begin{cases} f_1(x) & x \in S_1 \\ f_2(x) & x \in S_2. \end{cases}$$

Then it is an easy exercise to check that f is strong ∎

Thus the direct sum of matroids is the "correct" direct sum in the category of matroids and strong maps.

If $f\colon M \to N$ is strong we define the *kernel* of f, ker f by, ker $f = \{x \in S\colon f(x)$ is a loop of $N'\}$.

Likewise the *image* of f, $f(M)$ is $N'|f(S)$ and the *order* of f, denoted by $v(f)$ is

$$\rho_1 M - \rho_2(f(M)).$$

The *defect* of the strong map f is $v(f) - \rho_1(\ker f)$.

Finally if $S_1 \subseteq S$, the matroid M_1 on S_1 is a *quotient* of M on S if the map $q\colon S \to S_1$ defined by

$$q(x) = \begin{cases} x & x \in S_1 \\ 0 & x \notin S_1 \end{cases}$$

is a strong map from M to M_1 such that ker $q = S\backslash S_1$.

THEOREM 2. *If the strong map* $q\colon (S, M) \to (S_1, M_1)$ *defines a quotient of defect* 0 *then* $M_1 = M . S_1$.

Proof. Let $K = \ker q = S\backslash S_1$. If $\hat{\rho}$ denotes the rank function of $M . S_1$, for $X \subseteq S$,

$$\hat{\rho}X = \rho(X \cup K) = \rho K$$

$$\geqslant \rho_1(q(X \cup K) - \rho_1(q(K)) = \rho_1 X.$$

But

$$\rho_1 S_1 - \rho_1 X \leqslant \rho S - \rho(X \cup K)$$

$$= \rho S - (\rho K + \hat{\rho}X) = \rho_1 S_1 - \hat{\rho}X$$

Hence $\hat{\rho}X = \rho_1 X$ for all $X \subseteq S$ which proves the theorem. ∎

Now if M is a matroid on S we let $s(M)$ be the *simplification* of M, that is all mutually parallel elements are identified and any loops of M are sent to the zero element. It is easy to check that any simplification is a strong map. The *projection* of M to $T \subseteq S$ is the composite map $sc(M) \to s(M . T)$.

As an immediate consequence of Theorem 2 we have

COROLLARY. *If $f: (S, M) \to (T, N)$ is a surjective strong map with defect 0 then f is (up to isomorphism) a projection.*

Now if μ_1, μ_2 are set functions with the same domain S we write $\mu_1 \succ \mu_2$ if whenever $Y \subseteq X \subseteq S$,

$$\mu_1 X - \mu_1 Y \geqslant \mu_2 X - \mu_2 Y,$$

in words, μ_1 *grows faster than* μ_2.

LEMMA 1. *If μ_1, μ_2 are submodular set functions and $\mu_1 \succ \mu_2$ then $\mu = min(\mu_1, \mu_2)$ is submodular and $\mu \succ \mu_2$.*

Proof. Left as an exercise (non-trivial).

Now we may prove what is essentially the factorization theorem mentioned earlier.

THEOREM 3. *If $T \subseteq S$ and $q: (S, M) \to (T, N)$ defines a quotient with defect $r_1 > 0$ then if M_1 is any matroid of rank r_1 on S_1, $(S \cap S_1 = \varnothing)$, there is a matroid M_0 on $S_0 = S \cup S_1$ such that $\rho(M_0) = \rho(M)$ and*

$$M_0 | S = M, \qquad M_0 | S_1 = M_1, \qquad M_0 \cdot T = N.$$

Proof. Let ρ, ρ', ρ_1 be the rank functions of M, N, M_1 respectively. For $A \subseteq S_0$, $A = X \cup Y$ where $X \subseteq S$, $Y \subseteq S_1$ let

$$\mu_1 A = \rho X + \rho_1 Y$$
$$\mu_2 A = \rho'(X \cap T) + r_1.$$

Then μ_1, μ_2 are clearly submodular, integervalued, increasing and non-negative and since for $Y \subseteq X \subseteq S$,

$$\rho'(X \cap T) - \rho'(Y \cap T) \leqslant \rho X - \rho Y$$

we have that μ_1 grows faster than μ_2 and hence by the above lemma

(a)　$\rho_0 = min(\mu_1, \mu_2)$ is submodular: $2^{S_0} \to \mathbb{Z}^+$,
(b)　ρ_0 grows faster than μ_2 so that ρ_0 is an increasing function.

Also for $A \subseteq S_0$, $\rho_0 A \leqslant \mu_1 A \leqslant |A|$, so ρ_0 is the rank function of a matroid M_0 on S_0. Now if $X \subseteq S$,

$$\mu_1 X = \rho X \leqslant \rho S + \rho'(X \cap T) - \rho' T$$
$$= \mu_2 X$$

so that

$$\rho_0 X = \mu_1 X = \rho X \quad \text{and} \quad M_0 | S = M.$$

For

$$Y \subseteq S_1, \mu_1 Y = \rho_1 Y \leqslant \rho_1 S_1 = r_1 \leqslant \mu_2 Y$$

so that

$$\rho_0 Y = \mu_1 Y = \rho_1 Y \quad \text{and} \quad M_0 | S_1 = M_1.$$

Also

$$\rho_0 S_0 \leqslant \mu_2 S_0 = \mu_2 (S \cup S_1) = \rho'(T) + r_1 = \rho S$$

and clearly $\rho_0 S \geqslant \rho S$, so that M_0 and M have the same rank.

Finally let $\hat{\rho}_0$ be the rank function of $M_0 . T$. Then for $X \subseteq T$,

$$\hat{\rho}_0 X = \rho_0 X - \rho_0 (S_1 \cup (S \backslash T)) = \rho X - r_1.$$

Also since $\rho_0 \succ \mu_2$,

$$\rho_0 X - \rho_0 \varnothing \geqslant \rho' X + r_1$$

so that

$$\rho' X \leqslant \rho_0 X - r_1 = \rho X - r_1 = \hat{\rho}_0 X.$$

But again since $\rho_0 \succ \mu_2$,

$$\rho_0 S_0 - \rho_0 X \geqslant \mu_2 S_0 - \mu_2 X = \rho' T - \rho' X$$

so that

$$\rho' X \geqslant \rho' T + \rho_0 X - \rho_0 S_0$$

$$= \rho_0 X + \rho' T - \rho S,$$

since M, M_0 have the same rank and hence

$$\rho' X \geqslant \rho X - r_1$$

which shows that ρ' and $\hat{\rho}_0$ are the same rank function and $M_0 . T = M'$. ∎

We complete this section with a proof of the factorization theorem.

THEOREM 4. *Let* $f: M \to N$ *be a strong map and let* N *be a simple matroid. Then* f *is the composition of an injection and a projection.*

Proof

Case 1. If f is a surjection the theorem follows from Theorem 2.

Case 2. In the general case we can assume M, N are on disjoint sets S, T and consider the direct sum $M + N$ on the set $S \cup T$, and let j, j' be the natural injections $j: M \to M + N, j': N \to M + N$. By Theorem 3 there is a strong map $g: M + N \to N$ such that the diagram below commutes

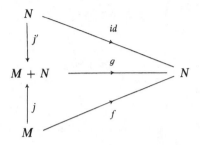

But g is surjective so by case 1, $g \simeq pi$ for some projection p and injection i so that

$$f \simeq pij$$

and ij is a projection which completes the proof. ∎

It is instructive to consider the above theorem geometrically. We illustrate what is going on in the following example.

Example 1. Let $M = M(K_4)$ with Euclidean representation as in Fig. 1 (a) and let N be the rank 2 matroid whose Euclidean representation is given in Fig. 1 (b).

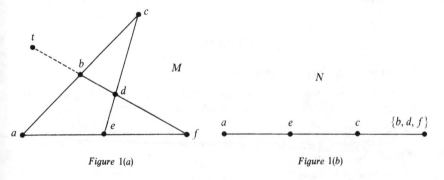

Figure 1(a)   Figure 1(b)

Then the identity strong map $M \to N$ is equivalent to adding on the point t collinear with b, d, f in Fig. 1(a) and projecting from t. Similarly the map $M \to N_1$ where N_1 is given in Fig. 2(b) is got by projecting from the point t_1 shown in Fig. 2(a).

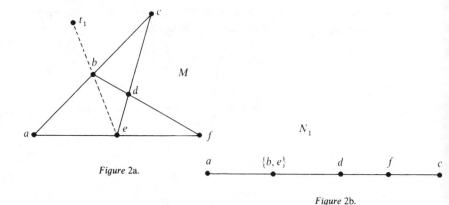

Figure 2a.

Figure 2b.

Clearly the strong map $M \to M$ is got by adding a point t not in the plane. The factorization theorem says that any strong map can be got by this process and hence the strong maps which reduce rank by at most one, correspond exactly to ways of adding the point t on to M to get a larger matroid M' which has M as a restriction and then projecting (essentially contracting) out this extra point.

This leads us naturally on to the theory of single element extensions of matroids, which we discuss in the next section.

We also give an alternative direct proof of the basic factorisation theorem in Section 4.

EXERCISES 2

1. Prove that the simplification operator is a strong map.

2. Prove lemma 1 that if μ_1, μ_2 are submodular set functions and μ_1 grows faster than μ_2 then $\mu = \min(\mu_1, \mu_2)$ is submodular and μ grows faster than μ_2.

3. Given matroids M on S, N on T, a *weak map* $\phi: M \to N$ is a map: $S \to T$ such that X independent in N implies that $\phi^{-1}X$ is independent in M. Prove that the following statements are equivalent

   (a) $\phi: M \to N$ is a weak map.
   (b) $A \subseteq S$, is a dependent set of $M \Rightarrow \phi A$ is dependent in N or $|\phi(A)| < |A|$.
   (c) if ρ, ρ' are the rank functions of M, N respectively then $\rho'\phi X \leqslant \rho X$ for all $X \subseteq S$.

4. Prove that if $T = S$ and N is such that $\mathscr{I}(M) \supseteq \mathscr{I}(N)$ then the identity map is a weak map which is not in general a strong map.

5. Let $f: S \to T$, and let M be a matroid on S. We saw in Chapter 8 that f defines a matroid $f(M)$ on T in which $X \subseteq T$ is independent in $f(M)$ if and only if it is the image of an independent set of M. Prove that $f: M \to f(M)$ is a weak map which in general is not strong.

3. SINGLE ELEMENT EXTENSIONS

If M is a matroid on S a matroid N on $S \cup e = S', e \notin S$, is a *single element extension* of M if $N \mid S = M$.

One obvious way of finding a single element extension of M is to define N by: the bases of N are the sets of the form $B \cup e$, where B is a base of M. This is a *trivial extension* of M and clearly exists for all M.

Note also that the trivial extension of M has rank strictly greater than M and it is easy to see that it is the only extension with this property.

We will henceforth demand that single element extensions be non-trivial.

Crapo [65] has characterized single element extensions of simple matroids and we outline his results here. The presentation is essentially that of Crapo–Rota [70] (Chapter 10) and therefore we will do no more than sketch the main ideas.

Throughout $\mathscr{L}$ will denote the geometric lattice of the simple matroid M on the set S (= atoms of $\mathscr{L}$). σ, ρ will denote respectively its closure function and rank function.

A pair of subsets A, B, of S is a *modular pair* if

$$\rho A + \rho B = \rho(A \cup B) + \rho(A \cap B).$$

A pair of elements x, y is a *modular pair* in a geometric lattice $\mathscr{L}$ if

$$\rho(x) + \rho(y) = \rho(x \vee y) + \rho(x \wedge y).$$

Clearly if A, B are a modular pair of subsets then σA, σB are a modular pair in $\mathscr{L}(M)$. However the converse is not true.

A *modular cut of $\mathscr{L}$* is a subset U such that

(a) If $x \in U$, and $x \leqslant y$ then $y \in U$.
(b) If $x, y \in U$ and x, y are a modular pair then $x \wedge y \in U$.

A *modular cut* of the matroid M on S is a collection U of subsets of S such that:

(a) A subset $A \in U$ if and only if $\sigma A \in U$.
(b) The collection of sets $\{A : A \in U, A$ a flat of $M\}$ is a modular cut of $\mathscr{L}(M)$.

The main result of this section is:

THEOREM 1. *Let U be a modular cut of the simple matroid M on S which does not* *contain any 1-flat of M. Then there is a unique extension N of M on* $S \cup e$ $(e \notin S)$ *such that U is exactly the set of flats A of M such that* $e \in \sigma' A$, *where* σ' *is the closure operator of N.*

Thus we can find the set of all possible single element extensions of a matroid M by finding all the modular cuts of M. On the surface at least this is a much more concrete (though still Herculean) task.

Sketch Proof of Theorem 1. Define a function $\tilde{\rho}$ on $S \cup e$ by defining for $A \subseteq S$,

$$\tilde{\rho}(A) = \rho A,$$
$$\tilde{\rho}(A \cup e) = \rho A + 1 \qquad \text{if } \sigma A \notin U,$$
$$\tilde{\rho}(A \cup e) = \rho A \qquad \text{if } \sigma A \in U.$$

It can be checked that since U is a modular cut $\tilde{\rho}$ is submodular, increasing, and hence is the rank function of a simple matroid $\tilde{M}$ on $S \cup e$.

It is obvious that $\tilde{M}|S = M$ and thus $\tilde{M}$ is the required single element extension. ∎

From this the reader can verify without too much difficulty the following proposition.

THEOREM 2. *Suppose* M_1 *on* $S \cup e$ *is an extension of the simple matroid M on S, which is determined by the modular cut U. Then the flats of* M_1 *are as follows:*

(i) *flats A of M which do not belong to U*

(ii) *sets of the form* $A \cup e$ *where A is a flat of M which belongs to the modular cut U*

(iii) *sets of the form* $A \cup e$ *where A is a flat of M,* $A \notin U$, *and A is not covered in* $\mathcal{L}(M)$ *by a flat* $B \in U$.

As Higgs [68] points out, the theory of single element extensions is closely related to the theory of strong maps.

We explain this relationship at the end of the next section.

EXERCISES 3

1. For M on S let a modular cut be *principal* if it consists of the collection of flats contained in a set A for some $A \subseteq S$. Show that if N is a single element extension of M corresponding to a principal modular cut then N can be induced from M by a bipartite graph. (Las Vergnas [75a])

2. Show that the modular cuts $U_1 = \varnothing$ and $U_2 = S$ correspond to single element extensions of M on S in which the extended element e is respectively a loop or a coloop of the corresponding single element extension.

4. STRONG MAPS INDUCED BY THE IDENTITY FUNCTION

Many of the ideas of the previous sections can be illustrated more clearly in the special case when M_1, M_2 are matroids on the same set and the identity map on S extends in the obvious way to a strong map between $\mathscr{L}(M_1)$ and $\mathscr{L}(M_2)$. Indeed for most purposes one need only consider strong maps between matroids on the same underlying set.

Throughout this section M_1, M_2 denote matroids on the same set S and ρ_i, σ_i ($i = 1, 2$) denote respectively the rank, and closure operator of M_i.

We write $M_1 \rightarrow M_2$ or $M_1 \rightarrow M_2$ is strong if every flat of M_2 is a flat of M_1. It is not difficult to show that this is equivalent to the identity map on S extending (in a canonical way) to a strong map between the geometric lattices $\mathscr{L}(M_1)$ and $\mathscr{L}(M_2)$.

A routine verification gives the following results.

THEOREM 1. *The following statements are equivalent:*

(i) $M_1 \rightarrow M_2$ *is a strong map.*
(ii)   *The σ_2-closure of any subset of S contains its σ_1-closure.*
(iii)   *For all subsets A, B of S with $A \subseteq B$*

$$\rho_1 B - \rho_1 A \geqslant \rho_2 B - \rho_2 A$$

Using this it is not difficult to prove:

THEOREM 2. *Let M_1, M_2 be matroids on S, then $M_1 \rightarrow M_2$ is strong if and only if $M_2^* \rightarrow M_1^*$ is strong.*

If M_1, M_2 are two matroids on S such that every flat of M_2 is a flat of M_1 we say M_2 is a *quotient* of M_1 and M_1 is a *lift* of M_2.

Equivalently, M_2 is a quotient of M_1 if and only if the identity map on the atoms of $\mathscr{L}(M_1)$ extends to a strong map from $\mathscr{L}(M_1)$ onto $\mathscr{L}(M_2)$.

Now if M_2 is a quotient of M_1 there may be several strong maps $\phi: M_1 \rightarrow M_2$ which have M_2 as their quotient. We therefore define the *canonical* map as that which takes each flat of M_1 into the minimal M_2-flat containing it. It is not difficult to prove that this canonical map is a strong map with quotient M_2.

Now if $M_1 \rightarrow M_2$ is a strong map we know by (1.3) that $\rho(M_2) \leqslant \rho(M_1)$ with equality if and only if $M_1 = M_2$.

The *order* of $M_1 \rightarrow M_2$ is $\rho(M_1) - \rho(M_2)$. Maps of order zero are called *trivial*. An *elementary* strong map is a strong map of order 1.

If $M_1 \rightarrow M_2$ is elementary, M_2 is an *elementary quotient* of M_1 and M_1 is an *elementary lift* of M_2.

More generally if $M_1 \to M_2$ is a strong map of order $n \geq 1$, an *elementary lift of M_2 in M_1* is an elementary lift of M_2 which is also a quotient of M_1; in other words a matroid M_3 on S is an elementary lift of M_2 in M_1 if:

(1) every flat of M_2 is a flat of M_3,
(2) every flat of M_3 is a flat of M_1,
(3) $\rho(M_3) = \rho(M_2) + 1$.

In this case $M_1 \to M_2$ factorizes as follows:

$$M_1 \to M_2 = M_1 \to M_3 \to M_2$$

where $M_1 \to M_3$ has order $n - 1$ and $M_3 \to M_2$ is elementary. The reader can verify that one possible M_3 consists of the matroid with flats

(a) all flats of M_2
(b) all flats X of M_1 satisfying $\rho_1 X = \rho_2(\sigma_2 X)$, where σ_2 is the closure function of M_2.

Induction on n shows that $M_1 \to M_2$ admits an *elementary factorization*

(4) $M_1 = N_0 \to N_1 \to N_2 \to \ldots \to N_{n-1} \to N_n = M_2$

where each $N_i \to N_{i+1}$, $0 \leq i \leq n - 1$ is elementary.

Analogously an *elementary quotient* of M_1 over M_2 is an elementary quotient of M_1 which is also a lift of M_2.

Thus in the elementary factorization (4) N_i is an elementary quotient of N_{i-1} over M_2, $1 \leq i \leq n$.

Now by Theorem 2, if $M_1 \to M_2$ is a strong map, $M_2^* \to M_1^*$ is a strong map. We call it the *dual map*. Since

$$\rho(M_1) - \rho(M_2) = \rho(M_2^*) - \rho(M_1^*)$$

dual maps have the same order. In particular

(5) $M_1 \to M_2$ is elementary if and only if $M_2^* \to M_1^*$ is elementary.

Moreover the elementary factorization (4) of $M_1 \to M_2$ leads to the elementary factorization.

(6) $M_2^* = N_n^* \to N_{n-1}^* \to \ldots \to N_1^* \to N_0^* = M_1^*$ of the dual map $M_2^* \to M_1^*$.

We call (6) the *dual factorization* of (4). Note also:

(7) M is an elementary quotient of N over P if and only if M^* is an elementary lift of N^* over P^*.

Consequently, every result on (elementary) quotients gives a dual result on (elementary) lifts.

We now relate the theory of strong maps with the theory of single element extensions, which we illustrated in Example 2.1.

Suppose $M \to N$ is strong and has order k. Let U be the collection of flats of M whose rank "drops the most", that is a flat A of M is a member of U if and only if

$$\rho A - \lambda A = k$$

where ρ, λ are the rank functions of M, N respectively.

LEMMA 1. *U is a modular cut of $\mathscr{L}(M)$.*

Proof. Routine verification.

Now let M_1 be the single element extension of M on $S \cup e_1$ ($e_1 \notin S$) generated by the modular cut U.

LEMMA 2. $M_1 . S = N$.

Proof. Routine application of the rank formula for single element extensions (see proof of Theorem 3.1).

Moreover this contraction map: $\mathscr{L}(M_1) \to \mathscr{L}(N)$ is a strong map of order $k - 1$, by a simple checking of the rank formulae.

Find the modular cut U_1 of flats of M_1 whose rank drops the most in this map $M_1 \to N$ and perform the associated single element extension by e_2, to get a matroid M_2 on $S \cup \{e_1, e_2\}$ such that the contraction map $M_2 \to N$ has order $k - 2$.

Continuing in this way we arrive at the situation where we have a k-element extension M_k of M on $S \cup \{e_1, \ldots, e_k\}$ such that

$$M_k . S = N$$

and clearly since we are performing single element extensions at each stage

$$M_k | S = M.$$

This gives us the factorization $M + N$ into

$$M \xrightarrow{i} M_k \xrightarrow{c} N$$

where i is an injection and c is a contraction.

Finally for those readers who just want a quick proof of the factorization theorem we note the following approach.

Sketch Proof of Factorization Theorem. Suppose that $M \to N$ is a strong map where again M has rank ρ, N has rank λ and S is the underlying set.

Let the order of $M \to N$ be k and let F be a set disjoint from S of cardinality k. Let $E = S \cup F$ and define $\mu: 2^E \to \mathbb{Z}^+$ by for all $X \subseteq E$,

$$\mu X = \min(\rho(X \cap S) + |X \backslash S|, \lambda(X \cap S) + \rho S - \lambda S)$$

The reader can check that μ is the rank function of a matroid on E (use Lemma 2.1). Call this matroid $H = H(M, N)$. It is very easy to check that

$$H \,|\, S = M, \quad H \,.\, S = N.$$

This gives the factorization theorem

where i is an injection, c a contraction. ∎

This is the slickest direct proof, it does not prove quite as much as Theorem 2.3 and does have the disadvantage that the function μ appears out of nowhere.

EXERCISES 4

1. Let M_1, M_2 be matroids on S. Prove that the following statements are equivalent:
    (a) the identity map is a weak map between M_1 and M_2.
    (b) each circuit of M_1 contains a circuit of M_2.
    (c) each independent set of M_2 is independent in M_1.

2. If M_1, M_2 above have the same rank prove that $M_1 \to M_2$ is a weak map if and only if $M_2^* \to M_1^*$ is a weak map.

5. THE SCUM THEOREM

We now use the theory of strong maps to prove a highly important result which has been aptly christened by Crapo and Rota [70] the "scum theorem". Loosely, it says that anything which occurs in a geometric lattice must also occur at the top of the lattice.

THEOREM 1. *Let $\mathcal{L}$ be a geometric lattice, let M_1 be any simple minor of its associated simple matroid $M(\mathcal{L})$. Then there exists an interval $[A, I]$ of $\mathcal{L}$ and a strong map from $\mathcal{L}(M_1)$ into $[A, I]$ which is an isomorphism.*

Before proving this theorem we give one application. We know from Theorem 10.2.1 that a matroid M is binary if and only if it does not contain as a minor the uniform matroid $U_{2,4}$. Thus by the scum theorem we know that M is binary if and only if the lattice of $U_{2,4}$ does not occur at the "top" of the lattice $\mathscr{L}(M)$. That is we have:

THEOREM 2. *A geometric lattice $\mathscr{L}$ is binary if and only if there do not exist copoints (or hyperplanes) H_1, H_2, H_3 such that $\rho(H_1 \cap H_2 \cap H_3) = \rho\mathscr{L} - 2$.*

Analogous theorems exist for a geometric lattice to be graphic, regular or planar.

Proof of the scum theorem:

Consider any minor M_1 of $M = M(\mathscr{L})$.

Let $\mathscr{L}(M_1)$ be isomorphic to the interval $[A, B]$ of $\mathscr{L}(M)$. Consider the interval $[A, I]$ of $\mathscr{L}$ and choose Z to be a minimal relative complement of B in this interval. (We know by Theorem 3.3.3 that Z exists.) Also we know that

$$\rho I - \rho Z = \rho B - \rho A,$$

so if we look at the strong map $\phi: [A, B] \to [Z, I]$ defined by

$$\phi(X) = X \vee Z \qquad (A \leqslant X \leqslant B),$$

then ϕ is a strong map which does not reduce the rank. Hence by (1.1) $[Z, I]$ must be isomorphic to $[A, B]$. Thus we have found an upper interval which is isomorphic to $[A, B]$ and have proved the theorem. ∎

It is worth emphasizing that the scum theorem does *not* say that if M_1 is a minor of a matroid M on S there exists a flat $A \leqslant I$ such that $\mathscr{L}(M_1)$ is isomorphic to the sublattice $[A, I]$. To illustrate this, consider the Steiner system $S(3, 6, 22)$ studied in Section 12.6.

If M denotes this matroid and H is any hyperplane we know that since M is paving, $M|H$ is the uniform matroid $U_{3,6}$. Now for any 1-flat x, the interval $[x, S]$ of $\mathscr{L}(M)$ is the projective plane $PG(2, 4)$ which is clearly not isomorphic to $U_{3,6}$. What the scum theorem does say is that $\exists$ a 1–1 order preserving map $U_{3,6} \to PG(2, 4)$. In other words $\mathscr{L}(PG(2, 4))$ contains $\mathscr{L}(U_{3,6})$.

EXERCISES 5

1. Prove the following "matroid version" of the scum theorem. Let M be a matroid on S and let $T \subseteq A \subseteq S$. Suppose that $M_1 = (M \cdot A)|T$ is a simple minor of S. Then there exists B with $T \subseteq B \subseteq A$ such that

   (a) $\rho((M \cdot B)|T) = \rho(M \cdot B)$.
   (b) the contraction map $c: M \cdot A \to M \cdot B$ induces an isomorphism between $(M \cdot A)|T$ and $(M \cdot B)|T$.

2. Exhibit the inclusion of $\mathscr{L}(U_{3,6})$ in $\mathscr{L}(PG(2, 4))$.

6. ERECTIONS OF MATROIDS

A natural and very interesting combinatorial question about a geometric lattice is whether or not it is possible to erect the associated geometry into one higher dimension while preserving its lower dimensional structure. This question has been extensively studied by Crapo in a series of papers [70, 70a, 71].

Given a matroid M of rank $r + 1 > 1$, those flats of rank $<r$ form a geometric lattice, the lattice of the truncation of M at r. If a matroid N is isomorphic to the truncation of a matroid M we say that M is an *erection* of N. There is no loss of generality in assuming that this isomorphism is the identity function on the elements of the ground set so that the geometric lattice $\mathscr{L}(M)$ results from the lattice $\mathscr{L}(N)$ of N by the insertion of a level of new copoints above the original copoints of $\mathscr{L}(N)$ and beneath the unit element of the lattice. The *trivial erection* of M is M itself.

Example. Let N be the uniform matroid $U_{r,n}$ of rank $r > 1$ on a set S of n elements. It is not difficult to prove that the erections of N are the paving matroids of rank $r + 1$ on S; that i! M is an erection of N if and only if its set $\mathscr{H}$ of hyperplanes is the set of blocks of an r-paving on S (see Section 2.3), that is

(1)   each block contains at least r points
(2)   each r-subset of S is contained in a unique block.

The main result of Crapo [70] is the extension of the basic idea illustrated by the above example to arbitrary (non-uniform) matroids.

If M is a matroid on S, and k is a non-negative integer we say that a subset $X \subset S$ is *k-closed* in M if and only if it contains the closures of all its j-element subsets, for all $j \leqslant k$. The *k-closure* of a set X is the smallest k-closed set containing the set X. (It is easy to see that k-closure is well defined since the intersection of k-closed sets is k-closed). We first prove the following easy lemma.

(3)   A subset D of S is k-closed in M if and only if $D \cap C$ is a flat for all k-flats C of M.

Proof. Let D be k-closed in M and let C be a k-flat. The intersection $D \cap C$ is contained in C and has rank $\leqslant k$. Let B be any independent subset of $D \cap C$ such that $\sigma B = \sigma(D \cap C)$. Since $|B| \leqslant k$, $\sigma B \subseteq D$. Since C is a flat, $\sigma B \subseteq C$, thus $\sigma(D \cap C) = \sigma(B) \subseteq C \cap D$, and $C \cap D$ is closed.

Assume conversely that D is a set of points such that $D \cap C$ is closed for all k flats C. For any k-element subset $E \subseteq D$, the closure σE is a flat of rank $\leqslant k$. Now $E \subseteq D \cap \sigma E \subseteq \sigma E$, $D \cap \sigma E$ is closed, and σE is the smallest closed set containing E. Thus $D \cap \sigma E = \sigma E$, $\sigma E \subseteq D$ and D is k-closed.

As an immediate corollary we have:

(4)   For any matroid, every flat of rank k is j-closed for $j \leqslant k$.

We now prove the main theorem of Crapo [70] which generalizes in a very natural way the ideas of Example 1.

THEOREM 1. *Let M be a simple matroid of rank r on S. A set $\mathscr{F}$ of subsets of S (called "blocks") is the set of hyperplanes of an erection of M if and only if:*

(a)   *each block spans M*
(b)   *each block is $(r - 1)$-closed*
(c)   *each base of M is contained in a unique block.*

Proof. If $\mathscr{F}$ is the set of copoints of an erection N of M then the copoints of N span in the truncation M. The intersection of any copoint of N with a coline of N (copoint of M) must be a flat of both N and M. Thus by (4) any member of $\mathscr{F}$ must be $(r - 1)$-closed. Any base B of M is of rank r in N and must span a copoint; having rank r in a matroid of rank $r + 1$ it cannot be contained in more than one copoint. Thus the conditions (a) to (c) are necessary.

Conversely let $\mathscr{F}$ satisfy (a) to (c). We prove that the set $\mathscr{L}$ of flats of M together with $\mathscr{F}$ form the flats of a geometric lattice of rank $r + 1$, which has $\mathscr{F}$ as its set of copoints.

To prove that $\mathscr{F} \cup \mathscr{L}$ is a lattice (with the inclusion ordering) we show that $\mathscr{L} \cup \mathscr{F}$ is the set of all intersections of blocks of $\mathscr{F}$, (where by convention we assume S is the empty intersection of blocks).

Suppose $|I| \geqslant 2$ and

$$A = \bigcap_{i \in I} F_i \qquad F_i \in \mathscr{F}.$$

Since no base of M is contained in more than one F_i $\rho A < r$ and A is contained in some copoint C of M. For any block F_i, $C \cap F_i$ is a flat of M because the blocks are $(r - 1)$-closed. Thus A can be written as the intersection of flats of M,

$$A = \bigcap_{i \in I} C \bigcap F_i$$

and thus A is a flat of M.

Thus the intersections of the members of $\mathscr{F}$ are contained in $\mathscr{L} \cup \mathscr{F}$.

Now take a copoint H of M and let B be a base of H.
Let $\{T_i : i \in J\}$ be the collection of bases of M which contain B and let $\{H_i, i \in J\}$ be the corresponding blocks which contain the bases T_i. Since the blocks are $(r - 1)$-closed

$$H = \sigma B \subseteq H_i \qquad (i \in J).$$

If there is more than one such block H_i, then as proven above $\bigcap_{i \in J} H_i$ is a proper flat of M which contains the copoint H, and thus must equal H.

If there is a single such block H_i, say H_1, then H_1 contains:

(a) all atoms dependent upon B, because H_1 is $(r - 1)$-closed;
(b) all atoms independent of B, because they, together with B, form the bases T_i all of which are in H_1.

Thus if $\mathscr{F}$ is not the trivial erection (yielding M itself), then all the flats of M are expressible as the intersections of blocks of $\mathscr{F}$. Thus $\mathscr{L} \cup \mathscr{F}$ with the inclusion ordering is a lattice, which it is clear has the Jordan–Dedekind chain condition and thus $\mathscr{F}$ must be its set of copoints.

To show that this lattice is geometric it suffices to show the submodularity of the height function for flats of rank $r - 1$.

Take C, D, any two copoints of M which contain a common coline E of M. Choose a base B for E and let p, q be atoms such that

$$E \vee p = C, \qquad E \vee q = D.$$

Then $B \cup p \cup q$ is a base for M and by hypothesis is in some block $F \in \mathscr{F}$. Since F is $(r - 1)$-closed, it contains both

$$C = \sigma(B \cup p) \quad \text{and} \quad D = \sigma(B \cup q).$$

Thus semimodularity is preserved in forming the lattice $\mathscr{L} \cup \mathscr{F}$ and $\mathscr{L} \cup \mathscr{F}$ is a geometric lattice with r-truncation $\mathscr{L}$. ∎

An interesting problem which is connected with the erection problem is finding what Crapo [71] calls the Dilworth completions of a geometric lattice (see Dilworth [44]).

Let $\mathscr{L}$ be a geometric lattice of rank $r > k$. Now let $Q_k(\mathscr{L})$ be the lattice obtained from $\mathscr{L}$ by identifying the flats of $\mathscr{L}$ of rank $\leqslant k$. In general $Q_k(\mathscr{L})$ will be a non-geometric lattice, called the *lower truncation* of $\mathscr{L}$ at k. The problem is to find a geometric lattice $D_k(\mathscr{L})$ of rank equal to the height of $Q_k(\mathscr{L})$ and which has $Q_k(\mathscr{L})$ as a sublattice. Any such geometric lattice $D_k(\mathscr{L})$ is obtained from $Q_k(\mathscr{L})$ by adding appropriate flats. It is called a *Dilworth completion* of $Q_k(\mathscr{L})$. A method of constructing Dilworth completions is given by Crapo [71], we refer the reader to this for details.

EXERCISES 6

1. Find the Dilworth completions of the lattices $Q_k(B_n)$ where B_n is the Boolean algebra with n atoms, and $1 \leqslant k \leqslant n$.

2. Prove that the uniform matroid $U_{3,7}$ has 171 distinct erections. (Crapo [70]).

3. Does the Vamos matroid have an erection?

○4. A flat F in a matroid is *essential* if the interval $[\varnothing, F]$ has a non-trivial erection. Otherwise the flat is *predictable*. It can be shown that the essential flats together with their ranks uniquely determine the matroid. Find a way of recognizing the essential flats of a matroid. (Crapo [71]).

5. Find a rank 3 matroid which has no erection.

7. THE AUTOMORPHISM GROUP OF A MATROID

We close this chapter with a brief account of the work done on the group of automorphisms of a matroid.

Let M be a matroid on S. A permutation $\pi: S \to S$ is an *automorphism* of M if πX is independent in M if and only if X is independent in M.

The (*automorphism*) *group* of M is the collection of automorphisms of M with the obvious operation of composition. It will be denoted by $A(M)$.

Let G be a graph. The *point automorphism group* of G, $A_p(G)$ is the collection of mappings $\pi: V(G) \to V(G)$ such that two vertices u, v are joined by an edge if and only if $\pi(u)$, $\pi(v)$ are adjacent in G.

The *cycle automorphism group* $A_c(G)$ is the collection of permutations $\phi: E(G) \to E(G)$ such that the set X of edges of G is a cycle if and only if ϕX is a cycle.

Note that $A_p(G)$ need not be isomorphic to $A_c(G)$.

Example. Let G be a simple cycle C_n of length n. Then $A_p(C_n)$ is the dihedral group of order n but $A_c(C_n)$ is the symmetric group of order n, S_n.

However the following lemma is obvious.

LEMMA. *For any graph G, the automorphism group of its cycle matroid $M(G)$ is the same as $A_c(G)$.*

Frucht [38] proved the following basic result.

THEOREM 1. *Given any finite group H there exists a finite graph G such that $A_p(G)$ is isomorphic to H.*

Proof. Harary [69] Chapter 14. ∎

Now from Theorem 6.1.1, we know that if G is 3-connected any cycle automorphism of G is induced by a vertex automorphism.

In other words, if G is 3-connected $A_p(G) \simeq A_c(G)$.
We now use the following result of Sabidussi [57].

(1) Given a positive integer k and finite group H there exists a finite k-connected graph G with $A_p(G) \simeq H$.

Combining these results we can prove our main theorem.

THEOREM 2. *Given any finite group H there exists a matroid M which has automorphism group $A(M)$ isomorphic to H.*

Proof. Take a finite group H, we can find a 3-connected graph G with $A_p(G) \simeq H$. By Whitney's Theorem 6.1.1 $A_p(G) \simeq A_c(G)$ which is isomorphic to $A(M(G))$. ∎

We now examine some specific finite groups and consider whether or not they are the automorphism group of any matroid on a given number of elements.
It is very easy to prove:

(2) If M is a matroid on an n-element set then it has automorphism group equal to the symmetric group S_n if and only if M is a uniform matroid $U_{r,n}$ for some r.

Similarly using the fact that the alternating group A_n is $(n/2)$-transitive and that $A(M) = A(M^*)$ it is easy to see:

(3) For $n > 2$, there is no matroid M on an n-element set with $A(M)$ the alternating group A_n.

Now for the cyclic group C_n of order n it is well known and easy to prove that for $n > 3$ there can be no graph G on n vertices (or n edges) with $A_p(G) = C_n$. The following result of Piff [72] emphasizes the fact that graphic matroids form only a small subclass of matroids.
Call a matroid on n elements *cyclic* if it has automorphism group C_n.

THEOREM 3. *For $3 \leqslant n < 8$ there does not exist a cyclic matroid on n elements. For $n \geqslant 8$ there exists a cyclic matroid on n elements.*

Proof. This is by construction. We sketch the basic ideas. For $n = 8$ let α be the permutation

$$\begin{pmatrix} 1 & 2 & 3 & 4 & 5 & 6 & 7 & 8 \\ 2 & 3 & 4 & 5 & 6 & 7 & 8 & 1 \end{pmatrix}$$

and let M_8 be the paving matroid of rank 4 on $\{1, 2, \ldots, 8\}$ which has as nontrivial hyperplanes $H = \{1, 2, 3, 5\}$, $\alpha H, \ldots, \alpha^7 H$. The reader can check that $A(M_8) = C_8$. For $n \geq 9$, take M_n to be the rank 3 matroid on $\{1, 2, \ldots, n\}$ with hyperplanes $H = \{1, 2, 4\}$ and all cyclic translations $\alpha^k H$. $(1 \leq k \leq n)$

Using the fact that $A(M) = A(M^*)$ it is routine to check that $A(M) \neq C_n$ for $3 \leq n \leq 7$. ∎

Piff [72] has made a more detailed study of properties of the automorphism group. We state briefly his main results.

(4) For any finite group H there exists a graphic, non-transversal connected matroid M with $A(M) \simeq H$.

(5) For any finite group H there exists a non-binary transversal matroid with automorphism group isomorphic to H.

These results show that it is impossible to characterize transversal or binary matroids by their automorphism groups.

From Theorem 2 we may also deduce very easily:

(6) For any finite group H there exists a geometric lattice whose automorphism group is isomorphic to H.

This last result complements a result of Birkhoff [46] who proved that for any finite group H there exists a distributive lattice with automorphism group H.

EXERCISES 7

1. Prove that the automorphism group of a matroid equals that of its dual matroid.

2. Prove that there exists no cyclic transversal matroid on a set S with $|S|$ a prime power greater than 2. (Piff [72])

3. Find a matroid which has the identity group as its automorphism group.

○4. Is it true that for any group H there is a matroid design with automorphism group isomorphic to H?

○5. Prove that for any group H there is a paving matroid with automorphism group isomorphic to H?

NOTES ON CHAPTER 17

The basic papers on strong maps are those of Crapo [67] and Higgs [68]. A most valuable survey of the area covered in Sections 1–6 is the unpublished survey paper of Crapo [71]. The treatment of the factorization theorems given here is a modification by A. W. Ingleton of the proof originally given in Crapo and Rota [70].

The content of Section 3 is based on the paper by Dowling and Kelly [72]. In another paper [74] they use the theory of strong maps to give an alternative characterization of transversal matroids. The scum theorem seems to have first appeared in Crapo and Rota [70], but they have pointed out to me that its proof is due to D. Higgs.

The theory of single element extensions and erections is mainly due to Crapo [65], [70], [70a], see also White [74], Kennedy [73], Brylawski [72a], Las Vergnas [75b].

Crapo [68a] has also made a study of cover preserving maps which are infimum preserving on modular pairs—he calls them *comaps* and uses it to piece together matroids in a way analogous to sticking graphs together at their vertices. See also Sachs [71], Graves [71], [71a].

The final section on groups and matroids is based on the paper of Harary, Piff and Welsh [72]—though the bulk of that paper deals with infinite structures. The main result, Theorem 2 has also been found independently by Gallant [70], [71] and implicitly by Doyen [68] though in this last paper the concept of matroid is not mentioned!

Las Vergnas [75a] has recently studied the relation between extensions of matroids and matroids induced by bipartite graphs (see Chapter 8). This is related to the work of Dowling and Kelly [74] mentioned above.

CHAPTER 18

Convex Polytopes Associated with Matroids

1. CONVEX POLYTOPES AND LINEAR PROGRAMMING

This section is a brief resume of those parts of convex set theory needed subsequently; for further details see Grünbaum [67] or Stoer and Witzgall [70].

If S is a finite set, we let $\mathbb{R}^S(\mathbb{R}^S_+)$ denote the space of real valued (non-negative) row vectors with coordinates indexed by S. For each $x \in \mathbb{R}^S$ and $e \in S$ denote the eth coordinate of x by $x(e)$. For $x, y \in \mathbb{R}^S$ we write $x \geqslant y$ if $x(e) \geqslant y(e)$ for $\forall e \in S$, and call y a *subvector* of x. This induces a partial order on $\mathbb{R}^S$ and $<, \leqslant, >$ are now defined in the obvious way. For $x \in \mathbb{R}^S$ and $A \subseteq S$ we define

$$x(A) = \sum_{e \in A} x(e)$$

and call the *modulus* $|x|$ of x the quantity

$$|x| = x(S) = \sum_{e \in S} |x(e)|.$$

For $x, y \in \mathbb{R}^S_+$ and $e \in S$, we define

$$(x \wedge y)(e) = \min(x(e), y(e))$$

$$(x \vee y)(e) = \max(x(e), y(e))$$

and thus define $x \wedge y$, $x \vee y$ in the obvious way.

A subset $X \subseteq \mathbb{R}^S$ is *convex* if $x, y \in X$ implies that for all $\lambda, 0 \leqslant \lambda \leqslant 1$,

$$\lambda x + (1 - \lambda)y \in X.$$

The *convex hull* of X, denoted by co(X) is the intersection of all convex sets containing X, or alternatively the smallest convex set containing X. A *hyperplane* of $\mathbb{R}^S$ is a subset $H \subseteq \mathbb{R}^S$ such that $\exists c \in \mathbb{R}^S \backslash \{0\}, d \in R$ such that

$$H = \{x \in \mathbb{R}^S : cx' = d\}.$$

Clearly H is uniquely determined by the pair (c, d) and is a maximal proper affine subspace of $\mathbb{R}^S$. Its *equation* is

$$cx' = d.$$

Associated with such a hyperplane are two *half spaces*

$$H^+ = \{x : cx' \geqslant d\},$$

$$H^- = \{x : cx' \leqslant d\}.$$

Clearly a half space is closed and convex. Thus the intersection of half spaces is closed and convex.

A *convex polytope* is a bounded region of $\mathbb{R}^S$ which can be expressed as the intersection of a finite number of half spaces.

Given a polytope $P \subseteq \mathbb{R}^S$ which is defined by

$$(\alpha) \qquad P = \{x \in \mathbb{R}^S : c_i x' \leqslant d_i : i \in T\}$$

we call the *face* associated with $J \subseteq T$, the subset $P(J)$ of P defined by

$$P(J) = \{x \in P : c_i x' = d_i, i \in J, c_i x' < d_i, i \in T \backslash J\}.$$

The *rank* of the face $P(J)$ is defined by $|S| - d(C^J)$ where $d(C^J)$ is the rank of the matrix C^J which has row vectors $\{c_i : i \in J\}$. A *vertex* of P is a face of zero rank. Thus to find the vertices of the convex polytope P we would have to solve the equations

$$c_i x' = d_i \qquad (i \in J)$$

for *all* $J \subseteq T$ such that $|J| = |S|$ and then check which of these solutions were members of P.

A basic theorem of convex polytopes is:

THEOREM 1. *A convex polytope P has only a finite number of vertices; moreover it can be expressed as the convex hull of its set of vertices.*

Conversely given any finite set X of points of $\mathbb{R}^S$ the convex hull of X is a convex polytope with vertex set a subset of X. This convex polytope can be expressed in the form (α) where the hyperplanes $c_i x' = d_i$ are its maximal proper faces.

Linear programming

We present the bare outlines of linear programming theory.

Let S, T be finite sets. Let $A = \{a_{ij} : i \in T, j \in S\}$ be a matrix with $a_{ij} \in \mathbb{R}$. Let $b \in \mathbb{R}^T$, $c \in \mathbb{R}^S$.

A (*primal*) *linear programme* is
$$\text{maximize } cx'$$

for $x \in \mathbb{R}^S$ satisfying

(1) $$Ax' \leqslant b'$$

(2) $$x \geqslant 0.$$

The *dual linear programme* is
$$\text{minimize } by'$$

for $y \in \mathbb{R}^T$ satisfying

(3) $$A'y' \geqslant c'$$

(4) $$y \geqslant 0.$$

A vector x satisfying (1) and (2) is a *feasible solution* to the primal problem. A vector y satisfying (3) and (4) is a *feasible dual solution*. A feasible primal solution which maximizes cx' is an *optimal primal* solution. An *optimal dual* solution is defined analogously.

A fundamental theorem of linear programming is:

THEOREM 2. *For any linear programming maximization problem exactly one of the following situations occurs.*

(i)   *There exists no feasible solution.*
(ii)   *For any $\alpha \in R$ there is a feasible solution x such that $cx' > \alpha$.*
(iii) *There is an optimal (feasible) solution.*

The next theorems explain the relationship between the solutions to the primal and dual problems.

THEOREM 3 (Weak Duality Theorem). *If x is a feasible primal solution and y is a feasible dual solution then*
$$cx' \leqslant by'.$$

THEOREM 4 (Strong Duality Theorem). *If there is a feasible primal solution and an upper bound for cx' over all feasible primal solutions x, then there is an optimal primal solution u and an optimal dual solution v and $cu' = bv'$.*

We now state two results which are intuitively obvious in 2 or 3 dimensions.

THEOREM 5.
(a)   *Let P be a convex polytope in $\mathbb{R}^S$. Then for any $c \in \mathbb{R}^S$ there exists a vertex v of P which maximizes cx' over P.*

(b)   *Let P be a convex polytope in $\mathbb{R}^S$ and let v by any vertex of P. Then there exists $c \in \mathbb{R}^S$ such that v is the unique member of P maximizing cx' over P.*

EXERCISES 1

1. Recall from Chapter 10 that a matrix A is totally unimodular if all its subdeterminants take the values 0, ± 1. If b is an integer vector prove that the vertices of the convex polytope

$$Ax \leqslant b$$
$$x \geqslant 0$$

are integer valued when A is totally unimodular.

2. With any polytope P in $\mathbb{R}^n$ we can associate a *face lattice* $\mathscr{L}(P)$ whose elements are the faces of P, ordered by inclusion.

   Prove that the Möbius function of the face lattice of a convex polytope takes alternately the values $+1$ and -1. More precisely prove that

$$\mu(o, F) = (-1)^{f-1}$$

where f is the dimension of the face F and o is the null face. (Rota [71])

2. POLYMATROIDS

A polymatroid can be defined in several equivalent ways. Here we give three definitions. Definition 1 is the original definition of Edmonds [70]; definitions 2 and 3 are essentially due to Dunstan [73] and Woodall [73], The main result of this section will be showing the equivalence of these three definitions.

Definition 1. A *polymatroid* $\mathbb{P}$ is a pair (S, P) where S, the *ground set* is a non-empty, finite set and P, the set of *independent vectors* of $\mathbb{P}$ is a non-empty compact subset of $\mathbb{R}^S_+$ such that:

(P1)   every subvector of an independent vector is independent,

(P2)   for every vector $a \in \mathbb{R}^S_+$, every maximal independent subvector x of a has the same modulus $r(a)$, the *vector rank* of a in $\mathbb{P}$.

In the above definition "maximal" has its obvious meaning that there exists no $y > x$ having the properties of x; since P is compact it is well defined.

Example. In two dimensions $(|S| = 2)$, the only possible types of polymatroid are the following shaded areas of R^2 in Fig. 1, and the degenerate cases.

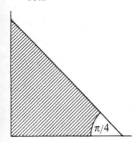 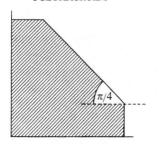 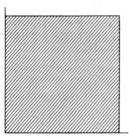

Figure 1

The most general polymatroid in 3 dimensions is of the following type

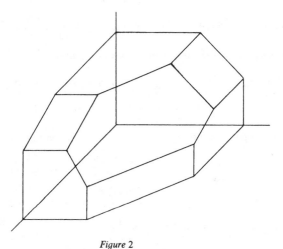

Figure 2

If $\mathbb{P} = (S, P)$ is a polymatroid, and $a \in \mathbb{R}_+^S$ we say a vector b is a *base* of a if it is a maximal independent subvector of a.

Definition 2. A *polymatroid* $\mathbb{P}$ is a pair (S, P) where S the *ground set* is a non-empty finite set and P, the set of *independent vectors* of $\mathbb{P}$ is a non-empty compact subset of $\mathbb{R}_+^S$ such that (P1) holds and also:

(P2)′   If $u, v, \in P$ and $|v| > |u|$, then there is a vector $w \in P$ such that

$$u < w \leqslant u \vee v$$

Axiom (P2)′ is the "exchange" axiom for polymatroids, it corresponds (though not exactly) to the exchange axiom for matroids.

Definition 3. A *polymatroid* $\mathbb{P}$ is a pair (S, ρ) where S, the *ground set* is a non-empty finite set and ρ, the *ground set rank function* is a function: $2^S \to \mathbb{R}^+$ satisfying

$$A \subseteq B \subseteq S \Rightarrow \rho A \leqslant \rho B$$

$$A, B, \subseteq S \Rightarrow \rho A + \rho B \geqslant \rho(A \cup B) + \rho(A \cap B)$$

$$\rho(\varnothing) = 0;$$

and the vectors $x \in \mathbb{R}_+^S$ such that $x(A) \leqslant \rho A$ for all $A \subseteq S$ are the *independent vectors* of $\mathbb{P}$. For each vector a in $\mathbb{R}_+^S$, the *vector rank* $r(a)$ of a is given by

$$r(a) = \min (a(X) + \rho(S \backslash X): X \subseteq S).$$

THEOREM 1. *The three definitions 1–3 are equivalent.*

We prove this in the following series of lemmas.

LEMMA 1. *The vector rank function r of Definition 1 satisfies for all $u, v \in \mathbb{R}^S$,*

$$r(u) + r(v) \geqslant r(u \vee v) + r(u \wedge v).$$

Proof. Let a be a base of $u \wedge v$. By (P2), $\exists b$, independent in P, satisfying

$$a \leqslant b \leqslant u \vee v$$

and

$$r(b) = |b| = r(u \vee v).$$

Since $a = b \wedge (u \wedge v)$ we get

$$a + b = b \wedge u + b \wedge v.$$

But $b \wedge u, b \wedge v$ are independent subvectors of u, v, respectively. Thus

$$r(u \wedge v) + r(u \vee v) = |a| + |b|$$

$$= |b \wedge u| + |b \wedge v|$$

$$\leqslant r(u) + r(v).$$

LEMMA 2. *Let P and ρ be defined as in Definition 3. Let $v \in P$ and define*

$$S(v) = \{X \subseteq S : v(X) = \rho(X)\}.$$

Then $S(v)$ is closed under unions and intersections.

Proof. Let X, $Y \in S(v)$. By the submodularity of ρ we have

$$\rho(X \cup Y) + \rho(X \cap Y) \leqslant \rho X + \rho Y$$
$$= v(X) + v(Y) = v(X \cap Y) + v(X \cup Y),$$
$$\leqslant \rho(X \cap Y) + \rho(X \cup Y),$$

since $v \in P$. Hence there must be equality throughout, so that, $X \cap Y$ and $X \cup Y \in S(v)$.

LEMMA 3. *Definition 2 $\Rightarrow$ Definition 1.*

Proof. Let P be defined by definition (P2)'. Let $z \in \mathbb{R}_+^S$ and let u, v be bases of z. If $|v| > |u|$, then by the exchange axiom, $\exists w \in P$ with

$$u < w \leqslant u \vee v \leqslant z$$

which contradicts the maximality of u. Thus all bases of z have the same modulus.

LEMMA 4. *Definition 1 $\Rightarrow$ Definition 3.*

Proof. Let $\mathbb{P} = (S, P)$ be defined by Definition 1. Define a vector $v \in \mathbb{R}^S$ to be the vector with e-coordinate r where $r = \max\{|x| : x \in P\}$. Then define $\rho : 2^S \to \mathbb{R}$ by

$$\rho X = v(X) \qquad (X \subseteq S).$$

The submodularity of ρ follows from that of r (Proved in Lemma 1 above). Clearly ρ is non-negative and increasing with $\rho(\varnothing) = 0$.

Now by the definition of ρ, it is clear that P (as defined in Definition 1) satisfies

$$(1) \qquad P \subseteq \{v \in \mathbb{R}_+^S : v(X) \leqslant \rho X \; \forall X \subseteq S\}.$$

Call the region defined in the right-hand side of (1) P_1. We will show $P_1 \subseteq P$ and thus complete the proof of Lemma 4.

Let $v \in P_1$. If $v \notin P$, choose u to be a base of v which maximizes $|N(u)|$ where

$$N(u) = \{e \in S : u(e) < v(e)\}.$$

Now,

$$u \leqslant w = (u + v)/2 \leqslant v$$

so that u is a base of w. Thus $r(v) = r(w)$ and every base of w is also one of v. For $X \subseteq S$ define $v_X(e) = v(e)$, $e \in X$ and $v_X(e) = 0$, $e \in S \backslash X$; then since

$$u(N(u)) < w(N(u)) \leqslant \rho(N(u)),$$

it follows that $u \wedge v_{N(u)}$ is not a base of $w \wedge v_{N(u)}$. Extend it to a base of $w \wedge v_{N(u)}$ and extend this in turn to a base $\hat{u}$ of w. Since $\hat{u}(N(u)) > u(N(u))$ and $|\hat{u}| = |u|$, $\hat{u}(e) < u(e)$ for some $e \notin N(u)$. Thus $\hat{u}$ is a base of v which violates the maximality of $|N(u)|$. This contradiction proves $v \in P$ which shows $P_1 = P$ and the lemma is proved.

LEMMA 5. *Definition 3 $\Rightarrow$ Definition 2.*

Proof

Let P and ρ be as in Definition 3. The only property in Definition 2 which is nontrivial to derive is the exchange axiom. So suppose there exist u and $v \in P$ for which the exchange axiom fails, with $|v| > |u|$.

Let $V = \{e \in S: v(e) > u(e)\}$, and let $e \in V$. Then there is a set $U_e \subseteq S$ such that $e \in U_e$ and

$$u(U_e) = \rho(U_e)$$

for otherwise we could increase $u(e)$ by say ε without violating the constraints $\{u(X) \leqslant \rho X, X \subseteq S\}$.

Let U be a maximal subset of S such that $u(U) = \rho(U)$. Then by Lemma 2,

$$u(U \cup U_e) = \rho(U \cup U_e)$$

and if $e \notin U$, this contradicts the maximality of U. Thus $e \in U$. But since this holds for arbitrary $e \in V$, $V \subseteq U$. But then

$$\rho U = u(U) < v(U)$$

which contradicts the supposition that $v \in P$. Hence the exchange axiom holds and the lemma is proved. ∎

Lemmas 3–5 prove Theorem 1. ∎

A consequence of Definition 2 is:

THEOREM 2. *The independent vectors of a polymatroid form a convex polyhedron in $\mathbb{R}_+^S$.*

We call this polyhedron the *independence polytope* of the polymatroid. If $\mathbb{P}$ is a polymatroid with ground set S or *on* S, we write $\mathbb{P} = (S, P, \rho)$ to indicate that $\mathbb{P}$ has independence polytope P and ground set rank function ρ.

For any compact subset $U \subseteq \mathbb{R}_+^S$ we say b is a *base* of U if it is a maximal subvector of U. An element is a *base* of the polymatroid $\mathbb{P} = (S, P, \rho)$ if it is a base of P. Using Theorem 1 the reader will be able to prove the following theorem, which gives axioms for a polymatroid in terms of its bases.

THEOREM 3. *If P is a compact subset of $\mathbb{R}_+^S$, P is the independence polytope* *of a polymatroid* $\mathbb{P} = (S, P, \rho)$ *if and only if the following conditions hold:*

(B1)   *if* $x \in P$ *and* $y \leqslant x$ *then* $y \in P$;

(B2)   *if* b, c *are bases of* P *and* d *is such that* $b \wedge c < d < b$ *there exists* e, *with* $d \wedge c < e \leqslant c$ *such that* $d \vee e$ *is a base of* P;

(B3)   *all the bases of* P *have the same modulus.*

These axioms illustrate the difference between matroids and polymatroids. First note that (B1) and (B2) alone would provide an analogue of the matroid base axioms. However consider the shaded region P of Fig. 1. P then satisfies (B1), (B2) but not (B3).

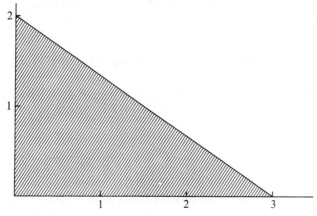

Figure 1

Similarly the shaded region P of Fig. 2 is an example of a set which satisfies (B1) and (B3) but not (B2).

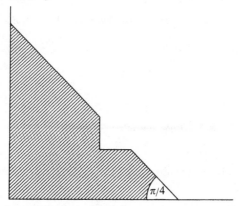

Figure 2

EXERCISES 2

1. Find a convex polytope which is not the independence polytope of a polymatroid but which satisfies (B2), (B3) of Theorem 3.

2. Show that in general the intersection of the independence polytopes of two polymatroids is a convex polytope which is not an independence polytope.

3. If P is the independence polytope of a polymatroid in $\mathbb{R}^n$ is its projection into $\mathbb{R}^{n-1}$ the independence polytope of a polymatroid?

°4. As far as I know there are no known necessary and sufficient conditions for a lattice to be the face lattice of a convex polytope. Are there any such conditions for a lattice to be the face lattice of an independence polytope?

3. POLYMATROIDS AND SUBMODULAR SET FUNCTIONS

From Definition 2.3 we know that if $\mathbb{P} = (S, P, \rho)$ is a polymatroid in $\mathbb{R}_+^S$, its independence polytope is the region of $\mathbb{R}_+^S$ defined by the intersection of the half spaces

(1) $x(A) \leqslant \rho A \qquad A \subseteq S.$

We also know that ρ is a non-decreasing submodular set function.

In this section we prove conversely that polymatroids can be obtained in this way from arbitrary non-negative submodular functions.

THEOREM 1. *Let $\mathscr{L}$ be a lattice of subsets of the finite set S ordered by inclusion and with the lattice operation of meet, set intersection. Then if $\mu: \mathscr{L} \to \mathbb{R}^+$ is submodular on $\mathscr{L}$, that is*

$$\mu(A \vee B) + \mu(A \wedge B) \leqslant \mu A + \mu B \qquad A, B \in \mathscr{L},$$

the region $P(S, \mu)$ of $\mathbb{R}_+^S$ defined by the intersection of the half spaces

$$x(A) \leqslant \mu A \qquad A \in \mathscr{L},$$

$$x \geqslant 0,$$

is the independence polytope of a polymatroid $\mathbb{P}$.

Example. Let $S = \{1, 2, 3\}$.

Let $\mathscr{L}$ be the lattice of subsets $\varnothing, \{1\}, \{2\}, \{3\}, \{1, 2, 3\}$; and let $\mu: \mathscr{L} \to \mathbb{R}^+$ be defined by

$$\mu(\varnothing) = 0, \quad \mu\{1\} = 3, \quad \mu\{2\} = 5, \quad \mu\{3\} = 4, \quad \mu\{1, 2, 3\} = 6.$$

Then μ is submodular on $\mathcal{L}$ and thus the polyhedron

$$
\begin{aligned}
x_1 && \leqslant 3 \\
x_2 && \leqslant 5 \\
x_3 &\leqslant 4 \\
x_1 + x_2 + x_3 &\leqslant 6 \\
x_i &\geqslant 0
\end{aligned}
$$

is a polymatroid in $\mathbb{R}^3$.

Proof of Theorem 1. We shall show that $P(S, \mu)$ satisfies the axioms (P1) and (P2) of Definition 1.

Clearly it is compact. Let $a \in \mathbb{R}^S_+$ and let x, y be bases of $P(S, \mu) \cap C(a)$, where $C(a) = \{z : z \leqslant a\}$. Suppose that $|x| > |y|$. Define

$$
\mathcal{L}(y) = \{A \in \mathcal{L} : y(A) = \mu A\}.
$$

Then if $A, B \in \mathcal{L}(y)$

$$
\begin{aligned}
\mu A + \mu B &= y(A) + y(B) \\
&= y(A \cap B) + y(A \cup B) \\
&\leqslant \mu(A \wedge B) + \mu(A \vee B)
\end{aligned}
$$

since $A \cup B \subseteq A \vee B$ and $A \cap B = A \wedge B$.

Since μ is submodular on $\mathcal{L}$, this implies that equality must hold in these inequalities, and hence

$$
\mu(A \wedge B) = y(A \wedge B), \qquad \mu(A \vee B) = y(A \vee B)
$$

and $A \wedge B, A \vee B \in \mathcal{L}(y)$. Thus if $D = \vee \{A : A \in \mathcal{L}(y)\}\ D \in \mathcal{L}(y)$ and if $S \in \mathcal{L}(y)$

$$
|y| = y(S) = \mu(S) \geqslant x(S) = |x| > |y|
$$

a contradiction. Hence $D \neq S$.

Since $x(D) \leqslant \mu(D) = y(D)$, there must exist $e \in S \backslash D$ with $x(e) > y(e)$.

Define $\qquad \gamma = \min \{\mu(A) - y(A) : A \notin \mathcal{L}(y)\}$

$$
\delta = \min \{\gamma, x(e) - y(e)\}.
$$

If z is a point of $\mathbb{R}^S_+$ such that

$$
z(e') = y(e') \ (e' \neq e), \quad z(e) = y(e) + \delta
$$

we assert that $z \in P(S, \mu)$. For $e \notin A$ for any $A \in \mathcal{L}(y)$

and so $$z(A) = y(A), \qquad A \in \mathscr{L}(y).$$

If $A \notin \mathscr{L}(y)$, then

$$z(A) \leqslant y(A) + \delta \leqslant \mu A.$$

Thus $z \in P(S, \mu)$ and since $z(e) \leqslant x(e)$, $z(e') \leqslant y(e')\, e' \neq e, z \leqslant a$ and thus $z \in P(S, \mu) \cap C(a)$.

This contradicts the maximality of y. Hence $|x| = |y|$ and $P(S, \mu)$ is the independence polytope of a polymatroid $\mathbb{P}$. ∎

COROLLARY 1. *The rank function of the polymatroid* $\mathbb{P} = P(S, \mu)$ *is given for all* $a \in \mathbb{R}_+^S$ *by*

$$r(a) = \min_{X \in \mathscr{L}} (\mu X + a(S \backslash X)).$$

Proof. Let $\alpha \in \mathbb{R}_+^S$ and let $x \in P(S, \mu) \cap C(a)$. Then for all $A \in \mathscr{L}$,

$$|x| = x(A) + x(S \backslash A) \leqslant \mu(A) + a(S \backslash A).$$

Hence, since $r(a) = |x|$ for some x,

(2) $$r(a) \leqslant \min_{A \in \mathscr{L}} (\mu A + a(S \backslash A)).$$

Suppose we have strict inequality in (2) and let $y \in P(S, \mu) \cap C(a)$ be such that $|y| = r(a)$.

Define $\mathscr{L}(y)$ as above, if $S \in \mathscr{L}(y)$, then $|y| = \mu S$ contradicting the strict inequality in (2). Hence $D \neq S$.

Now $\forall X \in \mathscr{L}$

$$y(S) < r(a) < \mu(X) + a(S \backslash X).$$

In particular

$$y(S) < \mu(D) + a(S \backslash D)$$

and so

$$y(S \backslash D) < a(S \backslash D).$$

Hence there must exist $e \in S \backslash D$ such that $y(e) < a(e)$.

If we augment y to z as in the proof of the theorem, then z will belong to $P(S, \mu)$ and $C(a)$ contradicting our choice of y. Hence we cannot have strict inequality in (2) and the result follows. ∎

For many purposes, given $\mathbb{P}$ by its independence polytope $P(S, \mu)$ it is useful to have a representation of $\mathbb{P}$ by an independence polytope $P(S, \mu')$ where μ' is defined on all subsets of S. This is given by the following result of Dunstan [73].

THEOREM 2. *Let $S, \mathscr{L}, \mu$ be as in Theorem 1 and let $S \in \mathscr{L}$. Then there exists a nondecreasing submodular function $\hat{\mu}: 2^S \to \mathbb{R}^+$ such that the independence polytope $P(S, \mu) = P(S, \hat{\mu})$. Moreover one such $\hat{\mu}$ is the set function defined for $A \subseteq S$ by*

$$\hat{\mu} A = \min (\mu B : B \supseteq A, B \in \mathscr{L}).$$

We leave the proof to the reader. Note that the additional constraint, $S \in \mathscr{L}$, in Theorem 1 ensures that $\hat{\mu}$ is well defined; also if $S \notin \mathscr{L}$ the necessary adjustments to the statements of Theorem 2 are quite straightforward.

EXERCISES 3

1. Let $\mu = 2^S \to \mathbb{R}^+$ be a submodular function. We say $A \subseteq S$ is μ-*closed* if for all sets B properly containing A, $\mu B > \mu A$. We say A is μ-*separable* if there is a proper partition (A_1, A_2) of A such that

$$\mu A = \mu A_1 + \mu A_2.$$

Let $\mathbb{P} = (P, S, \mu)$ be a polymatroid where μ is a non-decreasing submodular set function: $2^S \to \mathbb{R}^n$. Prove that a minimal set of inequalities which define $\mathbb{P}$ is $x_i \geqslant 0$, and

$$\sum_{i \in A} x_i \leqslant \mu A$$

where A runs through the class of μ-closed, μ-inseparable subsets. (Edmonds [70])

2. Let $\mu_1 : 2^S \to \mathbb{R}$ and $\mu_2 : 2^S \to \mathbb{R}$ be non-negative non-decreasing submodular functions which are not equal. Prove that the corresponding polymatroids are distinct.

$\circ$3. Find conditions on μ_1, μ_2 in the above exercise such that the polymatroids having μ_1, μ_2 as their ground set rank functions have combinatorially equivalent face lattices.

4. VERTICES OF POLYMATROIDS

In general, finding the vertices of a convex polyhedron which is presented as the intersection of a collection of half-spaces is very time consuming. It involves solving a large number of sets of linear equations.

For those polytopes P which are the independence polytopes of polymatroids, however, the following theorem of Edmonds [70] enables one to "read off" the vertices of P from its half-space presentation.

Suppose $S = \{1, 2, \ldots, n\}$ and $\mathbb{P} = (P, S, \rho)$ is a polymatroid where ρ is non-decreasing, non-negative and submodular.

For any permutation $\pi = (i_1, \ldots, i_n)$ of S define $A_\pi^1 = \{i_1\}$, $A_\pi^2 = \{i_1, i_2\}, \ldots, A_\pi^n = \{i_1, \ldots, i_n\}$.

THEOREM 1. *The vertices of the independence polytope P of $\mathbb{P}(S, P, \rho)$ are all points $v = v(k, \pi) \in \mathbb{R}_+^S$ where $v = (v_1, \ldots, v_n)$ and*

$$v_{i_1} = \rho(A_\pi^1)$$
$$v_{i_2} = \rho(A_\pi^2) - \rho(A_\pi^1),$$
$$v_{i_3} = \rho(A_\pi^3) - \rho(A_\pi^2),$$
$$\vdots \qquad \vdots \qquad \vdots$$
$$v_{i_k} = \rho(A_\pi^k) - \rho(A_\pi^{k-1}),$$
$$v_{i_{k+1}} = v_{i_{k+2}} = \ldots = v_{i_n} = 0,$$

and k ranges over the integers 0 to n and π ranges over all permutations of S.

COROLLARY. *If the ground set rank function ρ of a polymatroid $\mathbb{P} = (P, S, \rho)$ is integer valued all the vertices of P are integer points of $\mathbb{R}_+^S$.*

Example. Let $\mathbb{P}$ be the polymatroid whose independence polytope is given by

$$
\begin{aligned}
x_1 & & & \leqslant 2 \\
& x_2 & & \leqslant 3 \\
& & x_3 & \leqslant 5 \\
x_1 + & x_2 & & \leqslant 4 \\
& x_2 + & x_3 & \leqslant 6 \\
x_1 + & & x_3 & \leqslant 6 \\
x_1 + & x_2 + & x_3 & \leqslant 7 \\
& & x_i & \geqslant 0
\end{aligned}
$$

Take $\pi: \{1, 2, 3\}$ to be the permutation $(1, 2, 3)$. Then

$$v(3, \pi) = (2, 4, 4)$$
$$v(2, \pi) = (2, 4, 0).$$

To prove Theorem 1 we use the ingenious method of Edmonds [70]. The basic idea of this method is to show that any linear function cx' takes its maximum value over P at one of the points $v(k, \pi)$. Since (see Theorem 1.5) we can always choose c so that cx' has its maximum at a prescribed vertex of P we will have shown that the vertices must be the points $v(k, \pi)$.

Thus we get Theorem 1 as a corollary of the following "linear programming" type theorem.

THEOREM 2. *Let* $S = \{1, \ldots, n\}$ *and let* $\mathbb{P} = (P\,S, \rho)$ *be a polymatroid. For* $c \in \mathbb{R}^S$ *let* $\pi = (i_1, \ldots, i_n)$ *be a permutation of* S *such that*

$$c_{i_1} \geqslant c_{i_2} \geqslant \ldots \geqslant c_{i_k} > 0 \geqslant c_{i_{k+1}} \geqslant \ldots \geqslant c_{i_n}$$

Then cx' *takes its maximum value over* P *at the point* $v(k, \pi)$.

Proof. Consider the linear programme

(α) maximize $c\,x'$

$$x \in P.$$

Its dual linear programme is the following programme in the variables $y(A): A \subseteq S; A \neq \varnothing$

(α^*) minimize $\displaystyle\sum_{A \in S} \rho(A) y(A)$

$$\sum_{A:\, j \in A} y(A) \geqslant c_j \qquad (1 \leqslant j \leqslant n)$$

$$y(A) \geqslant 0.$$

It is routine to check:

(a) $v(k, \pi)$ is a feasible point of the primal programme (α).

(b)   The point $\hat{y}$ given by

$$\hat{y}(A_\pi^k) = c_{i_k}$$

$$\hat{y}(A_\pi^j) = c_{i_j} - c_{i_{j+1}} \qquad (1 \leqslant j \leqslant k-1)$$

$$\hat{y}(A^j) = 0 \qquad \text{otherwise}$$

is a feasible solution to the minimization problem (α^*).

(c)   That if v_i is the ith component of $v(k, \pi)$

$$\sum_{i=1}^{n} c_i v_i = \sum_{j=1}^{k-1} (c_{i_j} - c_{i_{j+1}}) \rho(A_\pi^j) + c_{i_k} \rho(A_\pi^k) = \sum_{A \in S} \hat{y}(A)\, \rho(A)$$

Hence we have feasible solutions to the dual linear programmes (α) and (α^*) for which the corresponding objective functions are equal. Thus by the basic Theorem 1.3 of linear programming duality theory the points $v(k, \pi)$ and $\hat{y}$ must be optimal. ∎

EXERCISES 4

1. Suppose P is the independence polytope of a polymatroid which is given in the form of the intersection of the half spaces

$$x(A) \leqslant \mu A$$
$$x \geqslant 0$$

where $A \in \mathscr{L}$ and $\mathscr{L}$ is a proper sublattice of 2^S. How would you modify Theorem 1 to read off the vertices of P.

2. If μ above is integer valued do all the vertices of the independence polytope have integer coordinates?

3. What is the maximum number of vertices of the independence polytope of a polymatroid in $\mathbb{R}^n$?

5. A NEW CLASS OF POLYTOPES WITH INTEGER VERTICES

An interesting subclass of polymatroids is defined as follows. A polymatroid $\mathbb{P} = (P, S, \rho)$ with vector rank function r is *integral* if it is a polymatroid and if (P2) of Definition 1 holds also when a and x are restricted to being integer valued. That is, for every integer valued vector a in $\mathbb{R}_+^S$ every maximal independent integer valued subvector x of a has the same modulus, which is equal to $r(a)$.

From the definition we get:

(1)   The (vector) rank function r of an integral polymatroid takes integer values on integer valued points of $\mathbb{R}_+^S$.

It is not too surprising that the converse is true. The following theorem, is not difficult to check.

THEOREM 1. *An integral polymatroid is a pair (S, P) satisfying* (P1), (P2), *of Definition 1 and also the condition:*

(P3)   *for a an integer valued vector of $\mathbb{R}_+^S$, $r(a)$ is an integer.*

THEOREM 2. *A polymatroid $\mathbb{P} = (P, S, \rho)$ is integral if and only if its ground set rank function ρ is integer valued.*

Proof. Consider the points v_X defined for all $X \subseteq S$ by

$$v_X(e) = \rho S \qquad e \in X$$
$$= 0 \qquad e \notin X.$$

Take N any integer greater than ρS. Then all the vertices of P are $\leqslant \hat{N} = (N, N, \ldots, N)$. But $\hat{N}$ is integer valued, and has rank equal to v_S. Hence the rank of v_S is an integer. But by Lemma 4 of Section 3,

$$r(v_S) = \rho S.$$

Hence ρS is an integer.

Also for any set X, since ρS is an integer, by Corollary 3.1 the rank of v_X is an integer. But again by Lemma 4 of Section 3,

$$r(v_X) = \rho X.$$

Now suppose that ρ is an integer valued function. We know that for any $a \in \mathbb{R}_+^S$

$$r(a) = \min \left(a(X) + \rho(S \backslash X) : X \subseteq S \right).$$

Hence if a is an integer vector, $a(X)$ is an integer and thus $r(a)$ is an integer.

Thus by Theorem 1, $\mathbb{P}$ is an integral polymatroid. ∎

As a corollary of Theorem 2 and Theorem 4.1 we get the following useful criterion.

THEOREM 4. *A polymatroid is integral if and only if all its vertices are integral.*

From its definition, if $\mathbb{P} = (S, P, \rho)$ is an integral polymatroid it is not difficult to check that the subsets of S with incidence vectors in P are the independent sets of a matroid on S. Conversely if we are given a matroid M on S with rank function ρ then ρ is the ground set rank function of an integral polymatroid $P(M)$ and the vertices of its independence polytope are precisely the incidence vectors of the independent sets of M.

We call this special type of polymatroid a *matroid polyhedron*.

Probably the deepest and most interesting property of integral polymatroids is contained in the intersection theorem of Edmonds [70].

THEOREM 5. *Let* $\mathbb{P}_1 = (S, P_1, \rho_1)$ *and* $\mathbb{P}_2 = (S, P_2, \rho_2)$ *be two integral polymatroids. Then the intersection of their independence polytopes* $P_1 \cap P_2$ *is a convex polytope all of whose vertices are integer valued.*

The proof of Theorem 5 hinges on showing that the convex polytope $P_1 \cap P_2$ can be represented in the form

$$Ax' \leqslant c$$
$$x \geqslant 0,$$

where A is a totally unimodular matrix and c is an integer vector. A well known property of such polytopes is that their vertices have integer coordinates. For an outline proof see Edmonds [70].

EXERCISES 5

1. If P is the matroid polyhedron determined by the matroid M describe how you would find the matroid polyhedra corresponding to restrictions, contractions or truncations of M.

2. Let $\Delta = (S, T; E)$ be a bipartite graph and let M_1, M_2 be matroids on E defined by $X \in \mathscr{I}(M_1)$, respectively $\mathscr{I}(M_2)$, if no two members of X have a common end point in S respectively T. Prove that the intersection of the matroid polyhedra, $P(M_1) \cap P(M_2)$ is the feasible region of the optimal assignment problem and hence always has integer vertices.

3. Prove that if $\mathbb{P}$ is an integral polymatroid which has only (0–1)-vertices then it is a matroid polyhedron.

4. Find three independence polytopes each of which has only integer vertices but which have an intersection with non-integer vertices.

6. FURTHER RESULTS ON POLYMATROIDS

In this section we outline the main results of McDiarmid [75c] and Woodall [73] who have extended familiar ideas of ordinary matroids to their continuous counterparts—polymatroids.

First note that matroid duality is easily extendable to polymatroids—though not uniquely.

Let $\mathbb{P} = (S, P, \rho)$ be a polymatroid with vector rank function r and let c be a vector in $\mathbb{R}_+^S$ which bounds P above.

Define a function $\rho^c : 2^v \to \mathbb{R}$ by, for each subset A of S

$$\rho^c A = c(A) + \rho(S \setminus A) - \rho S.$$

Then ρ^c is non-negative, increasing and submodular on 2^S and so is the ground set rank function of a polymatroid $\mathbb{P}^c = (S, P^c, \rho^c)$. We call this the c-dual of $\mathbb{P}$.

It is easy to check, for instance, that if we define a *spanning vector* of $\mathbb{P}$ to be a vector y with $r(y) = r(c)$ $(= \rho S)$ then we have:

(1) x is an independent vector of $\mathbb{P}^c$ if and only if $c-x$ is a spanning vector of $\mathbb{P}$.

Note also that if $\mathbb{P}$ is integral and c is integer valued then $\mathbb{P}^c$ is integral.

Now we will show how Rado's theorem on transversals has a continuous version which could be regarded as a polymatroid extension of the supply and demand theorem.

If $\Delta = (S, T; E)$ is a bipartite graph and A is a subset of the vertex set $S \cup T$ then we let

$$\alpha A = \{(u, v) \in E : u \in A, v \in V \backslash A\}.$$
$$\beta(A) = \alpha(V \backslash A).$$

Now let x be a vector in $\mathbb{R}_+^S$ and y be a vector in $\mathbb{R}_+^T$. It is helpful to regard x as a vector of supplies ($x(s)$ being the amount of supply at source $s \in S$) and to regard y as a vector of demands.

We say x can be *transported* to y via f for some vector $f \in \mathbb{R}_+^E$ if

(a) $\forall s \in S, \quad f(\alpha(s)) = x(s)$
(b) $\forall t \in T, \quad f(\beta(t)) = y(t).$

Alternatively we say f transports x to y.

Then the well-known theorem of supply and demand (see Berge [58a] or Ford and Fulkerson [62]) can be stated in the following form:

THEOREM 1. *If x, y are vectors of $\mathbb{R}_+^S$, $\mathbb{R}_+^T$ respectively, x can be transported to y if and only if for all $A \subseteq S$*

$$y(\partial A) \geqslant x(A).$$

(Recall $\partial A = \{v \in V(\Delta) : (u, v) \in E(\Delta)\}$ for some $u \in A$)

The reader will recognize this as just a continuous version of Hall's theorem. It is therefore natural to expect a "continuous analogue" of Rado's Theorem 7.2.2. This has been done by McDiarmid [75c] who proves:

THEOREM 2. *Let $\Delta = (S, T; E)$ be a bipartite graph and let $\mathbb{P} = (T, P, \rho)$ be a polymatroid on T. If $x \in \mathbb{R}_+^S$ then x can be transported onto some y independent in $\mathbb{P}$ if and only if for all $A \subseteq S$,*

$$\rho(\partial A) \leqslant x(A).$$

Furthermore if $\mathbb{P}$ is integral and x is integer valued we may insist that the transport vector f is integer valued.

Using this it is now possible to generalize the methods of Chapter 8 to obtain a very natural notion of *polymatroid sum*.

If S is a finite set, the *sum* of the subsets $P_1, \ldots, P_n$ of $\mathbb{R}_+^S$ is the subset $P_1 \vee P_2 \vee \ldots \vee P_n$ of $\mathbb{R}_+^S$ defined for each vector x of $\mathbb{R}_+^S$ by: $x \in P_1 \vee \ldots$

$\vee\ P_n$ if and only if there exist vectors $x_i (1 \leqslant i \leqslant n),\ x_i \in P_i,$ such that

$$x = \sum_{i=1}^{n} x_i.$$

THEOREM 3. *For* $i = 1, 2, \dots, n,$ *let* $\mathbb{P}_i = (S, P_i, \rho_i)$ *be a polymatroid on a set S. Then* $P = P_1 \vee \dots \vee P_n$ *is the independence polytope of a polymatroid* $\mathbb{P}$ *on S. Furthermore* $\mathbb{P}$ *has ground set rank function*

$$\rho = \sum_{i=1}^{n} \rho_i.$$

Also if each $\mathbb{P}_i$ *is integral then* $\mathbb{P}$ *is integral and for each integer valued vector x in P there exist integer valued vectors* x_i *in* $P_i\ (1 \leqslant i \leqslant n)$ *such that*

$$x = \sum_{i=1}^{n} x_i.$$

Using c-duality (with c any large enough vector) the ideas behind the proof of the intersection theorem for matroids can now be carried through to give the following intersection theorem for polymatroids.

THEOREM 4. *Let* $\mathbb{P}_1 = (S, P_1, \rho_1)$ *and* $\mathbb{P}_2 = (S, P_1, \rho_2)$ *be polymatroids on S and let* $k \in \mathbb{R}^+.$ *Then there exists a vector of* $\mathbb{R}_+^S$ *independent in both* $\mathbb{P}_1$ *and* $\mathbb{P}_2$ *and with modulus at least k if and only if for all subsets* $A \subseteq S$

$$\rho_1 A + \rho_2(S \backslash A) \geqslant k.$$

Furthermore if $\mathbb{P}_1,\ \mathbb{P}_2$ *are both integral we may insist that the independent vector x be integral.*

Provided we do not demand that our common independent vector be integral the intersection problem for polymatroids is completely solved. Woodall [73] proves the following common base theorem for polymatroids.

THEOREM 5. *Let* $\mathbb{P}_i = (S, P_i, \rho_i),\ (1 \leqslant i \leqslant k)$ *be polymatroids in* $\mathbb{R}^S.$ *Then they have a common independent vector of modulus* $t > 0$ *if and only if, for any set* $\{v_1, \dots, v_k\}$ *of subvectors of* $e = (1, 1, \dots, 1)$ *in* $\mathbb{R}^S$ *such that* $\Sigma_{i=1}^k v_i = e,$ *we have*

$$(2) \qquad \sum_{i=1}^{k} \lambda_i(v_i) \geqslant t$$

where $\lambda_i,$ *the free-rank function of* $\mathbb{P}_i$ *is defined by*

$$\lambda_i(v) = \max\{vw' : w \in P_i\}.$$

The conditions (2) are obviously necessary.

Using Theorem 5 we have an obvious new set of necessary conditions for the problem of deciding whether or not 3 matroids M_1, M_2, M_3 on S have a common independent set of cardinality t. By Theorem 5.4 the corresponding matroid polyhedra $\mathbb{P}(M_1)$, $\mathbb{P}(M_2)$, $\mathbb{P}(M_3)$ must have a common independent vector of modulus t.

However the following example shows that these new conditions although necessary are still not sufficient.

Example 1. Let $S = \{1, 2, 3, 4\}$ and let $\mathscr{A}$, $\mathscr{B}$, $\mathscr{C}$ be the families of subsets:
$$\mathscr{A} = \{1, 2\}, \ \{3, 4\}$$
$$\mathscr{B} = \{1, 3\}, \ \{2, 4\}$$
$$\mathscr{C} = \{1, 4\}, \ \{2, 3\}.$$
If $\mathbb{P}_1$ is the matroid polyhedron of $M[\mathscr{A}]$, $\mathbb{P}_2$ of $M[\mathscr{B}]$, $\mathbb{P}_3$ of $M[\mathscr{C}]$ then it can be checked that $\mathbb{P}_1$, $\mathbb{P}_2$, $\mathbb{P}_3$ have a common independent vector of modulus 2 namely $(\frac{1}{2}, \frac{1}{2}, \frac{1}{2}, \frac{1}{2})$. However $M[\mathscr{A}]$, $M[\mathscr{B}]$, $M[\mathscr{C}]$ do not have a common independent set of cardinality 2.

Another example of Woodall [73] shows that the conditions implied by (2) are in fact stronger than the obvious necessary conditions of Section 8.5 that the matroids M_1, M_2, M_3 on S have a common independent set of cardinality t.

Example 2. Let $S = \{1, 2, 3\}$ and let
$$\mathscr{A} = \{1\}, \ \{2, 3\}$$
$$\mathscr{B} = \{2\}, \ \{1, 3\}$$
$$\mathscr{C} = \{3\}, \ \{1, 2\}.$$
Then with the same notation as in Example 1, $\mathbb{P}_1$, $\mathbb{P}_2$, $\mathbb{P}_3$ do not have a common independent vector of modulus 2 even though their rank functions ρ_1, ρ_2, ρ_3 satisfy

$$\rho_1 X + \rho_2 Y + \rho_3 Z \geqslant 2$$

for any partition (X, Y, Z) of S.

EXERCISES 6

1. If $\mathbb{P}_1$ and $\mathbb{P}_2$ are matroid polyhedra in $\mathbb{R}^S$ with vertices the incidence vectors of the independent sets of matroids M_1, M_2 on S, prove that $\mathbb{P}_1 \vee \mathbb{P}_2$ is the matroid polyhedron corresponding to $M_1 \vee M_2$.

2. Show that when $\mathbb{P}$ is a matroid polyhedron, Rado's theorem for polymatroids reduces to Rado's theorem about independent transversals.

3. Use the theory of this section to find necessary (but not sufficient) conditions for 3 families of sets to have a common transversal. (Woodall [73])

4. Show that no polymatroid is indecomposable with respect to the sum operation. In particular show that the matroid polyhedron of the Fano matroid is the sum of two non-integral polymatroids.

7. SUPERMATROIDS

In this section we consider a generalization of matroids and integral polymatroids to structures on arbitrary finite partially ordered sets, which we call supermatroids. It turns out that a matroid on a set S is just the special case of a supermatroid on the ordered set of all subsets of S, and that integral polymatroids are essentially supermatroids on a special class of finite distributive sublattices of $\mathbb{R}^n$. The content of this section is taken from Dunstan, Ingleton and Welsh [72] to which we refer for all proofs.

Let $\mathbb{P} = (P, \leqslant)$ be a finite poset with a zero element. A collection $\mathscr{D}$ of elements of P is a *supermatroid* on P if the following conditions hold:

(A1) $o \in \mathscr{D}$
(A2)   If $x \in \mathscr{D}$ and $y \leqslant x$ then $y \in \mathscr{D}$
(A3)   For any $a \in P$ if

$$P(a) = \{x : x \in P, x \leqslant a\}$$

then all maximal elements of $P(a) \cap \mathscr{D}$ have the same height in P.

The elements of $\mathscr{D}$ are called the *independent* elements of the supermatroid, otherwise an element is *dependent*.

For any $a \in P$, the common height of maximal elements of $P(a) \cap \mathscr{D}$ is called the *rank* of a in $\mathscr{D}$ and is denoted by $\rho(a)$. A *base* of $\mathscr{D}$ is a maximal independent element of P; a *circuit* is a minimal dependent element of P; an element x is *spanning* in $\mathscr{D}$ if there exists a base b of $\mathscr{D}$ with $b \leqslant x$. An element a of P is *closed* in $\mathscr{D}$ if either a is a maximal element of P or if for all $c > a$, $\rho(c) > \rho(a)$.

Clearly a supermatroid is uniquely determined by a knowledge of its rank function or by a listing of (a) its bases (b) its circuits or (c) its spanning sets. However the many different concise sets of matroid axiom systems in terms of these concepts do not seem to extend to supermatroids.

We now sketch the main theorems for supermatroids.

First it is natural to see whether the attractive concept of duality can be extended to supermatroids.

If $(P, \leqslant)$ is a poset we write $P^* = (P^*, \overset{*}{\leqslant})$ for the dual poset having the same set of elements but with

$$x \overset{*}{\leqslant} y \Leftrightarrow y \leqslant x.$$

If $\mathcal{Q}$ is a supermatroid on P let $\mathcal{Q}^*$ be the set of spanning elements of Q.

THEOREM 1. *If P is a finite modular lattice and $\mathcal{Q}$ is a supermatroid on P, the set $\mathcal{Q}^*$ is a supermatroid on P^*.*

We call $\mathcal{Q}^*$ the *dual* of $\mathcal{Q}$, and it is easy to check that its rank function ρ^* is given by

$$\rho^*(a) = h(I) - h(a) - \rho(I) + \rho(a) \qquad (a \in P^*)$$

where h, ρ are the height and rank functions of P, $\mathcal{Q}$ respectively. This we note is the exact analogue of (2.1.5).

The strong connection between matroids and submodular set functions however seems to be essentially a phenomena of distributive lattices.

THEOREM 2. *If $\mathcal{Q}$ is a supermatroid on a distributive lattice $\mathcal{L}$ then its rank function ρ is submodular, that is for $x, y \in \mathcal{L}$,*

$$\rho(x \vee y) + \rho(x \wedge y) \leqslant \rho(x) + \rho(y).$$

Moreover if $\mathcal{L}$ is a non-distributive lattice there exists a supermatroid $\mathcal{Q}$ on $\mathcal{L}$ whose rank function is not submodular.

We call a supermatroid *strong* if its rank function ρ satisfies the further conditions:

(R1)   if b covers a then
$$\rho(a) \leqslant \rho(b) \leqslant \rho(a) + 1,$$
(R2)   if b, b' both cover a and c covers b, b' then
$$\rho(a) = \rho(b) = \rho(b') \Rightarrow \rho(a) = \rho(c).$$

For strong supermatroids we can prove results much closer to their matroid analogues. For example we have:

THEOREM 3. *If $\mathcal{Q}$ is a strong supermatroid on a semimodular lattice:*
(a)   *its rank function is submodular*
(b)   *the lattice of closed elements of $\mathcal{Q}$ under the induced ordering is a semimodular lattice.*

We illustrate the need to introduce the concept of strong supermatroid in the following example which typifies the sort of trouble that arises for arbitrary supermatroids even on the relatively well behaved class of modular lattices.

Example. Consider the modular lattice $\mathcal{D}_5$ with $Q = \{o, b\}$

$$\rho(c) + \rho(d) = 0 < \rho(c \vee d) + \rho(c \wedge d) = 1,$$
$$\rho(o) = \rho(c) = \rho(d) < \rho(e).$$

$\mathscr{D}_5$:

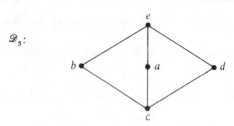

EXERCISES 7

1. Prove that if $\mathscr{D}$ is a supermatroid on a modular lattice $(\mathscr{D}^*)^* = \mathscr{D}$.

2. Prove that if x, y are elements of P with $y < x$ and y independent in the super-matroid $\mathscr{D}$ then if x has rank t in $\mathscr{D}$ there exists z with $y \leqslant z \leqslant x$ such that z is independent in $\mathscr{D}$ and $\rho(z) = t$.

3. Find a supermatroid on a geometric lattice whose rank function fails to satisfy the condition: b covers a implies

$$\rho(a) \leqslant \rho(b) \leqslant \rho(a) + 1$$

4. Prove that every supermatroid on a distributive lattice is strong.

NOTES ON CHAPTER 18

The origin of polymatroids is the paper of Edmonds [70], though it is somewhat lacking in proofs. Section 2, showing the equivalence of the various axiom systems for polymatroids is based on Edmonds [70], Dunstan [72] and an expanded unpublished version of Woodall [73]. The examples of Fig. 1 and 2 are due to Dunstan [72].

Theorems 3.1 and 4.1 are due to Edmonds [70]. Lawler [75] also gives a proof that the intersection of two matroid polyhedron has integer vertices.

Sections 6 and 7 are abbreviated versions of the papers by McDiarmid [75c], Woodall [73] and Dunstan, Ingleton and Welsh [72]. See also Finkbeiner [51], and Las Vergnas [73].

Integer polymatroids have also been studied under the name of *hyper-matroids* by Helgason [74] who uses them to find the weak chromatic polynomial of a hypergraph; see also Helgason [70].

CHAPTER 19

Combinatorial Optimization

1. THE GREEDY ALGORITHM

The most well-known algorithmic property of matroids is their intimate relationship with what has been termed the "greedy algorithm". Loosely speaking the greedy algorithm makes maximum improvements in an objective function at each stage and never back tracks. The main theorem of this section is that this approach optimizes the objective function if and only if the underlying structure has a matroid character. We now make these vague ideas precise. The basic idea for graphs is a well known result of W. Kruskal. The extension to matroids was first carried out by Rado [57].

Let S be a finite set and let $\mathscr{F}$ be a collection of subsets of S with the property that if $A \in \mathscr{F}$, $B \subset A$ then $B \in \mathscr{F}$. Let $w: S \to \mathbb{R}^+$ be a weight function and extend $w: 2^S \to \mathbb{R}^+$ in the obvious way

$$w(A) = \sum_{e \in A} w(e) \qquad (A \subseteq S).$$

Suppose we are asked to find a subset A_0 of S with the following properties:

(i) $A_0 \in \mathscr{F}$.
(ii) $w(A_0)$ is a maximum over all members of $\mathscr{F}$.

We call this *problem* $(\mathscr{F}, w)$.

The *greedy algorithm* for problem $(\mathscr{F}, w)$ is an automatic routine for selecting a member of $\mathscr{F}$. It proceeds as follows:

Step 1(a) Choose a member x_1 such that $\{x_1\} \in \mathscr{F}$ and such that $w(x_1) \geqslant w(x)$ for all x with $\{x\} \in \mathscr{F}$.

Step 1(b) If no such x exists, stop.

Step 2(a) Choose a member x_2 such that $\{x_1, x_2\} \in \mathscr{F}$ and such that $w(x_2) \geqslant w(x)$ for all $x \neq x_1$ with $\{x_1, x\} \in \mathscr{F}$.

Step 2(b) If no such x_2 exists, stop.

Step k(a) Choose a member x_k distinct from $x_1, \ldots, x_{k-1}$, such that

$\{x_1, x_2, \ldots, x_{k-1}, x_k\} \in \mathscr{F}$ and such that $w(x_k)$ is a maximum over all such x.

Step k(b)   If no such x_k exists, stop.

At the termination of the greedy algorithm we will have a subset X of S with $X \in \mathscr{F}$. We say the greedy algorithm *works* if $w(X) \geqslant w(A) \; \forall A \in \mathscr{F}$.

THEOREM 1. *If $\mathscr{F}$ is the collection of independent sets of a matroid on S, the greedy algorithm works for the optimization problem $(\mathscr{F}, w)$.*

Before proving the theorem we give an application.

Example 1. Suppose the edges of the connected graph G are each assigned a weight $w(e)$. Then since the subforests of G are the independent sets of a matroid on $E(G)$ the greedy algorithm will select a spanning tree of maximum weight.

Thus in Fig. 1(a) the greedy algorithm will select in order the edges e_7, e_5, e_1, e_9, e_4 giving the spanning tree shown in Fig. 1(b), of maximum weight 36.

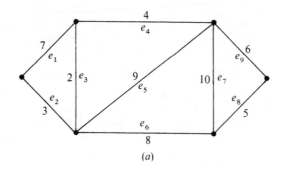

(a)

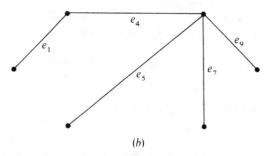

(b)

Figure 1

Proof of Theorem 1. Let B be the base of M selected by the greedy algorithm and suppose that T is a base of M with $w(T) > w(B)$ with the further property that $|T \cap B|$ is maximum. Let x_k be the first element selected by the greedy algorithm which belongs to $B \backslash T$. Then there exists a circuit C such that

$$x_k \in C \subseteq T \cup x_k.$$

Let $y \in C \backslash B$, so that $(T \backslash y) \cup x_k$ is a base of M. Since T has maximum weight $w(y) \leqslant w(x_k)$ and $w(y)$ cannot be strictly greater than $w(x_k)$ or the greedy algorithm would have selected y not x_k. Hence $w(y) = w(x_k)$ and hence $T_1 = (T \backslash y) \cup x_k$ also has maximum weight. But T_1 has more elements in common with B than T, which contradicts the choice of T and proves the theorem.■

It is probably more interesting that the converse of Theorem 1 is also true.

THEOREM 2. *Let $\mathscr{F}$ be a collection of subsets of S with the property that $A \in \mathscr{F}$, $B \subseteq A \Rightarrow B \in \mathscr{F}$. Then the greedy algorithm works for $(\mathscr{F}, w)$ for all non-negative weight functions w only if $\mathscr{F}$ is the collection of independent sets of a matroid on S.*

Proof. We have only to show that if $A = \{a_1, \ldots, a_k\} \in \mathscr{F}$ and $B = \{b_1, \ldots, b_{k+1}\} \in \mathscr{F}$ there exists a $b_i \notin A$ and $A \cup b_i \in \mathscr{F}$. To see this define a weight function w on S by

$$w(a_i) = u \qquad 1 \leqslant i \leqslant k,$$

$$w(b_i) = v \qquad b_i \in B \backslash A,$$

$$w(e) = 0 \qquad e \in S \backslash (A \cup B),$$

where $u > v > 0$. Then the greedy algorithm selects the elements $a_1, \ldots, a_k$ in order. If no b_i exists with $\{b_i, a_1, \ldots, a_k\}$ a member of $\mathscr{F}$ the algorithm will henceforth select members of $S \backslash (A \cup B)$ and will stop having selected a member of $\mathscr{F}$ of weight $w(A)$.

If however $|B \cap A| = t$, we see that

$$w(A) = ku, \qquad w(B) = tu + (k + 1 - t)v.$$

Choose u, v such that $w(A) < w(B)$ and $u > v$. This is clearly possible and then the greedy algorithm will not have slected a member of $\mathscr{F}$ of maximum weight. This is a contradiction and proves the theorem.■

Notice that Theorem 2 gives the following useful characterisation of matroids.

THEOREM 3. *A non-empty collection $\mathscr{F}$ of subsets of S is the set of independent sets of a matroid on S if and only if*

(a) $X \in \mathscr{F}, \quad Y \subseteq X \Rightarrow Y \in \mathscr{F}$,

(b)   *for all non-negative weight functions $w: S \to \mathbb{R}$, the greedy algorithm selects a member A of $\mathscr{F}$ with*

$$\sum_{e \in A} w(e) \geqslant \sum_{e \in B} w(e)$$

for all members B of $\mathscr{F}$.

EXERCISES 1

1. Let M be a matroid on S and w be a non-negative weight function on S. Prove that if B_0 is the base of M with maximum weight no element of B_0 is the minimum weight edge of any circuit of M unless w is a constant function.

2. Prove that the spanning tree of a graph $G = (V, E)$ which has minimum weight with respect to a non-negative weight function on E can be found in $O(|V|^2)$ computations.

3. Let M be a matroid on S and let $w: S \to \mathbb{R}^+$. Let $\mathscr{B}$ be the collection of bases of M and $\mathscr{C}^*$ be the collection of cocircuits of M. Prove

$$\min_{B \in \mathscr{B}} \max_{e \in B} w(e) = \max_{C^* \in \mathscr{C}^*} \min_{e \in C^*} w(e).$$

(*Note*: This is a special case of a more general result about blocking systems see Edmonds and Fulkerson [70]).

2. A GREEDY ALGORITHM FOR A CLASS OF LINEAR PROGRAMMES

In this section we present a "greedy algorithm" for linear programming problems and characterise those linear programmes for which it works. Many of the arguments will be only sketched. The interested reader can refer to Dunstan and Welsh [74].

Throughout $S = \{1, 2, \ldots, n\}$, U will denote a compact subset of $\mathbb{R}^S$ and $c = (c_1, \ldots, c_n)$ a point of $\mathbb{R}^S = \mathbb{R}^n$.

The *greedy algorithm* is an automatic routine for finding a point of $\mathbb{R}^n$ which maximizes the linear form cx' over U.

For any c and any U, the greedy algorithm selects a unique point $G(c, U) = (g_1, g_2, \ldots, g_n) \in U$ by the following method.

Given $c \in \mathbb{R}^n$ let $\pi(c) = (i_1, i_2, \ldots, i_n)$ be the unique permutation of $(1, 2, \ldots, n)$ which satisfies

$$|c_{i_1}| \geqslant |c_{i_2}| \geqslant \ldots \geqslant |c_{i_n}|$$
$$|c_{i_j}| = |c_{i_k}| \text{ and } j < k \Rightarrow i_j < i_k.$$

Example. If $c = (-4, 7, 4, 1, -5)$ then $\pi(c)$ is the permutation $(2, 5, 1, 3, 4)$. For any subset $\{i_1, i_2, \ldots, i_k\} \subseteq S$ we write

$$(x_{i_1}, x_{i_2}, \ldots, x_{i_k}) \hat{\in} U$$

if there exists $u \in U$ such that $y_{i_j} = x_{i_j}$ $(1 \leqslant j \leqslant k)$.

Now suppose that $\pi(c) = (i_1, i_2, \ldots, i_n)$. Choose g_{i_1} by

$$g_{i_1} = \begin{cases} \max \{x_{i_1} : (x_{i_1}) \hat{\in} U\} \text{ if } c_{i_1} > 0 \\ \min \{x_{i_1} : (x_{i_1}) \hat{\in} U\} \text{ if } c_{i_1} \leqslant 0. \end{cases}$$

Having chosen $g_{i_1}, g_{i_2}, \ldots, g_{i_{p-1}}$ we choose g_{i_p} by

$$g_{i_p} = \begin{cases} \max \{x_{i_p} : (g_{i_1}, \ldots, g_{i_{p-1}}, x_{i_p}) \hat{\in} U \text{ if } c_{i_p} > 0 \\ \min \{x_{i_p} : (g_{i_1}, \ldots, g_{i_{p-1}}, x_{i_p}) \hat{\in} U \text{ if } c_{i_p} \leqslant 0, \end{cases}$$

and continue recursively until $p = n$.

It is clear that since U is compact, we end up with a unique point $G(c, U) \in U$ which is essentially obtained by being as "greedy" as we can at each stage of the algorithm.

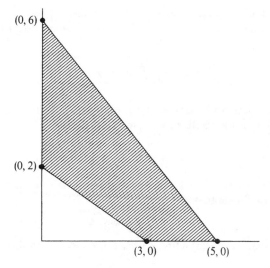

Figure 1

Example. Let $U \subseteq \mathbb{R}^2$ be the shaded area of Fig. 1. Let $c = (3, 5)$. Then $\pi(c) = (2, 1)$ and $g_{i_1} = 6, g_{i_2} = 0 \Rightarrow G(c, U) = (0, 6)$. Let $c' = (-4, -4)$. Then $\pi(c') = (1, 2)$ and $g_{i_1} = 0, g_{i_2} = 2 \Rightarrow G(c', U) = (0, 2)$.

We define $g(c, U)$ to be the value of cx' got by using the greedy algorithm, that is

$$g(c, U) = cG(c, U).'$$

Also let

$$m(c, U) = \max_{x \in U} cx'$$

Then we can define the *greedy cone* U^* of U by

$$U^* = \{c \in R^n : g(c, U) = m(c, U)\}.$$

In other words the greedy cone of a set U is the set of c for which our greedy algorithm gives a true optimum.

THEOREM 1. *If P is the independence polytype of a polymatroid its greedy cone P^* is $\mathbb{R}^n$.*

In other words the greedy algorithm gives a true optimum of cx' over the independence polytope of any polymatroid for any c.

To prove this we first need the following easy lemma.

LEMMA. *If $c = (c_1, c_2, \ldots, c_k, 0, \ldots, 0) \in \mathbb{R}^n$ where $c_1 \geqslant c_2 \geqslant \ldots \geqslant c_k > 0$ and $u = (u_1, u_2, \ldots, u_k, 0, 0, \ldots, 0) = G(c, P)$, then for any vector $v = (v_1, \ldots, v_n) \in P$,*

$$\sum_{i=1}^{k} u_i \geqslant \sum_{i=1}^{k} v_i.$$

Proof. Suppose N is so large that $N_k = (N, N, \ldots, N, 0, \ldots, 0)$ does not belong to P. (N_k has N at its ith coordinate $i \leqslant k$). If

$$\sum_{i=1}^{k} u_i < \sum_{i=1}^{k} v_i \leqslant r(N_k)$$

where r is the vector rank function of the polymatroid, then we must be able to find $w \in P$ such that

$$u < w \leqslant N_k$$

Hence there must exist a first integer t, $1 \leqslant t \leqslant k$, such that $w_t > u_t$.

Clearly, this contradicts the choice of u by the greedy algorithm

Proof of Theorem 1. Since the polytope $P \subseteq \mathbb{R}_+^n$ and $x \in P \Rightarrow y \in P$ for all $y \leqslant x$, it is easy to see that if any c_i is non-positive, the ith coordinate g_i of $G(c, P)$ will be zero. Hence it is easy to see that if c^+ is defined by

$$c_i^+ = \begin{cases} c_i & \text{if} \quad c_i > 0 \\ 0 & \text{if} \quad c_i \leqslant 0, \end{cases}$$

then $G(c^+, P) = G(c, P), g(c^+, P) = g(c, P)$, and $m(c^+, P) = m(c, P)$.

Thus the theorem will be proved if we prove it for all $c \geqslant 0$. Let c be such that

$$c_1 \geqslant c_2 \geqslant \ldots \geqslant c_p > 0 = c_{p+1} = \ldots = c_n.$$

We will use induction on p, the number of strictly positive c_i. If $p = 0$ or 1, clearly $g(c, P) = m(c, P)$.

Suppose therefore that $g(c, P) = m(c, P)$ for all c with fewer than p positive components, and consider the problem of maximizing over P the linear form

$$c_1 x_1 + \ldots + c_p x_p, \qquad c_i > 0.$$

Let $G(c, P) = (u_1, u_2, \ldots, u_p, 0, \ldots, 0)$ and suppose

$$g(c, P) = \sum_{i=1}^{p} c_i u_i < \sum_{i=1}^{p} c_i v_i$$

for some $(v_1, v_2, \ldots, v_p, 0, 0, \ldots, 0) \in P$.

Then since $c_p > 0$ we can write (using the previous lemma),

$$\sum_{i=1}^{p} \frac{c_i}{c_p} v_i > \sum_{i=1}^{p} \frac{c_i}{c_p} u_i, \qquad \sum_{i=1}^{p} v_i \leqslant \sum_{i=1}^{p} u_i.$$

Subtracting we get

$$\sum_{i=1}^{p-1} \left(\frac{c_i}{c_p} - 1 \right) v_i > \sum_{i=1}^{p-1} \left(\frac{c_i}{c_p} - 1 \right) u_i$$

and letting $c_i' = (c_i/c_p - 1)$ we have $c_1' \geqslant c_2' \geqslant \ldots \geqslant c_{p-1}' \geqslant 0$.

Now consider the problem of maximizing over P

$$\sum_{i=1}^{p-1} c_i' x_i.$$

By the induction hypothesis the greedy algorithm will work and since it will pick $(u_1, u_2, \ldots, u_{p-1}, 0, \ldots, 0)$ if $c_{p-1}' > 0$ or $(u_1, u_2, \ldots, u_t, 0, \ldots, 0)$ if $c_{t+1}' = 0$ and $c_t' > 0$ for some $t + 1 \leqslant p - 1$, we have a contradiction. Hence by induction the greedy algorithm works for all c and $P^* = \mathbb{R}^n$. ∎

However it is easy to see that the converse of Theorem 1 is false. In Fig. 2

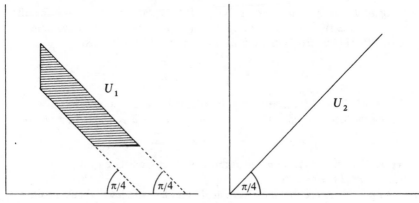

Figure 2

the sets U_1, U_2 are subsets of $\mathbb{R}^2$ such that $U_i^* = \mathbb{R}^2$ but the U_i are not the independence polytopes of polymatroids.

The second theorem of Dunstan and Welsh [74] characterizes those U for which the greedy algorithm works.

For $A \subseteq \mathbb{R}_+^n$ we define the *hereditary closure* $H(A)$ of A by

$$H(A) = \{x : x \geqslant 0, x \leqslant y \text{ for some } y \in A\}.$$

$A + y$ denotes the *translate* of A by the vector y, that is

$$A + y = \{x : x = z + y, z \in A\}.$$

and co (A) denotes the convex hull of A.

THEOREM 2. *The greedy cone U^* of a compact subset U of $\mathbb{R}^n$ contains $\mathbb{R}_n^+$ if and only if there exists a translate $U + y$ of U such that $H(\text{co } (U + y))$ is a polymatroid.*

EXERCISES 2

1. Prove that for any U, U^* is a cone.

2. In $\mathbb{R}^2$, let $U = \{x : x_1^2 + x_2^2 \leqslant 1\}$. Prove that the greedy cone U^* is the union of the x_1 and x_2 axes.

3. In $\mathbb{R}^2$, let U be defined by

$$x \in U \Leftrightarrow x \geqslant 0, \qquad 3x_1 + x_2 \leqslant 3.$$

Prove that the corresponding greedy cone U^* is $\mathbb{R}^2 \backslash T$ where T is that subset of the positive quadrant bounded by the lines $x_1 = x_2$ and $x_1 = 3x_2$.

3. PARTITIONING AND INTERSECTION ALGORITHMS

In this section we shall describe an algorithm for efficiently partitioning a set into k disjoint sets E_i such that for each i, E_i is independent in a given matroid M_i, $1 \leq i \leq k$. An original algorithm for this problem was presented by Edmonds [67b], the treatment here is a modification and extension of his method due to Knuth [73], a similar method has been discovered independently by Greene and Magnanti [74].

Let $M_1, M_2, \ldots, M_k$ be matroids on S. We shall be seeking a partition of S into $(E_1, \ldots, E_k)$ where each E_i is independent in M_i, $1 \leq i \leq k$. We shall imagine that the elements of S are being coloured with k colours so that E_j consists of the elements of colour j.

The natural approach is to start with every element uncoloured and to successively paint them. Suppose therefore that we have painted certain elements and that X_j is the set of elements having colour j, so that X_j is independent in M_j. Let $X_0 = S \setminus (X_1 \cup \ldots \cup X_k)$ be the set of unpainted elements.

If x is an element not of colour j we can paint it colour j if $x \cup X_j$ is independent in M_j. On the other hand if $x \cup X_j$ is dependent there is a unique circuit $C \subseteq x \cup X_j$ and we can paint x with colour j if the colour of any element $y \in C \cap X_j$ is scraped off, and y is painted another colour.

We denote such a sequence of repaintings, by, say,

$$x \rightarrow y \rightarrow z \rightarrow O_t$$

which means: "Paint x with the present colour of y, repaint y with the present colour of z, then repaint z with colour t."

Thus we can write, for $y \in X_j$ and $x \notin X_j$,

$$x \rightarrow y \Leftrightarrow (X_j \cup x) \setminus y \in \mathscr{I}(M_j)$$

and if $x \notin X_t$

$$x \rightarrow O_t \Leftrightarrow x \cup X_t \in \mathscr{I}(M_t),$$

and O_t is a special symbol distinct from the elements of S. Thus this arrow notation defines a digraph on the vertx set $S \cup \{O_1, O_2, \ldots, O_k\}$, and we write

$$x \xrightarrow{*} y \Leftrightarrow \text{there is a path } x = x_0 \rightarrow x_1 \rightarrow \ldots \rightarrow x_m = y$$

The key to the painting (partitioning) algorithm is the following lemma.

LEMMA 1. *Suppose there is a path $x = x_0 \rightarrow x_1 \rightarrow \ldots \rightarrow x_m = O_r$, $x_i \not\rightarrow x_j$ for $j > i + 1$. Then if x_i is painted the colour of x_{i+1} for $0 \leq i \leq m$ the resulting elements of colour j are independent in M_j, for $1 \leq j \leq k$.*

Proof. The result is trivial when $m = 1$. If $m > 1$, consider the state of the elements after the mth step of the repainting.

Let x_{m-1} have colour s, and let

$$X'_r = X_r \cup x_{m-1}$$
$$X'_s = X_s \backslash x_{m-1}$$
$$X'_j = X_j \qquad j \neq r, s.$$

Let $\overset{\wedge}{\rightarrow}$ denote the arrow relation in the digraph determined by this set of X'_j. Let $x'_i = x_i$ for $0 \leqslant i \leqslant m$ and also let $x'_{m-1} = O_s$. The lemma will follow by induction on m if we prove that

(1) $x'_o \overset{\wedge}{\rightarrow} x'_1 \overset{\wedge}{\rightarrow} \ldots \overset{\wedge}{\rightarrow} x'_{m-1}$

and

(2) $x'_i \overset{\wedge}{\rightarrow} x'_j$ for $j > i + 1$.

Trivially $x'_i \overset{\wedge}{\rightarrow} x'_{i+1}$ except when x'_{i+1} has colour r. In this case, $i + 1 < m - 1$, and we must show that the set $A = (X'_r \backslash x_{i+1}) \cup x_i$ is independent in M_r.

Suppose A is dependent. Then A contains a single circuit C and since both $A \backslash x_i$ and $A \backslash x_{m-1}$ are independent, C must contain x_i and x_{m-1}. This would imply $x_i \rightarrow x_{m-1}$ contrary to hypothesis, and (1) holds.

Suppose that $x'_i \overset{\wedge}{\rightarrow} x'_j$ for $j > i + 1$. Then there is an immediate contradiction unless x'_j has colour s and $j < m - 1$. But in this case

$$D = (X'_s \backslash x'_j) \cup x'_i \in \mathscr{I}(M_s)$$
$$E = (X_s \backslash x_j) \cup x_i \notin \mathscr{I}(M_s)$$

and since $x'_i = x_i$, $x'_j = x_j$, there is a circuit C of M_s satisfying, $C \subseteq E$, and since $E \backslash D = x_{m-1}$ we must have both x_i and x_{m-1} in the circuit C. So $x_i \rightarrow x_{m-1}$ and this contradiction completes the proof of the lemma.

This lemma is the key to proving that the following algorithm is an efficient partitioning algorithm. Suppose we have the problem: given matroids $(M_i: 1 \leqslant i \leqslant k)$ on S, does there exist a collection of disjoint subsets $E_1, \ldots, E_k$ of S such that $|E_i| = n_i$ $(1 \leqslant i \leqslant k)$ and E_i is independent in M_i?

Consider the following algorithm.

Algorithm
(1) Start with $X_1 = X_2 = \ldots = X_k = \emptyset$.
(2) Construct the digraph $G(\mathscr{X})$ on the vertex set $S \cup O_1 \cup \ldots \cup O_k$ associated with the painting (partitioning) $\mathscr{X} = (X_1, X_2, \ldots, X_k)$.

(3)   For the unique t defined by $|X_1| = n_1, \ldots, |X_{t-1}| = n_{t-1}$, $|X_t| < n_t$ attempt to find a path $y \overset{+}{\to} O_t$ in $G(\mathcal{X})$ from some uncoloured y to O_t, which satisfies the conditions of Lemma 1.

(4)   If such a path exists repaint the elements of S corresponding to the elements of this path and change $\mathcal{X} = (X_1, \ldots, X_t)$ accordingly. Then move to (2).

(5)   If no such path exists then we either have

$$|X_1| = n_1, \ldots, |X_k| = n_k$$

and we have our required painting or the algorithm has stopped because there is no path in $G(\mathcal{X})$ and $\mathcal{X}$ is an incomplete painting, and we known from the theorem below that no painting of the sort we required exists.

THEOREM 1. *The above algorithm "works" in the sense that if it stops before* $|X_1| = n_1, \ldots, |X_k| = n_k$ *there is no partitioning of S into independent sets of these cardinalities.*

Proof. Suppose the algorithm stops with

$$|X_1| = n_1, \qquad |X_2| = n_2, \ldots, |X_{r-1}| = n_{r-1}, \qquad |X_r| < n_r.$$

We may assume that M_i has rank n_i $(1 \leqslant i \leqslant k)$ for otherwise we can just truncate M_i to level n_i without affecting the existence of a partitioning.

If $G = G(\mathcal{X})$ and X_0 is the set of unpainted elements of S, then we know there is no path in G, $y \overset{+}{\to} O_r$ for any $y \in X_0$. Now define $A_i, B_i, 1 \leqslant i \leqslant r$, by

$$B_i = \{x \in X_i : x \overset{+}{\to} O_r\}$$

$$A_i = X_i \backslash B_i, \qquad A = A_1 \cup A_2 \cup \ldots \cup A_r \cup X_0.$$

We assert

$$\rho_i(A) = |A_i| \qquad 1 \leqslant i \leqslant r-1,$$

where ρ_i is the rank function of M_i. Clearly $\rho_1(A) \geqslant |A_1|$, suppose $\rho_1(A) > |A_1|$. Then $\exists y \in A_2 \cup \ldots \cup A_r \cup X_0$ with y independent of A_1 in M_1. Suppose $y \in A_2$ then since $|X_1| = n_1 = \rho(M_1)$, $y \to z$ for some $z \in B_1$ and $B_1 \neq \varnothing$ since $X_1 = A_1 \cup B_1 = n_1 = \rho(M_1)$. But there is a path $z \overset{+}{\to} O_r$ and hence a path $y \to z \overset{+}{\to} O_r$ and thus $y \in B_2$ not A_2. Similarly if $y \in A_i$ $(3 \geqslant i \geqslant r-1)$. If $y \in X_0$ there can be no path $y \to z \overset{+}{\to} O_r$ or we would have continued our repainting. Hence $\rho_1(A) = |A_1|$ and similarly $\rho_i(A) = |A_i| 1 \leqslant i \leqslant r-1$.

Similarly if $x \in X_0 \cup A_1 \cup \ldots \cup A_{r-1}$ a similar argument shows that x depends on A_r in M_r and thus $\rho_r(A) = |A_r|$. Hence we have

$$\sum_{i=1}^{r} \rho_i(A) + |S \backslash A| = \sum_{i=1}^{r} |A_i| + \sum_{i=1}^{r} |B_i|$$

$$= \sum_{i=1}^{r} |X_i| < \sum_{i=1}^{r} n_i.$$

But a necessary condition that a partitioning of the required size exist is that the union $M_1 \vee \ldots \vee M_r$ has rank $n_1 + \ldots + n_r$. By (8.3.2) a necessary and sufficient condition for this is that for all $Z \subseteq S$

$$\sum_{i=1}^{r} \rho_i(Z) + |S \backslash Z| \geqslant \sum_{i=1}^{r} n_i$$

and this is not true when $A = Z$ so that the stopping of the algorithm before a painting is completed shows the non-existence of a painting of the cardinalities required and completes the proof of the theorem. ∎

The algorithm is "fast" in the sense that it involves making at most $n^3 + n^2 k$ tests of independence in the given matroids where n is the cardinality of S. Whether there exists an algorithm involving fewer such tests remains an open problem.

Matroid intersection algorithms

The matroid intersection problem is the problem of finding a subset $X \subseteq S$ of cardinal k which is independent in two matroids M_1 and M_2 on S.

Application: The matching problem. Let $\Delta = (T_1, T_2; E)$ be a bipartite graph.

One version of the matching problem is finding a subset of k edges of G which form a matching in Δ. It is clearly equivalent to finding when M_1, M_2 have a common independent set of cardinal k, where M_i is the matroid on E in which a set of edges is independent if it is incident with distinct vertices of T_i.

We show how the partitioning algorithm can be modified to give an efficient algorithm for the matroid intersection problem.

Method. Let M_1^k, M_2^k be the truncations of M_1, M_2 at level k. Then M_1^k, M_2^k have a common independent set X of cardinal k if and only if S can be partitioned into two sets I_1, I_2 which are independent respectively in M_1^k and $(M_2^k)^*$. Hence applying the partitioning algorithm with these two matroids gives us an algorithm for finding our common independent set of cardinality k.

Thus we have got a good algorithm for finding a set of given cardinality which is independent in 2 matroids.

For three matroids however the problem is much more difficult and no non-exhaustive algorithm is known. As a measure of the difficulty of this problem recall from Section 8.5 that if had such an algorithm we would have a good algorithm for finding a Hamiltonian path in a direct graph and this is well known to be a intractable problem.

An interesting extension of the matroid intersection algorithm has been proved by Aigner and Dowling [71a].

Let $\Delta = (S, T; E)$ be a bipartite graph and let M_1, M_2 be matroids on S, T respectively.

Recall that an *independent matching* in $(M_1 \Delta, M_2)$ is a set U of edges $\{(s_i, t_i): s_i \in A, t_i \in B, A \subseteq S, B \subseteq T\}$ of Δ such that:

(a) U is a matching of Δ;
(b) A is independent in M_1;
(c) B is independent in M_2.

Given an independent matching $U \subseteq E(\Delta)$ an *augmenting chain* of U with respect to (M_1, Δ, M_2) is a sequence

$$(a'_0, b'_1)(b_1, a_1), (a'_1, b'_2), \ldots, (b_n, a_n), (a'_n, b'_{n+1})$$

of $2n + 1$ ($n \geq 0$) distinct ordered pairs such that:

(i) $(a_i, b_i) \in U \qquad 1 \leq i \leq n$

   $(a'_i, b'_{i+1}) \in E(\Delta) \backslash U \qquad 0 \leq i \leq n$,

(ii) $a_0 \in S \backslash \sigma_1 A, \qquad b'_{n+1} \in T \backslash \sigma_2 B$

   where σ_i is the closure operator of M_i,

(iii) $a'_i \in \sigma A, \qquad a'_i \notin \sigma_1 \left[\left(A \backslash \bigcup_{j=1}^{i} a_j \right) \cup \bigcup_{j=1}^{i-1} a'_j \right]$

   $b'_i \in \sigma A, \qquad b'_i \notin \sigma_2 \left[\left(B \backslash \bigcup_{j=1}^{i} b_j \right) \cup \bigcup_{j=1}^{i-1} b'_j \right]$

   for $1 \leq i \leq n$.

Note that if M_1, M_2 are the free matroids this last condition implies that $a'_j = a_i$, $b'_i = b_i$ ($1 \leq i \leq n$), and hence our definition reduces to the standard definition of an augmenting chain of a bipartite graph (see for example Berge [58]).

The main result of Aigner and Dowling [71a] is the following matroid generalization of the assignment problem.

THEOREM 2. *An independent matching U is of maximum cardinality in (M_1, Δ, M_2) if and only if it has no augmenting chain.*

An interesting extension of the intersection problem has been formulated by Lawler [71].

Let M be a matroid on a set S of even cardinality and let π be a partition of S into disjoint blocks each of size 2. Each block of π contains exactly two elements e and $\bar{e}$; we call e the *mate* of $\bar{e}$ and vice versa. A set $X \subseteq S$ is called a *parity set* if for each e, $e \in X$ if and only if its mate $\bar{e} \in X$. The *matroid parity problem* is to find an independent parity set with a maximum number of elements. We now consider some examples of matroid parity problems.

Example 1. Consider the problem of finding a maximum matching in a graph G (see Section 14.6).

Replace each edge e_i of G by a pair of edges $f_i, \bar{f}_i$ with a new vertex between them. Let $G' = (V', E')$ be the new graph so obtained. Let T be the set of subsets $A \subseteq E'$ such that no two edges of T are incident with the same vertex of G' unless it is one of the new vertices created by subdivision. Then T is the collection of independent sets of a matroid M on E' and if $f_i, \bar{f}_i$ are mates the problem of finding a maximum matching in G is the matroid parity problem for M with this partition of E'.

Example 2. Let M_1, M_2 be matroids on S and suppose that we wish to find a set X independent in both M_1 and M_2 and having maximum cardinality.

To represent this as a matroid parity problem take $\bar{S}$ a set of cardinal equal to S but disjoint from S and let M_2' be an isomorphic copy of M_2 on $\bar{S}$. For $e \in S$ let $\bar{e} = \phi(e)$ where ϕ is an isomorphism $M_2 \to M_2'$. Now let $N = M_1 + M_2'$ on $S \cup \bar{S}$ and we see that there is a $1 - 1$ correspondence between independent parity sets of N and common independent sets of M_1 and M_2.

If we now partition S into blocks each of size k and call a set X a *k-parity set* if it either contains a block or does not intersect it, the *k-parity problem* for M is to find an independent k-parity subset of S of maximum cardinality.

EXERCISES 3

1. Find a "fast" algorithm for the 2-parity problem.

2. Show how the problem of finding a set of maximum cardinality which is independent in k matroids can be reduced to a k-parity problem. (Lawler [71]).

3. Show that a k-parity problem can be reduced to the problem of finding a set of maximum cardinality which is independent in 3 matroids. (Lawler [71]).

4. Use the intersection theorem to prove that if M is a simple matroid on $2n$ elements $\{x_i : 1 \leqslant i \leqslant n\} \cup \{x_i' : 1 \leqslant i \leqslant n\}$ and in which $\{x_1, \ldots, x_n\}$ is a base then provided M has no circuits of size k or less then M has at least 2^k bases B such that for each i, B either contains x_i or x_i'. (Magnanti [74]).

5. Let M be a matroid of rank r on $S = \{b_1, \ldots, b_r, e_1, \ldots, e_q\}$ where $B = \{b_1, \ldots, b_r\}$ is a basis of M. Let $A(B)$ be the $r \times q$ 0–1 matrix in which $a_{ij} = 1$ if and only if b_i belongs to the fundamental circuit of e_j in B. Define $t(A(B))$ to be the term rank of $A(B)$, $p(B) = \rho(S \backslash B)$. Prove that

$$\min_B t(A(B)) = \max_B p(B).$$

($p(B)$ is the *pseudo-combivalence* rank of B). (Maurer [75]).

4. LEHMAN'S SOLUTION OF THE SHANNON SWITCHING GAME

Closely connected with the partitioning algorithm is an attractive game on graphs invented by C. Shannon, and solved by Lehman [64]. Lehman's arguments are in terms of a corresponding game played on matroids.

Let G be an undirected graph with two distinguished vertices u, v. Game $\langle G; u, v \rangle$ is played between two players A (the *join player*) and B (the *cut player*). Of the two players, A's goal is to construct a path in G which connects u and v. Player B's goal is to obstruct all routes which connect u and v. In his turn B can delete one of the edges of G, while player A can make an edge invulnerable to deletion. The game ends when one of the players reaches his goal.

It is clear that if a player can win by playing second he can win, *a fortiori*, playing first.

Thus game $\langle G; u, v \rangle$ has exactly one of the following properties.

(i)   Player B plays first and player A can win against all possible strategies of B.

(ii) Player A plays first and player B can win against all possible strategies of A.

(iii) The player who plays first, regardless of identity can win against all possible strategies of the other player.

Games having properties (i), (ii), (ii) we call *join*, *cut*, or *neutral* games respectively.

Simple examples of such games are shown in Fig. 1.

Less trivial examples shown in Fig. 2 illustrate the general principle:

(1)   A neutral game is converted into a join game by inserting a new edge between the distinguished vertices u and v.

Join game   Cut game   Neutral Game

Figure 1

It turns out to be just as easy to analyse games where any subset U of vertices of G is taken as the distinguished set and the goal of player A is to tag a set of edges $\{e_1, \ldots, e_p\} \subseteq G$ such that the subgraph determined by $\{e_1, \ldots, e_p\}$ contains a connected subgraph containing U.

To interpret this game in matroid terms, adjoin to G a set of new edges which form a connected graph K containing precisely the terminals as vertices. Relative to the cycle matroid of the new graph $G \cup K$ the goal of A is to tag a set of edges of G which span the set of edges in K.

Thus the matroid generalization can be formulated as follows. Let M be a matroid on S and let N, K be nonempty subsets of S. In the game $\langle M; N, K \rangle$ players A and B take turns tagging different elements of N with player B going first. Player A wins if he tags a set of elements which span K. (A set

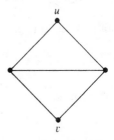

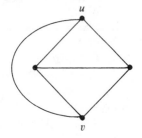

Neutral Game   Join Game

Figure 2

X_1 *spans* X_2 if $X_2 \subseteq \sigma X_1$ in the matroid M). Call $\langle M; N, K \rangle$ a *join game* if A can win against any strategy of player B, *cut games* and *neutral games* are defined similarly.

Lehman's main result (explicitly for the case where K is a single element) and generalized by Edmonds [65a] to arbitrary K is

THEOREM 1. $\langle M; N, K \rangle$ *is a join game if and only if N contains two disjoint sets, A_0 and B_0, which span each other and which span K.*

Proof. Call two (or more) sets which span each other *cospanning*. Given two disjoint cospanning sets A_0, B_0 of N which span K we will provide a winning strategy for the join player. We may assume A_0, B_0 are independent.

Let M_0 be the restriction of M to $S_0 = \sigma(A_0 \cup B_0)$. If the cut player tags an element not in $A_0 \cup B_0$ we can pretend that at the same time he also tags some element of $A_0 \cup B_0$. Clearly the join player would not be taking an illegal advantage by pretending this. Hence suppose the cut player first tags element $a_0 \in A_0$. There exists (at least one) $b_0 \in B_0$ such that $(A_0 \backslash a_0) \cup b_0$ is a base of M_0. The join player should tag b_0.

Consider $M_1 = M_0 . (S_0 \backslash b_0)$. It is easy to prove:

(2) $A_1 = A_0 \backslash a_0$, $B_1 = B_0 \backslash b_0$ are cospanning in M_1.

Since it is the cut player's turn again the situation of A_1 and B_1 relative to M_1 is as it was for A_0 and B_0 except that M_1 is on a set of one fewer elements.

Assume there is a strategy for the succeeding turns whereby the join player can tag a set of elements which contains a base T of M_1. Now use the following easy lemma.

LEMMA. *If M is a matroid on S, $X \subseteq S$, then the elements in a base of $M | (S \backslash X)$ together with the elements in a base of $M . X$ form a base of M.*

Thus the set $T \cup b_0$ which the join player will have tagged is a base of M_0 and hence must span K.

When B_0 contains only one element b_0, then b_0 itself spans M_0 and K. Hence by induction on the number of elements in $\sigma(A_0 \cup B_0)$ we have a winning strategy for the join player. This proves that the existence of two disjoint cospanning sets guarantees the join player a win. The harder converse is proved in Lehman [64]. ∎

Note also that the above proof gives the join player a winning strategy for the game. In an actual playing of the game he would of course have to locate the sets A_0, B_0 in advance.

As a corollary we state the version of Theorem 1 for graphs.

Call two trees *coextensive* if they have the same set of vertices. Then for the game $\langle G; u, v \rangle$ Theorem 1 reduces to

THEOREM 2. *The Shannon switching game $\langle G; u, v \rangle$ is a join game if and only if G contains two edge disjoint coextensive trees which have among their vertices the distinguished vertices u and v.*

The main theorem can be used to classify neutral and cut games for the general matroid game $\langle M; N, K \rangle$. We shall merely state the results for the game $\langle G; u, v \rangle$.

THEOREM 3. *Consider the game $\langle G; u, v \rangle$. Let G_1 be obtained from G by inserting a new edge between u and v. Then*

(a) $\langle G; u, v \rangle$ is neutral $\Leftrightarrow$ $\langle G; u, v \rangle$ is not a join game but $\langle G_1; u, v \rangle$ is.

(b) $\langle G; u, v \rangle$ is a cut game $\Leftrightarrow$ $\langle G_1; u, v \rangle$ is not a join game.

In general there is no easily describable strategy for the cut player going second in cut games or going first in neutral games. When G is a planar graph such a strategy does exist,

Suppose $\langle G_1; u, v \rangle$ is a cut game and the graph G_1 formed by adding a new edge e_0 to G joining u and v is a planar graph. Let G_1^* be any planar dual of G_1 and let e_0^* be the edge of G_1^* corresponding to e_0. Let u^*, v^* be the endpoints of e_0^*. Then we have:

(3) $\langle G; u, v \rangle$ is a cut game if and only if $\langle G_1^*; u^*, v^* \rangle$ is a join game.

Moreover the winning strategy described in the proof of Theorem 1 for player A on $\langle G_1^*; u^*, v^* \rangle$ gives in the obvious way a winning strategy for player B in $\langle G; u, v \rangle$.

EXERCISES 4

1. Prove that $\langle G; u, v \rangle$ is a cut game if and only if G has two edge disjoint coextensive cotrees which have u and v among their vertices.

2. Prove that $\langle G; u, v \rangle$ is a neutral game if and only if both (a) and (b) hold:
    (a) there exist edge disjoint coextensive trees T_1, T_2 and $(u, v) \in T_1 \backslash T_2$,
    (b) there exist edge-disjoint cotrees T_1^*, T_2^* which are cospanning in the cocycle matroid of G and $(u, v) \in T_1^* \backslash T_2^*$. (Bruno and Weinberg [71]).

5. AN EXTENSION OF NETWORK FLOW THEORY

Many problems in operations research and combinatorial optimization theory can be formulated in terms of finding maximum flows in capacity constrained networks. An account of both the theory and its applications is given in the book by Ford and Fulkerson [62]. In this section we show how the basic max-flow min-cut theorem of network flow theory has a natural extension to matroids.

Suppose that we are given an undirected graph G and two distinguished vertices u, v, to be called the *source* and *sink* respectively. If $c_i \geqslant 0$ is the *capacity* of the edge e_i and represents the amount of flow which the edge e_i can support, the maximum feasible flow from u to v in the capacitated graph can be found by the following method.

Insert an additional distinguished edge e joining the vertices u, v.

Let $C_1, \ldots, C_p$ be the cycles of G passing through the edge e. The value of the maximum flow from u to v is the maximum value of $u_1 + \ldots + u_p$ subject to the constraints that the flow along e_i is not more than c_i and $u_i \geqslant 0$ represents the *flow* or *circulation* around the cycle C_i.

Analogously we can formulate the maximum flow problem for matroids. Let M be a matroid on S and let e be a distinguished element of S. Let $S = \{e, e_1, \ldots e_n\}$, and let $C_1, \ldots, C_p$ be the circuits of M which contain e. Let $c_i \geqslant 0$ be a real number ($1 \leqslant i \leqslant n$) which we call the *capacity* of e_i.

Define the $n \times p$ matrix $D = (d_{ij})$ by

$$d_{ij} = \begin{cases} 1 & e_i \in C_j, \\ 0 & e_i \notin C_j. \end{cases}$$

Then a (*feasible*) *e-flow* in M is a vector $(u_1, \ldots, u_p)$ satisfying

$$(1) \quad \sum_{j=1}^{p} d_{ij} u_j \leqslant c_i \quad (1 \leqslant i \leqslant n),\ u_i \geqslant 0\ (1 \leqslant i \leqslant p).$$

The vector $u = (u_1, \ldots, u_p)$ is a *maximum e-flow* if it maximizes

$$u_1 + \ldots + u_p$$

subject to the constraints (1) and Σu_i is then called the *value* of the maximum e-flow.

It is not difficult to show that when M is the cycle matroid of an (undirected) graph G with distinguished edge e, a maximum e-flow in M will be a maximum flow between the endpoints of e in the graph $G \backslash e$.

Now let C^* be any cocircuit of M which contains e. We define the *capacity* $c(C^*)$ of C^* by

$$c(C^*) = \sum_{i:\, e_i \in C^*} c_i$$

We say that M has the *max-flow min-cut* (MFMC) property if for any element e which is not a loop and any set of real capacities $c_i \geqslant 0$, the maximum value of an e-flow equals

$$\min c(C^*)$$

where the minimum is taken over all cocircuits C^* containing e. We call this minimum the *min capacity of e*.

LEMMA. *In any matroid the maximum value of an e-flow is less than or equal to the min capacity of e.*

Proof. Let $C^* = \{e, e_1, \ldots, e_t\}$ be a cocircuit of M, and suppose that

$$c(C^*) = c_1 + c_2 + \ldots + c_t < u_1 + \ldots + u_p,$$

for some feasible flow u.

By (1)

$$(d_{11}u_1 + \ldots + d_{1p}u_p) + \ldots + (d_{t1}u_1 + \ldots + d_{tp}u_p) < u_1 + \ldots + u_p.$$

Rearranging we get

$$(2) \quad (d_{11} + d_{21} + \ldots d_{t1})u_1 + \ldots + (d_{1p} + \ldots + d_{tp})u_p < u_1 + \ldots + u_p.$$

Now a circuit and cocircuit of M cannot have exactly one element in common, and since $C_i \cap C^* \ni e$, we must have

$$C_i \cap \{e_1, \ldots, e_t\} \neq \varnothing \ (1 \leqslant i \leqslant p).$$

Hence for each i, $1 \leqslant i \leqslant p$,

$$d_{1i} + \ldots + d_{ti} \geqslant 1.$$

This contradicts (2) and proves the lemma.

For general matroids we cannot say much more, it is easy to find examples where the maximum e-flow has a value strictly less than the min capacity.

However for regular matroids, the natural analogue of the maxflow–mincut theorem of networks remains true.

THEOREM 1. *In any regular matroid the maximum value of an e-flow equals the min capacity of e.*

Since the cycle matroid of a graph is regular this gives as a corollary the main theorem of Ford and Fulkerson [56].

The proof of Theorem 1 is not easy. We refer to Minty [66] or Fulkerson [68].

Moreover because of the unimodular property of vector representations of regular matroids if we let the $c_i = 1$ for each element e_i we can then get the following extension of Menger's theorem to regular matroids.

THEOREM 2. *Let M be a regular matroid on S and let e not be a loop of S, then if C^* is a cocircuit of M which contains e and is of minimum cardinality then*

$$|C^*| - 1 = p$$

where p is the maximum number of circuits of M which contain e but are otherwise pairwise disjoint.

Note that even without the hypothesis that M is regular we have $|C^*| - 1 \geqslant p$, for otherwise there will exist a circuit C_i with $|C_i \cap C^*| = 1$ which is a contradiction.

A consequence of Menger's theorem for regular matroids is the following theorem about graphs.

If u, v are distinct vertices of a graph G we say that a set A of edges of G is a *uv-cocycle* if A is a minimal set of edges such that in the graph $G \backslash A$ there is no path joining u and v.

COROLLARY 1. *Let u, v be distinct vertices of a connected graph G. Then the maximum number of pairwise disjoint uv-cocycles equals the length of the shortest path from u to v in the graph G.*

Proof. Let G' be the graph obtained from G by adding an additional edge e_0 joining u, v. Now the cocycle matroid $M^*(G')$ is regular and applying Theorem 2 to the matroid $M^*(G')$ with $c(e) = 1 \; \forall \, e$, we notice that the cardinality of the smallest cocircuit of $M^*(G')$ is just one more than the length of the shortest path from u to v in G. ∎

It is interesting to note that although Corollary 1 is the exact matroid dual of Menger's theorem and should therefore have a proof of the same order of difficulty it is in fact very much easier to find an ad hoc proof of Corollary 1 than to supply an ad hoc proof of Menger's theorem (see Exercise 6).

Yet another consequence of the Menger's theorem is the length–width inequality for regular matroids.

Let M be a matroid on S and let e be a distinguished element of S. Let $\lambda(e)$, the *length* of e and $w(e)$ the *width* of e be defined by: $\lambda(e) + 1$ is the cardinality of the smallest circuit of M containing e, $w(e) + 1$ is the cardinality of the smallest cocircuit of M containing e. If e is a loop we define $w(e) = 0$, if e is a coloop $\lambda(e) = 0$.

We say that the *length width inequality* holds for the matroid M on S if for each element $e \in S$,

$$(3) \qquad \lambda(e)\,w(e) \leqslant |S| - 1.$$

THEOREM 3. *If M is a regular matroid the length–width inequality is satisfied.*

Proof. When e is a loop or coloop $\lambda(e)\,w(e) = 0$ and (3) holds. Since M is regular M^* is regular and hence by Menger's theorem for regular matroids if e is not a loop or coloop of M^* there exist circuits $C_1, \ldots, C_{w(e)}$ which are pairwise disjoint apart from their common element e. But $|C_i| \geqslant \lambda(e) + 1$ and since the $C_i \backslash e$ are disjoint

$$|S \backslash e| \geqslant |C_1 \backslash e| + \ldots + |C_{w(e)} \backslash e|$$

which gives

$$|S| - 1 \geqslant w(e)\lambda(e)$$

and proves the theorem. ∎

Note that there exist many non-regular matroids for which the length–width inequality holds.

Example. Let M be the Fano matroid. Then for any element e,

$$\lambda(e) = 2, \quad w(e) = 3$$

and we have equality in the length–width inequality.

Recently Seymour [75] has proved a remarkable characterisation of those matroids for which an integer max-flow min-cut theorem holds. We briefly describe his main results.

If M is a matroid on $S \cup e$ we say that (M, e) has the *integer max flow min cut property* (($Z^+ -$ MFMC) property) if when the capacity c_i are restricted to being non-negative integers there exists a non-negative integer flow $(u_1, \ldots, u_p)$ which equals the min capacity of e.

It is clear that:

(4) If (M, e) has the $Z^+ -$ MFMC property then (M, e) has the (MFMC)– property.

However the dual of the Fano matroid is an example of a matroid which shows that the converse of (4) is false.

Seymour's main result is the following complete characterisation of those matroids for which the integer max flow min cut theorem is true.

THEOREM 4. *Let M be a connected matroid on $S \cup e$. Then (M, e) has the integer max flow min cut property if and only if M is binary and there does not exist* *a minor M_1 of M on a subset $T \cup e$ of $S \cup e$ such that M_1 is isomorphic to the dual of the Fano matroid.*

The proof of this is quite involved.

EXERCISES 5

1. Show that the max-flow min-cut property is not true for the non-binary matroid $U_{2,4}$.

2. Give an example to show that some non-regular matroids have the MFMC property.

3. Prove that the affine geometry AG(3, 2) does not have the max-flow min-cut property. (Seymour [75]).

4. Let M be binary on $T \cup e$, $e \notin T$ and let $M \mid T$ be graphic. Prove that with respect to the edge e, M satisfies the max-flow min-cut theorem. (Seymour [75]).

5. Prove that (M, e) has the MFMC property if and only if (M^*, e) has the MFMC property. (Fulkerson [70]).

6. Supply a direct *ad-hoc* proof of Corollary 1.

∘7. Characterize those matroids which have the MFMC property.

6. SOME INTRACTABLE PROBLEMS

We close with a brief discussion of the limitations of matroids as a tool in discrete optimization. This is probably best illustrated by two specific problems which at one time it seemed might be solved by the techniques of matroid theory.

Covering problems
    We say that a family of sets $\mathscr{A} = (A_i : i \in I)$ is a *cover* for a set S if $A(I) \supseteq S$. The covering problem is the following.

(P1)   If $\mathscr{A} = (A_i : i \in I)$ is a cover of S find a subcover $(A_i : i \in J)$, $J \subseteq I$, of S which has the property that $|J|$ is a minimum.

By the principle of point-set duality (P1) is equivalent to the dual problem.

(P1)*   Given a family $\mathscr{A} = (A_i : i \in I)$ of subsets of S find a system of representatives $(x_i; i \in I)$ of $\mathscr{A}$, such that $|\{x_i : i \in I\}|$ is a minimum.

Problem (P1)* is just a special case of the very basic matroid problem.

(P2)   Given a matroid M on S and a family $\mathcal{A}$ of subsets of S find necessary and sufficient conditions for $\mathcal{A}$ to have a system of representatives $(x_i : i \in I)$ such that the set $\{x_i : i \in I\}$ is independent in M.

Fundamental problems of discrete optimisation are to find efficient (non-exhaustive) algorithms for these three problems.

To see that any algorithm for (P2) can be modified to give an algorithm for (P1)* we just take M to be the uniform matroid of rank k, and then let k run through the integers.

In view of Rado's Theorem it is surprising that there exist no Hall-type conditions for the existence of such a set of representatives in problem (P2). At the moment I know of no solution to this even in special cases such as when the A_i are pairwise disjoint sets.

For a more general discussion of covering problems we refer to Lawler [66].

The Timetabling Problem

An important but apparently intractable problem of operations research is the timetabling problem. Although it has applications to many scheduling problems it is best described in terms of teachers and classes, and meetings to be arranged between them.

Let $\mathcal{T} = (T_i : 1 \leqslant i \leqslant m)$ be a collection of teachers and let $\mathcal{C} = (C_i : 1 \leqslant i \leqslant n)$ be a collection of classes. Suppose we are given:

(i)   an $m \times n$ *requirements matrix* $R = (r_{ij})$ where r_{ij} is an integer denoting the number of times teacher T_i must meet with class C_j;

(ii)   a collection of *preassignments* (i, j, k), $(1 \leqslant i \leqslant m, 1 \leqslant j \leqslant n, 1 \leqslant k \leqslant d)$ where the preassignment (i, j, k) means that teacher T_i must meet class C_j on day k.

The *timetabling problem* is:

(a)   to find necessary and sufficient conditions for deciding whether the given set of requirements and preassignments can be met in a schedule of less than or equal to d days.

(b)   if (a) is possible, find an algorithm for constructing a timetable.

Regarded as a problem of graph theory the timetabling problem can be rewritten in the following form.

Given a bipartite graph $\Delta = (\mathcal{T}, \mathcal{C}; E)$ where the vertex T_i is joined to C_j by r_{ij} edges suppose that some of the edges of Δ are coloured using up to d-colours, (if (i, j, k) is a preassignment then colour one of the edges (i, j) with colour k).

(a)   Test whether this partial edge colouring of Δ can be extended to a full edge colouring of Δ in which no two incident edges have a common colour.

(b) Find an algorithm for extending the given partial edge colouring to a full colouring in which no two incident edges have the same colour.

An easy extension of König's theorem, see Berge [58] is the following result.

THEOREM 1. *Let Δ be a bipartite graph. Then there exists a matching in Δ which is incident with every vertex of maximum degree.*

A consequence of this is:

COROLLARY. *If a bipartite graph Δ has maximum vertex degree d then Δ is the union of d matchings.*

When these results are applied to the timetabling problem we have

THEOREM 2. *When there are no preassignments a timetable is possible in d_0 days where d_0 is the maximum vertex degree of the bipartite graph $\Delta = (\mathscr{F}, \mathscr{C}; E)$.*

Clearly d_0 is best possible and moreover easy algorithms (either the Hungarian algorithm for the assignment problem, or repetitions of the Ford–Fulkerson maximum flow algorithm) will give an optimum timetabling for this particular problem.

When preassignments are present however the problem seems to be extremely difficult.

Viewed as a problem in transversal theory the timetabling problem is equivalent to the following problem.

Given two families $\mathscr{A} = (A_i : i \in I)$ and $\mathscr{B} = (B_i : i \in I)$ of subsets of S where

$$A_i \cap A_j = B_i \cap B_j = \varnothing \qquad (i \neq j)$$

and also a family $(P_i : 1 \leqslant i \leqslant d)$ of subsets of S find necessary and sufficient conditions for $\mathscr{A}$ and $\mathscr{B}$ to have d pairwise disjoint common transversals $(T_i : 1 \leqslant i \leqslant d)$ with $P_i \subseteq T_i$, $1 \leqslant i \leqslant d$. Fulkerson [71], and de Sousa [71] have found necessary and sufficient conditions for two families $\mathscr{A} = (A_i : i \in I)$ and $\mathscr{B} = (B_i : i \in I)$ to have d disjoint common transversals, namely that for all subsets J, K of I,

$$|A(J) \cap B(K)| \geqslant d(|J| + |K| - |I|).$$

However there seems to be no way of extending their techniques to ensure that these common transversals contain prescribed subsets.

Roughly speaking we say that a problem is *hard* if all presently known algorithms for it require in the worst case a computing time which is at least an exponential function of the amount of input data.

All of the problems discussed above are hard in this sense. Examples of graph problems which are not hard are:

(a) computing maximum flows in capacitated networks
(b) finding maximum weight spanning trees of graphs
(c) computing shortest routes in networks
(d) finding a common transversal of two families of sets
(e) finding a transversal of a family of sets which is independent in a given matroid.

This concept of "hard problem" has been made much more precise by recent work of Cook [71] and Karp [72]. In his very interesting paper Karp exhibits a logical equivalence among twenty-one classic computationally difficult (hard) problems which suggests strongly that they are all intractable in the sense that there is no algorithm for their solution which is a polynomial function of the amount of data needed to describe them.

Both the timetabling and cover problem are intractable in this sense. It would seem likely that many of the more difficult unsettled problems about matroids have algorithmic formulations which are also intractable in this sense. In particular, the key problem of finding an algorithm which will decide whether or not three matroids have a common independent set of prescribed cardinality must surely fall into this category.

EXERCISES 6

1. Find an algorithm which computes the shortest routes between each pair of vertices in an n-vertex graph in $O(n^3)$ computations.

○2. Prove that there is no algorithm for problem 1 which needs $O(n^\alpha)$ computations for $\alpha < 3$.

NOTES ON CHAPTER 19

That the greedy algorithm is an efficient algorithm for selecting a spanning tree of a graph of maximum weight is well known, see for example the references in Ore [62]. Rado [57] was the first to recognize its extension to matroids and later independently Edmonds [67c], Gale [68] and Welsh [68a] reproved this same theorem. Klee [71] has extended it to infinite sets. Theorem 1.2 is essentially due to Gale [68]. Section 2 is based on Dunstan and Welsh [73].

The partitioning algorithm presented in Section 3 is due to Knuth [73]. A similar algorithm was discovered by Greene and Magnanti [74] and both are based on the original method of Edmonds [67b]. However the particular simplicity and elegance introduced to the presentation by interpreting it in terms of paintings is due to Knuth. Various algorithms (of the same order of complexity) exist for the intersection problem, in particular Lawler [70], [75] and Aigner and Dowling [71] have methods which, via matroid duality, are equivalent to the partitioning algorithm for two matroids. The discussion of parity problems is based on Lawler [71]. For a more detailed discussion of these algorithms see the forthcoming book by Lawler [75].

The treatment of the Shannon switching game given in Section 4 is based on the expository paper of Brualdi [74b]. For other treatments of this problem, in addition to the fundamental paper of Lehman [64], see Edmonds [65a], Bruno and Weinberg [71].

Section 5 is based on Minty [66], Fulkerson [68], [70] and Seymour [75], [75a], [75b]. A different extension of Menger's theorem to general non-binary matroids has been carried out by Tutte [65a]. However this is in terms of the connectivity function introduced in Section 5.6 and though in this way we get an extension of Menger's theorem to arbitrary non-regular matroids the extension is not as natural in that it is far from obvious that when M is graphic, Menger's graph theorem is a special case of Tutte's extension of his theorem.

Section 6 is included because of the recent interest in the work on the complexity of computational algorithms. Although there is now a quite extensive literature on the computational complexity of graph algorithms little is known at the moment about the complexity of matroid algorithms.

In addition to the algorithms mentioned in this chapter we should also note that M. Hall [56] has an algorithm for finding a transversal of a family of sets, and Hautus [70] has an extension which finds a transversal independent in a given matroid. Clearly these are special cases of any algorithm for finding maximal sets which are independent in two matroids.

Although not explicitly about matroids, the papers of Fulkerson [68], [70], [72] are very interesting extensions of some of the ideas of this chapter, in particular of the ideas of generalised network flows. Related to this is the paper of Rockafellar [69] which relates various graph or matroid theorems with results in convex analysis. He also suggests a theory of oriented matroids, recently developed axiomatically by Las Vergnas [75] and Bland [75].

A *pivotal system* or *combivalence class* of matrices is the set of matrices obtainable from a given matrix A by the standard pivot exchange procedure of linear programming. The study of such a class is the basis of the combinatorial theory of linear programming developed by Tucker [63]. Iri [68] proved the interesting result that the maximum rank for matrices in a pivotal system equals the minimum term rank, and that moreover some matrix in the system has both. Maurer [75a] has generalized this result to matroids extending an unpublished matroid argument of D. R. Fulkerson. See also Bruno and Weinberg [71]. This study of pivotal systems is related to Maurer's study of *basis graphs* of matroids [73–73b, 75]. These are graphs in which a vertex corresponds to each basis and two vertices are joined only if it is possible to move between the corresponding bases in a single exchange operation. Maurer completely characterises those graphs which can arise as basis graphs.

CHAPTER 20

Infinite Matroids

1. PRE-INDEPENDENCE SPACES

The aim of this chapter is to indicate directions in which the matroid theory developed for finite sets in the earlier chapters can be extended to infinite sets. In many ways the theory developed so far for infinite sets is very incomplete. The basic problem seems to be that there are many non-equivalent ways in which the axioms of finite matroids can be extended when the underlying set is infinite. We will indicate some of these and hope that this will give some flavour of the uncertainty which seems to surround the subject.

The simplest (or most naive) approach is of course to add nothing to the finite axioms. Accordingly we define a *pre-independence space* (*pi-space*) to be a set S and a non-empty family $\mathscr{I}$ of subsets of S called *independent* sets satisfying

(I1)   (Hereditary axiom). If $A \in \mathscr{I}$, and $B \subseteq A$ then $B \in \mathscr{I}$

(I2)   (Finite exchange axiom). If A, B are finite members of $\mathscr{I}$ with $|A| = |B| + 1$ there exists $a \in A \backslash B$ such that $B \cup a \in \mathscr{I}$.

Thus when S is finite a pi-space on S is just a matroid. A set is *dependent* if it is not independent.

As an example of all that is wrong with pi-spaces consider the following example.

Example 1. Let S be an infinite set and let $\mathscr{I}$ consist of all finite subsets of S.

This illustrates the remark that a pi-space need have no bases and no circuits.

To get around this first difficulty it is natural to propose the additional axiom

(m)   (Maximal condition). If X is an independent subset of S there exists a maximal independent subset of S which contains X.

385

We call a pi-space satisfying (m) an *mpi-space*.

Although we can now define a *base* of a set A to be a maximal independent subset of A consider the following example noted by Dlab [65].

Example 2. Let α_1, α_2 be any two infinite cardinal numbers. Take S_1, S_2 to be disjoint sets, $|S_1| = \alpha_1, |S_2| = \alpha_2$ and let $\mathscr{I}$ be the mpi-space defined on $S_1 \cup S_2$ as follows. The maximal members of $\mathscr{I}$ are all sets of the form

$$(S_1 \backslash F_1) \cup F_2, (S_2 \backslash F_2) \cup F_1$$

where F_1, F_2 satisfy

$$F_1 \subseteq S_1, \qquad F_2 \subseteq S_2, \qquad |F_1| = |F_2| < \infty.$$

It is easy to check that this is an mpi-structure and has bases of cardinalities α_1 and α_2.

Dlab [65] has generalized this idea and proved the following result.

(1)   Given any set $\{\alpha_i : i \in I\}$ of distinct cardinals there exists an mpi-space for which the set of cardinalities of bases is exactly the set $\{\alpha_i : i \in I\}$.

With a pi-space $(S, \mathscr{I})$ we associate a *rank function* ρ taking values from $\{0, 1, 2, \ldots, \infty\}$ and defined for $X \subseteq S$ by

$$\rho X = \sup \{|Y| : Y \in \mathscr{I}, Y \subset \subset X\}$$

where $Y \subset \subset X$ indicates that Y is a finite subset of X.

It is easily shown that for $X, Y \subseteq S$,

$$\rho X + \rho Y \geqslant \rho(X \cup Y) + \rho(X \cap Y).$$

For further properties of the rank function of pre-independence spaces we refer to Mirsky [71].

2. INDEPENDENCE SPACES

We first state two basic results of set theory. For proofs and a discussion of the relation between these theorems we refer to Mirsky's book [71 Chapter 1].

A collection $\mathscr{E}$ of subsets of a set S has *finite character* if a set X belongs to $\mathscr{E}$ if and only if all its finite subsets belong to $\mathscr{E}$.

THEOREM 1 (Tukey's lemma). *if a collection $\mathscr{E}$ of subsets has finite character then every member of $\mathscr{E}$ is a subset of some member maximal with respect to inclusion. In particular $\mathscr{E}$ possesses a maximal member.*

The proof of this is an easy deduction from Zorn's lemma. Conversely it can be shown that it implies Zorn's lemma, so that the axiom of choice, Zorn's lemma, and Tukey's lemma are all equivalent to each other.

Given a family $\mathscr{A}$ of subsets $(A_i : i \in I)$ of S, a *choice function* θ for $\mathscr{A}$ is a map $\theta : I \to S$ such that $\theta(i) \in A_i \; \forall \, i \in I$. In other words, a choice function is a system of representatives of the family $\mathscr{A}$.

The *axiom of choice* asserts that if the collection of choice functions of $\mathscr{A}$ is empty then there exists some $i \in I$ such that $A_i = \varnothing$.

Rado's selection principle is a fundamental tool of infinite combinatorics and can be stated as follows.

THEOREM 2 (Rado's selection principle). *Let* $\mathscr{A} = (A_i : i \in I)$ *be a family of finite subsets of* S. *Let* $\mathscr{F}(I)$ *be the collection of finite subsets of* I *and for each* $J \in \mathscr{F}(I)$ *let* θ_J *be a choice function for the family* $(A_i : i \in J)$. *Then there exists a choice function* θ *for* $\mathscr{A}$ *such that for each* $J \in \mathscr{F}(I)$, *there exists* $K \in \mathscr{F}(I)$ *with* $J \subseteq K$ *and* $\theta | J = \theta_K | J$.

A very short proof of this, based on Tychonoff's theorem that the product of compact topological spaces is compact can be found in Mirsky [71 Chapter 4]. This also contains some interesting applications of the principle to diverse branches of mathematics.

An *independence space* is a pi-space $(S, \mathscr{I})$ which satisfies the further condition:

(FC) (Finite character) If $A \subseteq S$ and every finite subset of A is a member of $\mathscr{I}$ then $A \in \mathscr{I}$.

From Tukey's lemma we see immediately that an independence space is an mpi-space and hence bases exist. Indeed more can be said, for if we define a set to be *spanning* if it contains a base we have the following result of Rado [49] or Robertson and Weston [58]).

THEOREM 3. *In an independence space all bases have the same cardinality. If* X *is independent and* Y *is spanning with* $X \subseteq Y$ *there exists a base* B *with* $X \subseteq B \subseteq Y$.

We now define a *hyperplane* to be a maximal set not containing any base and a *circuit* to be a minimal dependent set. Clearly these concepts are well defined and it is a routine exercise to prove:

THEOREM 4. *A collection* $\mathscr{C}$ *of incomparable subsets of* S *is the set of circuits of an independence space on* S *if and only if*

(C1) $C \in \mathscr{C} \Rightarrow C \neq \varnothing, |C| < \infty,$

(C2)   *If C_1, C_2 are distinct members of $\mathscr{C}$ and $x \in C_1 \cap C_2$ there exists $C_3 \in \mathscr{C}$, such that $C_3 \subseteq (C_1 \cup C_2) \backslash x$.*

In the same way as in the finite case we can now define the *closure operator* σ of an independence space $(S, \mathscr{I})$ by $x \in \sigma A$ if $x \in A$ or if there exists a circuit C with $x \in C \subseteq A \cup x$. A set X is *closed* or a *flat* if $\sigma X = X$ and it is easy to check that σ satisfies the axioms (S1)–(S4) of Section 1.7.

From these the reader familiar with universal algebra and in particular with Cohn [65] will recognize an independence space as a *transitive abstract dependence relation* or equivalently as a set S equipped with an *algebraic closure operator* $\sigma: 2^S \to 2^S$ which in addition to the usual axioms (S1)–(S4) of Section 1.7 also satisfies the algebraic condition:

(S5)   If $a \in \sigma X$ for some $X \subseteq S$ then $a \in \sigma(X_f)$ for some finite subset X_f of X.

The relationship between independent sets and closure operators spelt out for matroids in Chapter 1 carries over to independence spaces without difficulty. Thus for an alternative proof of Theorem 3 see Cohn [65, p. 253].

Suppose now we look at the collection of closed sets or flats of an independence space.

Since the closure operator σ of an independence space satisfies (S1)–(S4) we have from a basic theorem of lattice theory (see Crawley and Dilworth [73] p. 25).

THEOREM 5. *The closed sets of an independence space form a complete semi-modular lattice under set inclusion, and each of its elements is a join of atoms.*

However in general we no longer have a geometric lattice since even when infinite a geometric lattice is defined to have only finite dimension (= height). Thus the theory of general geometric lattices corresponds to the theory of rank-finite independence spaces.

Maeda and Maeda [70] call a lattice in which each element is the join of atoms, *atomistic* and go on to define an atomistic lattice $\mathscr{L}$ to be *compactly atomistic* if $\mathscr{L}$ also satisfies:

(CA)   If p is an atom of $\mathscr{L}$ and A is a set of atoms of $\mathscr{L}$ such that $p \leqslant \vee (q: q \in A)$ then there exists a finite subset $\{q_1, \ldots, q_n\}$ of A such that

$$p \leqslant q_1 \vee \ldots \vee q_n.$$

Finally they define a *matroid lattice* to be a complete, compactly atomistic, semimodular lattice.

Thus we have seen that essentially the same structure keeps on cropping up; it appears again as the finitary matroid of Klee see section 5 below. This is probably because the underlying model which has been abstracted in each case has had an algebraic nature—either linear or algebraic dependence over a field. Thus on an infinite set any "matroid" not of finite character is not linearly representable or algebraically representable—a consequence of the definitions of linear and algebraic dependence.

EXERCISES 2

1. If $T \subseteq S$ and $\mathscr{I}$ is an independence space on S show that the restriction and contraction of $\mathscr{I}$ to T as defined for matroids in Chapter 4 remains valid in the sense that the resulting structures are still independence spaces.

2. Let $(S, \mathscr{I})$ be an independence space. Call $X \supseteq S$ *codependent* if it has non-empty intersection with every base of $\mathscr{I}$. Call C^* a *cocircuit* if it is minimal codependent. Prove that the hyperplanes of $(S, \mathscr{I})$ are the complements of the cocircuits.

3. INFINITE TRANSVERSAL THEORY

The area of combinatorics in which matroid theory has been most fruitfully extended to infinite sets seems to be transversal theory. Here we will indicate the main ideas of this theory, referring the reader to the book of Mirsky [71] for further details.

The cornerstone of transversal theory are the Hall–Rado theorems and we first look at two infinite versions of these:

THEOREM 1. *Let $(S, \mathscr{I})$ be a preindependence space with rank function ρ. The finite family $(A_i : 1 \leqslant i \leqslant n)$ of subsets of S possesses an independent transversal if and only if for all $J \subseteq \{1, \ldots, n\}$,*

$$(R) \qquad \qquad \rho(A(J)) \geqslant |J|.$$

Proof. When each A_i is finite the reduction argument used in the proof of Theorem 7.2.1. proves the theorem. In the general case the condition (R) implies that for each $J \subseteq \{1, \ldots, n\}$ $A(J)$ contains an independent set C_J say, with $|C_J| = |J|$. Let

$$C = \bigcup_{J \supseteq \{1, \ldots, n\}} C_J, \qquad A_i^* = A_i \cap C, \qquad (1 \leqslant i \leqslant n).$$

Then the A_i^* are finite subsets of C and it is easy to see

$$\rho(A^*(J)) = \rho(A(J) \cap C) \geqslant \rho(C_J) = |J|.$$

Hence by the theorem already proved for finite A_i we know that $(A_i^*:$

$1 \leqslant i \leqslant n$) has an independent transversal. This transversal is clearly an independent transversal of $(A_i : 1 \leqslant i \leqslant n)$. ■

THEOREM 2. *Let* $(S, \mathcal{I})$ *be an independence space with rank function* ρ. *Let* $\mathcal{A} = (A_i : i \in I)$ *be a family of finite subsets of* S. *The following statements are equivalent.*

(a) $\rho(A(J)) \geqslant |J|$ *for every finite subset* J *of* I.
(b)   *Every finite subfamily of* $\mathcal{A}$ *has an independent transversal.*
(c) $\mathcal{A}$ *has an independent transversal.*

Proof. From the finite case of Rado's theorem we know (a) $\Rightarrow$ (b) and it is trivial that (c) $\Rightarrow$ (a). We prove (b) $\Rightarrow$ (a) by Rado's selection principle.

Let $J \subset\subset I$. The subfamily $\mathcal{A}_J$ has an independent transversal by (b) and so there is an injective choice function θ_J of $\mathcal{A}_J$ with $\theta_J(J) \in \mathcal{I}$.

Let θ be a choice function of $\mathcal{A}$ chosen in accordance with Rado's selection principle. It is easy to check that θ must be injective, and hence $\theta(I)$ is a transversal of $\mathcal{A}$.

Let $Y \subset\subset \theta(I)$, say $Y = \theta(J)$ where $J \subset\subset I$. Then there exists $K \subset\subset I$ containing J with $\theta_K(i) = \theta(i)$ $(i \in J)$.

But since $\theta_K(K) \in \mathcal{I}$, $\theta_K(J) \in \mathcal{I}$, and hence $\theta(J) \in \mathcal{I}$. Thus every finite subset of $\theta(I)$ is independent. Since $\mathcal{I}$ has finite character, $\theta(I)$ must be independent and is our required independent transversal. ■

The more stringent conditions of Theorem 2 as compared with Theorem 1 are necessary as the following examples show.

Example 1. Let $S = \{1, 2, \dots\}$. Let $\mathcal{I} = 2^S$. Then the family $\mathcal{A}$ defined by

$$A_0 = S, \qquad A_1 = \{1\}, \qquad A_2 = \{2\}, \dots$$

satisfies (a) and (b) but has no independent transversal.

Example 2. Let $S = \{1, 2, \dots\}$. Let $\mathcal{I}$ be the preindependence structure defined by $X \in \mathcal{I}$, only if X is finite. The family $\mathcal{A} = (A_i : 1 \leqslant i \leqslant \infty)$ defined by

$$A_i = \{i\} \qquad (i = 1, 2, \dots),$$

satisfies (a) and (b) but has no independent transversal.

The other key theorem of transversal theory Theorem 7.3.1 has also only a partial extension to the infinite case.

THEOREM 3. *Let* $\mathcal{A}$ *be a family of subsets of a set* S. *Then the collection of partial transversals of* $\mathcal{A}$ *is a preindependence space on* S. *If each element* x *of*

S belongs to only a finite number of the A_i, then these partial transversals form
an independence space.

Proof. Mirsky [71, p. 102, 103]. ∎

The following example shows the need for some restriction before we can
guarantee that the partial transversals form an independence space,

Example 3. Let $S = \{1, 2, 3, \ldots\}$ and let $A_i = \{1, i + 1\}: 1 \leqslant i < \infty$, then
the set of PTS of $\mathscr{A}$ is the collection of proper subsets of $\{1, 2, 3, \ldots\}$.

What is surprising in view of the amount of effort expended on transversal
theory in recent years is that the following basic problems are essentially
untouched in their full generality.

Problem 1. Find necessary and sufficient conditions for a family of sets to
have a transversal.

Problem 2. Find necessary and sufficient conditions for two families of sets to
have a common system of representatives (or a common transversal).

Problem 1 seems to be exceptionally difficult. The present state of the problem can best be seen by studying the recent paper of Damerell and Milner
[74].

Closely related to finite transversal theory is the idea of matroid union.
The following theorems of Pym and Perfect [70] illustrate further this
relationship.

Let $(\mathscr{I}_j: j \in I)$ be a collection of preindependence spaces on the set S with
rank functions ρ_i. Let

$$\bigvee_{j \in I} \mathscr{I}_j = \{X: X = \bigcup_{i \in I} A_i, A_i \in \mathscr{I}_i\}$$

and call this collection of sets the *union* of the preindependence spaces $(\mathscr{I}_j: j \in I)$.

THEOREM 4. *The union of a collection* $(\mathscr{I}_j: j \in I)$ *of preindependence spaces is a
preindependence space with rank function* ρ *given for* $Y \subseteq S$ *by*

$$\rho Y = \inf_{U \subseteq Y} \left(\sum_{i \in I} \rho_i U + |Y \backslash U| \right).$$

If each $\mathscr{I}_j$ *is an independence space on S and for each* $x \in S$, $\{x\}$ *is independent
in only finitely many* $\mathscr{I}_j$ *then* $\bigvee \mathscr{I}_j$ *is an independence space on S.*

(Note that we set $\sum_{i \in I} \rho_i U = \infty$ if $\rho_i U > 0$ for infinitely many *i*).

Notice the correspondence between "$\{x\}$ independent in only finitely many

$\mathscr{I}_j$" and the corresponding condition for families of sets (Theorem 3) that "each element x belongs to only finitely many A_i". It is of course easy to prove Theorem 3 from Theorem 4.

As in the finite case it is easy to obtain from Theorem 4 a set of necessary and sufficient conditions for a set S to be contained in the union of k disjoint independent sets or to have k disjoint independent sets. However what seem to be very difficult problems (unlike the finite case where they reduce to the above problem) are the following:

Problem 3. When does an independence space have a pair of disjoint bases?

Problem 4. When do two independence spaces have a common base?

In both cases we know of no clear set of necessary conditions which (as often seems to happen in the finite case) might turn out to be sufficient and the first step in both of these problems is to "guess" some form of an answer.

Notice that an answer to problem 4 would give a partial solution to problem 2 for the case where we are looking for two restricted families of sets to have a common transversal.

EXERCISES 3

1. Let $(\mathscr{I}_j: 1 \leqslant j \leqslant n)$ be a family of rank finite independence spaces on S and let the rank function of $\mathscr{I}_j$ be $\rho_j (1 \leqslant j \leqslant n)$. Prove that there exists a family $(B_i: 1 \leqslant i \leqslant n)$ of pairwise disjoint sets, where B_j is a base of $\mathscr{I}_j$, $1 \leqslant j \leqslant n$, if and only if for all finite subsets E of S

$$|E| \geqslant \sum_{i=1}^{n} (\rho_i S - \rho_i (S \backslash E)).$$

2. Let B be a distinguished subset of the vertex set of the infinite graph G. Let $S \subseteq V(G)$ and let $Z \in \mathscr{I}$ if $Z \subseteq S$ and there exist vertex disjoint paths for which the set of initial vertices is Z and the set of end vertices is contained in B. Prove that $\mathscr{I}$ is a pi-space on S and that if for each $z \in S$ the number of paths from z to B is finite then $\mathscr{I}$ is an independence space. (Pym and Perfect [70])

4. DUALITY IN INDEPENDENCE SPACES

Suppose we mimic the approach to duality in matroid theory on finite sets and define B^* to be a base of a "dual structure" if $S \backslash B^*$ is a base of the primal.

In preindependence spaces this is a non-starter since bases may not be defined.

In independence spaces this process is defined but has the serious disadvantage that it can take one outside the class of independence structures.

Example 1. Take S to be an infinite set and let k be a finite integer. Let $\mathscr{I}$ be $\{X : X \subseteq S, |X| \leqslant k\}$. Then $\mathscr{B}(\mathscr{I}) = \{B : |B| = k\}$ and hence if we define B^* to be a base if $S \backslash B^* \in \mathscr{B}(\mathscr{I})$ we see for example that $S \notin \mathscr{I}^*$ but every finite subset of S does.

We therefore reject this approach, and look at an alternative approach proposed by Las Vergnas [70c].

Given an independence space $(S, \mathscr{I})$ we define its *dual space* $(S, \mathscr{I}^*)$ by; X is independent in $\mathscr{I}^*$ if and only if for all finite subsets Y of X, $S \backslash Y$ is a spanning subset of $(S, \mathscr{I})$.

An alternative characterisation of $(S, \mathscr{I}^*)$ is given by:

(1) C^* is a circuit of $(S, \mathscr{I}^*)$ if and only if $C^* \subset\subset S$ and $S \backslash C^*$ is a hyperplane of the space $(S, \mathscr{I})$.

It is routine to prove that $(S, \mathscr{I}^*)$ as defined is an independence space. The price paid however is that in general

$$(S, \mathscr{I}^{**}) \neq (S, \mathscr{I}).$$

Example 2. If S is an infinite set, let $\mathscr{I} = \{X : |X| \leqslant k\}$ where k is some positive integer. Then $\mathscr{I}^* = 2^S$, so that $\mathscr{I}^{**} = \{\varnothing\}$ which is not equal to $\mathscr{I}$.

Suppose now we look at the interpretation of this duality in graphs—the original motivation of (finite) matroid duality.

If $G = (V, E)$ is an infinite graph, that is a graph with vertex set V, edge set E both possibly infinite, let $\mathscr{I}$ be the collection of subsets X of E which do not contain any (necessarily finite) cycle of G. Then since all minimal dependent sets ($=$cycles) are finite we trivially have.

THEOREM 1. *The cycle matroid of an infinite graph is an independence space.*

The situation with cocycles is not quite so straightforward. This is because a cocycle (minimal cutset) of an infinite graph can be infinite.

Example 3. Let G be a graph with two vertices a, b, and with an infinite number of edges joining a and b.

Define a *bond* of an infinite graph G to be a finite cocycle, and let $\mathbb{B}(G)$ be the collection of sets $X \subseteq E(G)$ such that X does not contain a bond.

THEOREM 2. *The set $\mathbb{B}(G)$ is an independence space on the edge set of a graph G.*

The (dual) relation between the cycle and bond matroids of an infinite graph is summarized in the following theorem—for a proof see Las Vergnas [70c].

THEOREM 3. *For any infinite graph G,*

$$(M(G))^* = \mathbb{B}(G).$$

But

$$(\mathbb{B}(G))^* = M(G)$$

if and only if between any two distinct vertices of G there do not exist an infinite number of edge disjoint paths.

Example 4. If G is the 2-vertex graph on (a, b) with edge set $E(G)$ of Example 3 then $\mathbb{B}(G)$ has as its independent sets all subsets of $E(G)$. Hence $(\mathbb{B}(G))^*$ has as its independent sets only the empty set, and is not isomorphic to the cycle matroid of G.

Despite the fact that $\mathscr{I}^{**} \neq \mathscr{I}$ in general, this duality proposed by Las Vergnas has some attractive features. For example he proves:

THEOREM 4. *If* $(S, \mathscr{I})$ *is an independence space, then*

$$(S, \mathscr{I}^{***}) = (S, \mathscr{I}^*).$$

This result is similar to one proved for a duality theory proposed by Brualdi and Scrimger [68] for a class of objects which they call *exchange systems*. The trouble with the Brualdi–Scrimger model is that in addition to $\mathscr{I}^{**} \neq \mathscr{I}$ the definition of an exchange system is so wide that the dual of an exchange system need not always exist.

In the next section we look at other approaches for duality which do not have these disadvantages.

EXERCISES 4

1. Call an independence space *connected* if it is not the direct sum of non-trivial spaces, (where direct sum is defined as in Section 5.2 for matroids). Prove that if $(S, \mathscr{I})$ is connected and $(S, \mathscr{I}^{**}) = (S, \mathscr{I})$ then S is a countable set. (Las Vergnas [70c])

2. Prove that a 2-connected graph with only finite cocycles is finite. (Las Vergnas [70c])

5. THE OPERATOR APPROACH TO DUALITY

Although the duality theory proposed in the last section has some attractive features it does suffer from the drawback that the dual of the dual is not in general the primal. In this section we follow the treatments of Higgs [69a] and Klee [71] and achieve this objective.

Sierpinski [52] described for Fréchet V-spaces the same duality as Whitney [35] had earlier described for matroids. We follow him and proceed as follows.

A *space* is a pair (S, Ω) where S is a set and $\Omega: 2^S \to 2^S$, satisfies

(1) $A \subseteq B \Rightarrow \Omega A \subseteq \Omega B$
(2) $x \in \Omega(A \backslash x)$ whenever $x \in \Omega A$.

The *dual* of the space (S, Ω) is defined to be the space (S, Ω^*) where

$$\Omega^* A = S \backslash \Omega(S \backslash A) \qquad (A \subseteq S).$$

It is easy to check that (S, Ω^*) is a space and that $(S, \Omega^{**}) = (S, \Omega)$.

Now Ω is an abstraction of the derived set operator of point set topology and associated with Ω we have an enlarging function $\langle \ \rangle_\Omega: 2^S \to 2^S$ given by

$$\langle A \rangle_\Omega = A \cup \Omega A \qquad (A \subseteq S).$$

Now in general $\langle \ \rangle_\Omega$ is not a closure operator.

However we *define* a space (S, Ω) to be an *H-matroid* if both $\langle \ \rangle_\Omega$ and $\langle \ \rangle_{\Omega^*}$ are closure operators.

THEOREM 1. *A space (S, Ω) is a H-matroid if and only if for all $A \subseteq S$,*

(i) $\Omega(A \cup \Omega A) \subseteq A \cup \Omega A$,
(ii) $\Omega(A \cap \Omega A) \supseteq A \cap \Omega A$.

Proof. Routine check of definitions. ∎

Klee's approach is far more exhaustive. He starts from a closure like operator and builds up an algebra of equivalent conditions. We give the outlines.

A function $f: 2^S \to 2^S$ is an *operator* if for all $A, B \subseteq S$,

(i) $A \subseteq fA$
(ii) $A \subseteq B \Rightarrow fA \subseteq fB$.

The *dual operator* f^* is the unique function $2^S \to 2^S$ which satisfies (i) and (ii) and also:

(*)   For any partition $\{p\} \cup X \cup Y$ of S with $\{p\}$ a singleton either $p \in fX$ or $p \in f^*Y$ but not both. That is

$$f^*(X) = X \cup \{x: x \notin f(S \backslash (X \cup x))\}.$$

For any operator f on S we can define the usual classes of subsets of S by:

(3) X is *dependent* if $\exists x \in X$ and $x \in f(X \backslash x)$, otherwise X is *independent*;

(4) X is *spanning* if $fX = S$;
(5) X is a *base* if it is spanning and independent;
(6) X is a *circuit* if it is minimal dependent;
(7) X is a *hyperplane* if it is maximal non-spanning.

A *K-matroid* is a set S and a operator f on S which satisfies:

(vwI) (*Very weakly idempotent*) If X is finite, Y is independent, and $X \subseteq fY$ then $f(X \cup Y) = fY$.
(vwE) (*Very weakly exchanging*) If X is finite and $p \in S$ is such that $p \cup Y$ is spanning, $p \in fY$, $p \notin f(Y \backslash X)$ then $x \in f(p \cup (X \backslash x))$ for some $x \in X$.

The conditions (vwI) and (vwE) are dual conditions in that if f satisfies (vwI), f^* satisfies (vwE) and therefore we have:

THEOREM 2. *The dual of a K-matroid is also a K-matroid.*

We first point out that both K-matroids and H-matroids are significantly wider classes of structure than independence spaces. For one thing, in general, they both have infinite circuits.

Example 1. Let G be an infinite graph and let f be the operator on $E(G)$ whose circuits are the usual (finite) cycles and also all 2-way (doubly infinite) paths. Then it can be checked that this uniquely determines a K-matroid. We call this the 2-*way path matroid* of G and denote it by $TWP(G)$.

Example 2. Let S be the edge set of an infinite graph G. Call C a circuit of M if $C \subseteq E(G)$ and satisfies the following four conditions.

(8) C is connected.
(9)   The vertices of G incident with C have degree 2, 3 or 4 in the subgraph C.
(10) If C is finite it has precisely one vertex of degree 4, none of degree 3; or precisely two of degree 3 and none of degree 4.
(11) If C is infinite it has no vertex of degree 4, at most one of degree 3, and contains at most two vertex disjoint singly infinite paths.

These conditions amount to saying that C is homeomorphic to one of the five graphs shown below, where an arrow indicates a singly infinite path.

Example 1 also shows that there exist K-matroids which are not H-matroids for if we take G_0 to be the infinite graph of Fig. 2(b).

If $\Omega(A)$ is defined to be the set of x in $E(G_0)$ such that $x \in P \subseteq A \cup x$ for some cycle or 2-way in finite path P in G then $(E(G), \Omega)$ is a space which in general is not a H-matroid.

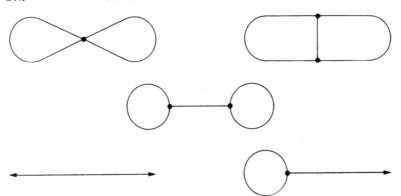

Figure 1

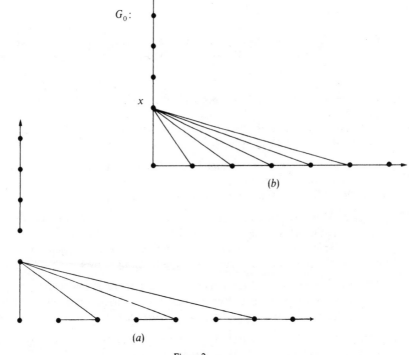

Figure 2

For in G_0 let A be the set of edges of the infinite subgraph shown in Fig. 2*a*, and let x be the edge of G_0 marked in Fig. 2*b*. Then letting σ denote the operator $\langle\ \rangle_\Omega$ the closure of A, σA in G_0 satisfies

$$x \notin \sigma A \text{ but } x \in \sigma\sigma A.$$

Thus σ is not a closure operator and thus $TWP(G_0)$ is not a H-matroid.

In general the 2-way path matroid of a graph is not a H-matroid—a characterization of those graphs which do not give H-matroids has been given by Higgs [69]. See also Bean [70].

As Klee remarks, the conditions (vwI) and (vwE) on an operator f are about the minimum requirements on f under which any interesting theory can be developed.

Progressively stronger requirements are:

(wI) (*Weakly idempotent*) If $x \in fY$ then $f(x \cup Y) = f(Y)$
(I)   (*Idempotent*) If $X \subseteq fY$ then $f(X \cup Y) = f(Y)$.

This last condition is equivalent to the condition that $f^2 = f$.

The dual conditions of these are:

(WE) (*Weakly exchange*) If $p \in fY$ and $p \notin f(Y \backslash x)$ then $x \in f(p \cup (Y \backslash x))$.
(E)   (*Exchange*) If $p \in f Y$ and $p \notin f(Y \backslash X)$ then $x \in f(p \cup (Y \backslash x))$ for some $x \in X$.

More precisely f satisfies WI (respectively I) if and only if its dual operator f^* satisfies WE (respectively E).

Closely connected with these different conditions are the following conditions on the circuits of an operator f:

(C)   If $p \in f Y$ there is a minimal $U \subseteq Y$ for which $p \in fU$ and U is independent.
(C_F) If $p \in f Y$ there is a finite $U \subseteq Y$ for which $p \in fU$.

The condition (C_F) is called the *finitary* condition and a *finitary matroid* is a K-matroid which satisfies C_F, wI and wE.

The condition C_F is clearly the condition that circuits of the Klee matroid exist and are finite. In other words, routine checking shows:

(12) The collection of independent sets of a finitary K-matroid forms an independence space and conversely the closure operator of an independence space forms a finitary K-matroid.

If f satisfies (C), respectively (C_F), its dual operator f^* must satisfy the conditions (H), respectively (H_F), stated below.

(H)   If $p \notin fY$ there is a maximal $V \supseteq Y$ for which $p \notin fV$ and $p \cup V$ is spanning.

(H_F)   If $p \notin fY$ there is a cofinite $V \supseteq Y$ for which $p \notin fV$.

Condition H_F demands that the hyperplanes should be cofinite, H demands the existence of hyperplanes.

The relationship between these different conditions is summarized very elegantly in the following theorem of Klee [71].

THEOREM 3. *The above conditions on operators are related as follows*

$$I \Rightarrow wI \Rightarrow vwI, \qquad E \Rightarrow wE \Rightarrow vwE,$$

$$wI \wedge H \Rightarrow I, \qquad wE \wedge C \Rightarrow E,$$

$$vwI \wedge C_F \Rightarrow I \wedge C, \qquad vwE \wedge H_F \Rightarrow E \wedge H,$$

$$wI \wedge H_F \Rightarrow C, \qquad wE \wedge C_F \Rightarrow H.$$

Notice to prove Theorem 3 it suffices to prove the identities on the left-hand side, since the right-hand relations then follow by duality.

We close by observing that K-matroids are not determined by either their collection of circuits or by their collection of independent sets.

Example 3. Let S be an infinite set and let f_1, f_2 be defined by

$$f_1 X = X \qquad \forall X \subseteq S$$

$$f_2 X = \begin{cases} X & X \text{ finite} \\ S & X \text{ infinite.} \end{cases}$$

Then (S, f_1) and (S, f_2) are distinct K-matroids, in fact they satisfy I, and wE but both have the same collection of circuits—namely the empty set.

Example 4. Let S be infinite and let (S_1, S_2) be a partition of S with S_1 and S_2 infinite. Define f_3 by

$$f_3 X = \begin{cases} X & |X| < \infty, \\ X \cup S_1 & |X \cap S_2| < \infty, |X \cap S_1| \geqslant \infty, \\ X \cap S_2 & |X \cap S_1| < \infty, |X \cap S_2| \geqslant \infty, \\ S & |X \cap S_1| \geqslant \infty, |X \cap S_2| \geqslant \infty. \end{cases}$$

Then the independent sets of the K-matroid (S, f_3) are the collection of finite subsets of S. Thus (S, f_3) has the same collection of independent sets as (S, f_2) but $f_3 \neq f_2$.

However provided some existence condition such as (C) is placed on the matroid we can characterize certain classes of K-matroids by their circuits. We have the following theorem of Klee [71].

THEOREM 4. *The circuits of any* (EC)-*K-matroid form an incomparable collection $\mathscr{C}$ of sets satisfying:*

(γ) *If C_1, C_2 are distinct circuits and $q \in C_1 \cap C_2$ there is a circuit C_3 with $C_3 \subseteq (C_1 \cup C_2) \backslash q$.*

Moreover for any collection $\mathscr{C}$ of incomparable sets satisfying (γ) there is a unique EC-matroid whose circuits are precisely the members of $\mathscr{C}$.

An example of Klee shows that unlike the finite situation the *weak circuit axiom* (γ) is not equivalent to the *strong circuit axiom* ($\hat{\gamma}$):

($\hat{\gamma}$) If C_1, C_2 are circuits with $p \in C_1 \backslash C_2$ and $q \in C_1 \cap C_2$ there is a circuit C_3 with $p \in C_3 \subseteq (C_1 \cup C_2) \backslash q$.

A second theorem of Klee reads exactly as Theorem 4 except that ($\hat{\gamma}$) and (wIEC) replace (γ) and (EC) respectively.

To conclude, these K-matroids of Klee seem to possess many of the properties which one would like an infinite extension of matroids to possess and have, no obvious disadvantages.

For a comprehensive discussion of the various properties of matroid operators, and for a proof of a transfinite extension of the greedy algorithm of Section 19.1 to finitary matroids ($=$ independence spaces) we refer to Klee [71].

EXERCISES 5

1. Are the cocycles of an infinite graph the circuits of a K-matroid?

NOTES ON CHAPTER 20

This chapter is meant to introduce the reader to infinite matroid theory. For a discussion of the interrelationship between the various axiom systems and structures, especially from the algebraic viewpoint, see Dlab [62]–[66a].

Mirsky's book [71] is the best source of information for the work done by Rado, Mirsky, Perfect, Brualdi and others on the relationship between infinite transversal theory and (pre)-independence spaces. Perfect [69], [71] and Brualdi [71a]–[71e], extend the linking theory developed in Chapters 8 and 13 to infinite sets. Some of the theory of submodular functions developed in Chapter 8 goes through to infinite sets—see Pym and Perfect [70]. Brualdi [70a], [71e] extends the covering and packing theorems of Chapter 8 to independence spaces but as pointed out, there are still some very basic open problems.

Sections 4 and 5 are based on the papers of Las Vergnas [70a], Higgs [69], [69a], Bean [70] and particularly Klee [71]. Example 5.2 was pointed out to me by A. W. Ingleton.

We have spent little time discussing the lattice theory approach to infinite matroids—this is mainly because the relevant lattice theory seems to be dominated by algebraic thinking—hence a predominance of finite character axioms.

Bibliography

AIGNER, M. and DOWLING, T. A.
70 Matching theorems for combinatorial geometries. *Bull. Amer. Math. Soc.* **76** (1970), 57–60.
71 An extension of the König–Egervary theorem (abstract). *Combinatorial Theory and its Applications* (Erdös, Renyi and Sos, eds), North-Holland (1971), 39–40.
71a Matching theory for combinatorial geometries. *Trans. Amer. Math. Soc.* **158** (1971), 231–245.

ASCHE, D. S.
66 Minimal dependent sets. *J. Australian Math. Soc.* **6** (1966), 259–262.

ATKIN, A. O. L.
72 Remark on a paper of Piff and Welsh. *J. Combinatorial Theory,* **13** (1972), 179–183.

BACLAWSKI, K.
75 Whitney numbers of geometric lattices. *Advances in Math.* **16** (1975), 125–138.

BAKER, K. A.
69 A generalisation of Sperner's lemma. *J. Combinatorial Theory,* **6** (1969), 224–226.

BASTERFIELD, J. G. and KELLY, L. M.
68 A characterisation of sets of *n* points which determine *n* hyperplanes. *Proc. Camb. Phil. Soc.* **64** (1968), 585–588.

BEAN, D. W. T.
69 Solution to a matroid problem posed by D. J. A. Welsh. *Canad. Math. Bull.* **12** (1969), 129–131.
70 Minimal edge sets meeting all circuits of a graph. *Proceedings of 1st Louisiana Conference on Combinatorics* (1970), 1–46.
71 Some results on a matroid conjecture of Harary and Welsh. *Proceedings of 2nd Louisiana Conference on Combinatorics* (1971), 147–167.
72 Refinements of a matroid *Proc. Third South Eastern Conf. Combinatorics, Graph Theory and Computing* (1972), 69–72.

BERGE, C.
58 Sur le couplage maximum d'un graphe. *C.R. Acad. Sciences* (Paris) **247** (1958), 258–259.
58a *Théorie des Graphes et ses Applications.* Dunod (Paris) (1958) translated as *The Theory of Graphs and its Applications.* Methuen (London) (1962).
69 The rank of a family of sets and some applications to graph theory. *Recent Progress in Combinatorics* (W. T. Tutte, ed.) (1969), 49–57.
73 *Graphs and Hypergraphs.* North Holland (1973).

BERLEKAMP, E. R.
68 *Algebraic Coding Theory.* McGraw-Hill (1968).

BIGGS, N.
71 *Finite Groups of Automorphisms.* Cambridge Univ. Press (1971).
74 *Algebraic Graph Theory.* Cambridge Univ. Press (1974).

BIRKHOFF, G.
35 Abstract linear dependence in lattices. *Amer. J. Math.* **57** (1935), 800–804.
46 On groups of automorphisms. *Rev. Un. Mat. Argentina,* 11 (1946), 155–157.
67 *Lattice Theory. A.M.S. Colloq. Publ.* (3rd ed.), Providence (1967).

BIRKHOFF, G. D.
12 A determinantal formula for the number of ways of colouring a map. *Annals of Math.* (2) **14** (1912), 42–46.

BIXBY, R. E.
72 Composition and decomposition of matroids and related topics. Thesis—Cornell University (1972).
74 *l*-matrices and a characterisation of binary matroids. *Discrete Math.* (to appear).
75 A composition for matroids. *J. Combinatorial Theory B* **18** (1975) 59–73.
75a On Reid's characterisation of the matroids representable over GF(3). *J. Combinatorial Th.* (to appear).

BLACKBURN, J. A., CRAPO, H. H. and HIGGS, D. A.
73 A catalogue of combinatorial geometries. *Math. Comput.* **27** (1973), 155–166.

BLAND, R. G.
75 Complementary orthogonal subspaces of R^n and orientability of matroids (to appear) (1975).

BLEICHER, M. N.
62 Complementation of independent subsets and subspaces of generalized vector spaces. *Fund. Math.* **51**, (1962), 1–7.

BLEICHER, M. N. and MARCZEWSKI, E.
62 Remarks on dependence relations and closure operators. *Colloq. Math.* **9** (1962), 209–212.

BLEICHER, M. N. and PRESTON, G. B.
61 Abstract linear dependence relations. *Publ. Math. Debrecen* **8** (1961), 55–63.

BOLLOBAS, B.
69 A lower bound for the number of non-isomorphic matroids. *J. Combinatorial Theory,* 7 (1969), 366–368.

BONDY, J. A.
72 Presentations of transversal matroids. *J. London Math. Soc.* (2), **5** (1972), 289–292.
72a Transversal matroids, base orderable matroids, and graphs. *Quart. J. Math.* (Oxford), **23** (1972), 81–89.

BONDY, J. A. and WELSH, D. J. A.
71 Some results on transversal matroids and constructions for identically self dual matroids. *Quart. J. Math.* (Oxford), **22** (1971), 435–451.

BROWN, T. J.
72 Deriving closure relations with the exchange property I & II. *Kansas State Univ. Technical Report Nos.* 35 & 36 (1972).
72a Finitary exchange relations and F-products. *Kansas State Univ. Technical Report No.* 37 (1972).

74   Transversal theory and F-products. *J. Combinatorial Theory A* **17** (1974), 290–299.

BRUALDI, R. A.

69   Comments on bases in dependence structures, *Bull. Australian Math. Soc.* **2** (1969), 161–169.

69a An extension of Banach's mapping theorem. *Proc. Amer. Math. Soc.* **20** (1969), 520–526.

69b A very general theorem on systems of distinct representatives. *Trans. Amer. Math. Soc.* **140** (1969), 149–160.

70   Common transversals and strong exchange systems. *J. Combinatorial Theory* (8), **3** (1970), 307–329.

70a Admissible mappings between dependence spaces. *Proc. London Math. Soc.* (3), **21** (1970), 296–312.

71   Matchings in arbitrary graphs. *Proc. Cambridge Phil. Soc.* (1971) **69**, 401–407.

71a Menger's theorem and matroids. *J. London Math. Soc.* **4** (1971), 46–50.

71b Generalized transversal theory. Springer Lecture Notes 211 (1971), 5–31.

71c Induced matroids. *Proc. Amer. Math. Soc.* **29** (1971), 213–221.

71d Strong transfinite version of König's duality theorem. *Monatshefte für Math.* **75** (1971), 106–110.

71e On families of finite independence structures. *Proc. London Math. Soc.* **22** (1971), 265–293.

71f Common partial transversals and integral matrices. *Trans. Amer. Math. Soc.* **155** (1971), 475–492.

71g A general theorem concerning common transversals. *Combinatorial Mathematics and its Applications* (D. J. A. Welsh, ed.). Academic Press (1971), 9–60.

74   On fundamental transversal matroids, *Proc. Amer. Math. Soc.* **45** (1974), 151–156.

74a Weighted join-semilattices and transversal matroids. *Trans. Amer. Math. Soc.* **191** (1974), 317–328.

75b Networks and the Shannon switching game, (Preprint) (1974) (to be published in *Delta*).

BRUALDI, R. A. and DINOLT, G. W.

72   Characterisation of transversal matroids and their presentations. *J. Combinatorial Theory,* **12** (1972), 268–286.

BRUALDI, R. A. and PYM, J. S.

71   A general linking theorem in directed graphs. *Studies in Pure Maths. Festschrift for R. Rado* (L. Mirsky, ed.) Academic Press (1971), 17–31.

BRUALDI, R. A. and SCRIMGER. E. B.

68   Exchange systems, matchings, and transversals. *J. Combinatorial Theory,* **5** (1968), 244–257.

BRUNO, J. and WEINBERG, L.

71   The principal minors of a matroid. *Linear Algebra and its Applications,* **4** (1971), 17–54.

BRUTER, C. P.

69   Vue d'ensemble sur la theorie des matroides. *Bull. Soc. Math. France,* **17** (1969), 1–48.

69   Sur différents problèmes posés en théorie des matroides. *Publ. Inst. de Stat. Univ. de Paris* **3** (1969), 1–96.

70 *Les Matroides*. Dunod (1970).
70a Applications des notions d'extrémalité et de stabilité à la théorie des matroides. *C.R. Acad. Sci.* **271** (1970), 542–546.
71 *Theorie des Matroides*. Springer Lecture Notes **211** (1971).
71a Deformation des matroides. *C.R. Acad. Sci.* **273** (1971), 8–9.
74 *Elements de la Théorie des Matroides*. Springer Lecture Notes **387** (1974).

BRYLAWSKI, T. H.
71 A combinatorial model for series-parallel networks. *Trans. Amer. Math. Soc.* **154** (1971), 1–22.
71a The Möbius function as a decomposition invariant. *Proc. Waterloo Conf. on Möbius Algebras* (1971), 143–148.
72 A decomposition for combinatorial geometries. *Trans. Amer. Math. Soc.* **171** (1972), 235–282.
72a Modular constructions for combinatorial geometries (1972) (preprint).
73 An affine representation for transversal geometries (1973) (preprint).
73a Some properties of basic families of subsets. *Discrete Math.* **6** (1973), 333–341.
74a Reconstructing combinatorial geometries. *Graphs and Combinatorics* (Springer Lecture Notes) **406** (1974), 227–235.
75 On the non-reconstructibility of combinatorial geometries (1975) (preprint).

BRYLAWSKI, T. H., KELLY, D. G. and LUCAS, T. D.
74 *Matroids and Combinatorial Geometries*. Lecture Note Series, University of North Carolina, Chapel Hill (1974).

BRYLAWSKI, T. H. and LUCAS, T. D.
75 Uniquely representable combinatorial geometries (preprint) (1975).

BUMCROT, R. J.
69 *Modern Projective Geometry*. Holt Reinhart & Wilson (1969).

CAMION, P.
68 Modules unimodulaires. *J. Combinatorial Theory*, **4** (1968), 301–362.

CARDY, S.
73 The proof of and generalisations to a conjecture by Baker and Essam. *Discrete Math.* **4** (1973), 101–122.

COHN, P. M.
65 *Universal Algebra*. Harper & Row (1965).

COOK, S. A.
71 The complexity of theorem proving procedures, *Proc. Third Annual ACM Symposium on Theory of Computing*. May (1971), 151–158.

CRAPO, H. H.
65 Single element extensions of matroids. *J. Res. Nat. Bur. Stand.* **69B** (1965), 57–65.
66 The Möbius function of a lattice. *J. Combinatorial Theory*, **1** (1966), 126–131.
67 Structure theory for geometric lattices. *Rend. Sem. Mat. Univ. Padova*, **38** (1967), 14–22.
67a Geometric duality. *Rend. Sem. Math. Univ. Padova*, **38** (1967), 23–26.
67b A higher invariant for matroids. *J. Combinatorial Theory* **2** (1967), 406–417.
68 Möbius inversion in lattices. *Archiv. der Math.* **19** (1968), 595–607.
68a The joining of exchange geometries. *J. Math. Mech.* **17** (1968), 837–852.
69 The Tutte polynomial. *Aequationes Math.* **3** (1969), 211–229.

69a Geometric duality and the Dilworth completion. *Calgary International Conference*. Gordon & Breach (1969), 37–47.
70   Erecting geometries. *Annals. N. Y. Acad. Sc.* **175** (1970), 89–92.
70a Erecting geometries. *Proceedings of 2nd Chapel Hill Conference on Combinatorial Math.* (1970), 74–99.
71   Constructions in combinatorial geometries. (N.S.F. Advanced Science Seminar in Combinatorial Theory) (*Notes,* Bowdoin College) (1971).
71a Orthogonal representations of combinatorial geometries. *Atti del Convegno di Geometria Combinatoria* (Perugia) (1971), 175–186.

CRAPO, H. H. and CHEUNG, A.
74   On relative position in extensions of combinatorial geometries (preprint) (1974).

CRAPO, H. H. and ROTA, G. C.
70   *On the Foundations of Combinatorial Theory: Combinatorial Geometries.* M.I.T. Press Cambridge, Mass. (1970).
71   Simplicial geometries. *Amer. Math. Soc. Symposium XIX*, (1971), 71–75.
71a Geometric lattices. *Trends in Lattice Theory* (H. Abbot, ed.) (1971), 127–165.

CRAWLEY, P. and DILWORTH, R. P.
73   *Algebraic Theory of Lattices.* Prentice Hall (1973).

DAMERELL, M. R. and MILNER, E. C.
74   Necessary and sufficient conditions for transversals of countable set systems. *J. Combinatorial Theory,* **A 17** (1974), 350–375.

DAVIES, J. and MCDIARMID, C. J. H.
75   Disjoint common partial transversals and exchange structures (preprint) (1975).

DE BRUIJN, N. G. and ERDÖS, P.
48   On a combinatorial problem. *Indigationes Math.* **10** (1948), 421–423.

DEMBOWSKI, P. and WAGNER, A.
60   Some characterisations of finite projective spaces. *Arch. Math.* **11**, (1960) 465–469.

DENNISTON, R. H. F.
75   Some new 5-designs. *Bull. London. Math. Soc.* (to appear).

DE SOUSA, J.
71   Disjoint common transversals. *Combinatorial Mathematics & its Applications* (D. J. A. Welsh, ed.), Academic Press (1971).

DE SOUSA, J. and WELSH, D. J. A.
72   A characterisation of binary transversal matroids. *J. Math. Analysis Appl.* **40** (1) (1972), 55–59.

DILWORTH, R. P.
41   Arithmetic theory of Birkhoff lattices. *Duke Math. J.* **8** (1941), 286–299.
41a Ideals in Birkhoff lattices *Trans. Amer. Math. Soc.* **49** (1941), 325–353.
44   Dependence relations in a semimodular lattice. *Duke Math. J.* **11** (1944), 575–587.
50   The structure of relatively complemented lattices. *Annal. Math.* **51** (1950), 348–359.
50a A decomposition theorem for partially ordered sets. *Annal. Math.* (2) **51** (1950), 161–166.
54   Proof of a conjecture on finite modular lattices. *Annal. Math.* **60** (1954), 359–364.
60   Some combinatorial problems on partially ordered sets. *Proc. Symposia Applied Maths.* **10** (Combinatorial Analysis) (Providence) (1960), 85–90.

DILWORTH, R. P. and GREENE, C.
71 A counter-example to the generalisation of Sperner's theorem. *J. Combinatorial Theory*, **10** (1971), 18–21.

DINOLT, G. W.
71 An extremal problem for non-separable matroids. *Theorie des Matroides* (C. P. Bruter, Ed.) Springer Lecture Notes in Mathematics **211** (1971), 31–50.

DIRAC, G. A.
52 A property of 4-chromatic graphs and some remarks on critical graphs. *J. London Math. Soc.* **27** (1952), 85–92.
57 A theorem of R. L. Brooks and a conjecture of H. Hadwiger. *Proc. London Math.* **7** (1957). 161–195.

DLAB, V.
62 General algebraic dependence relations. *Publ. Math. Debrecen* **9** (1962), 324–355.
62a On the dependence relation over abelian groups. *Publ. Math. Debrecen* **9** (1962), 75–80.
65 Axiomatic treatment of bases in arbitrary sets. *Czech. Math. J.* **15** (1965), 554–563.
65a The role of the "finite character property" in the theory of dependence. *Comment. Math. Univ. Carolinae*, **6** (1) (1965), 97–104.
66 Algebraic dependence structures. *Zeit. Math. Logik und Grundlagen*, **12** (1966), 345–377.
66a General algebraic dependence structures and some applications. *Coll. Math.* **14** (1966), 265–273.
66b Dependence over modules. *Czech. Math. J.* **16** (1966), 137–157.
69 Rank theory of modules. *Fund. Math. LXIV* (1969), 313–324.

DOOB, M.
73 An interrelation between line graphs, eigenvalues and matroids. *J. Combinatorial Theory*, **15** (1973), 40–50.

DOWLING, T. A.
71 Codes, packings and the critical problem. *Atti del Convegno di Geometria Combinatoria e sue Applicazioni*, (Perugia) (1971), 210–224.
73 A class of geometric lattices based on finite groups. *J. Combinatorial Theory*, **13**, (1973), 61–87.
73a A q-analog of the partition lattice, *A Survey of Combinatorial Theory*. North Holland (1973), 101–115.
75 A note on complementing permutations in finite lattices *J. Combinatorial Theory* (to appear) (1975).

DOWLING, T. A. and KELLY, D. G.
72 Elementary strong maps between combinatorial geometries. *Institute of Statistics Series North Carolina* (1972). **821**
74 Elementary strong maps and transversal geometries. *Discrete Math.* **7** (1974), 209–225.

DOWLING, T. A. and WILSON, R. M.
74 The slimmest geometric lattices. *Trans. Amer. Math. Soc.* **196** (1974), 203–215.
75 Whitney number inequalities for geometric lattices. *Proc. Amer. Math. Soc.* **47** (1975), 504–512.

DOYEN, J.
68 Constructions groupales d'espaces linéaires finis. *Bull. Acad. Royale Belgique*, **2** (1968), 144–156.

DOYEN, J. and HUBAUT, X.
71   Finite regular locally projective spaces. *Math. Zeit,* **119** (1971), 83–88.

DUBREUIL-JACOTIN, M. L., LESIEUR, L. and CROISOT, P.
53   Lecons sur la théorie des treillis, des structures algébriques ordonnées et des treillis géométrique. Gauthin-Villars (Paris) (1953).

DUCHAMP, A.
72   Sur certains matroides définis à partir des graphes. *C.R. Acad. Sci.* **274** (1972), 9–11.
74   Une caractérisation des matroides binaires par une propriété de coloration. *C.R. Acad. Sci.* **278** (1974), 1163.

DUFFIN, R.
65   Topology of series-parallel networks. *J. Math. Anal. Appl.* **10** (1965), 303–318.

DUNSTAN, F. D. J.
73   Some aspects of combinatorial theory (D. Phil. thesis) Oxford (1973).

DUNSTAN, F. D. J., INGLETON, A. W. and WELSH, D. J. A.
72   Supermatroids *Combinatorics* (Inst. Math. & its Appl.) (D. J. A. Welsh and D. R. Woodall, eds.) (1972), 72–123.

DUNSTAN, F. D. J. and WELSH, D. J. A.
73   A greedy algorithm for solving a certain class of linear programmes. *Math. Programming.* **5** (1973), 338–353.

EDMONDS, J.
65   Minimum partition of a matroid into independent subsets, *J. Res. Nat. Bur. Stand.* **69B** (1965), 67–72.
65a Lehman's switching game and a theorem of Tutte and Nash-Williams. *J. Res. Nat. Bur. Stand.* **69B** (1965), 73–77.
65b Paths, trees and flowers. *Canad. J. Math.* **17** (1965), 449–467.
65c On the surface duality of linear graphs. *J. Res. Nat. Bur. Stand.* **69B** (1965), 121–123.
65d Maximum matching and a polyhedron with 0–1 vertices. *J. Res. Nat. Bur. Stand.* **69B** (1965), 125–130.
67   Systems of distinct representatives and linear algebra. *J. Res. Nat. Bur. Stand.* **71B** (1967), 241–245.
67a Optimum branchings. *Lectures in Appl. Math.* **11** *(Mathematics of the Decision Sciences)* (1967), 346–361.
67b Matroid partition, *Lectures in Appl. Math.* **11** *(Mathematics of the Decision Sciences)* (1967), 335–346.
67c Matroids and the greedy algorithm *International Symposium on Programming* (Princeton) (1967).
70   Submodular functions, matroids and certain polyhedra. *Proc. Int. Conf. on Combinatorics (Calgary),* Gordon and Breach (New York) (1970), 69–87.
71   Matroids and the greedy algorithm. *Math. Programming,* **1** (1971), 127–136.

EDMONDS, J. and FULKERSON, D. R.
65   Transversals and matroid partition. *J. Res. Nat. Bur. Stand.* **69B** (1965), 147–153.
70   Bottleneck extrema. *J. Combinatorial Theory,* **8** (1970), 299–306.

EDMONDS, J. and ROTA, G. C.
66   Submodular set functions (Abstract) *Waterloo Combinatorics Conference* (1966).

ENTRINGER, R. C. and JACKSON, D. E.
69 A direct proof of the duality theorem of König. *Archiv. der Math.* **20** (1969), 666–667.

FINKBEINER, D. T.
51 A general dependence relation for lattices. *Proc. Amer. Math. Soc.* **2** (1951), 756–659.
60 A semimodular imbedding of lattices. *Canad. J. Math.* (1960) 582–591.

FOLKMAN, J.
66 The homology groups of a lattice. *J. Math. Mech.* **15** (4) (1966), 631–636.

FORD, L. R. and FULKERSON, D. R.
58 Network flow and systems of representatives. *Canad. J. Math.* **10** (1958), 78–84.
62 *Flows in Networks.* Princeton University Press (1962).

FORTUIN, C. M., KASTELEYN, P. W. and GINIBRE, J.
71 Correlation inequalities on some partially ordered sets. *Commun. Math. Phys.* **22** (1971), 89–103.

FOURNIER, J-C.
70 Sur la représentation sur un corps des matroides à sept et huit éléments. *C.R. Acad. Sci.* (Paris) *Ser.* **270** (1970), 810–813.
71 Une propriété de connexité caractéristique des matroides graphiques. *C.R. Acad. Sci.* (Paris), **272** (1971), 1092–1093.
73 Orthogonalité généralisée entre matroides et application à la représentation des graphes sur les surfaces. *C.R. Acad. Sci.* (Paris), **276** (1973), 835–838.
74 Une relation de séparation entre cocircuits d'un matroide. *J. Combinatorial Theory,* **16** (1974), 181–190.

FRUCHT, R.
28 Herstellung von Graphen mit vorgegebener abstrakten Gruppe. *Compositio Math.* **6** (1938), 239–250.
48 On the construction of partially ordered systems with a given group of automorphisms. *Rev. Un. Mat. Argentina,* **13** (1948), 12–18.

FULKERSON, D. R.
64 The maximum number of disjoint permutations contained in a matrix of zeros and ones. *Canad. J. Math.* **16** (1964), 729–735.
68 Networks, frames and blocking systems. *Mathematics of the Decision Sciences, (Lectures in Applied Math.), Amer. Math. Soc.* **11** (1968), 303–335.
70 Blocking polyhedra. *Graph Theory and Its Applications* (B. Harris, ed.), Academic Press (1970), 93–112.
71 Disjoint common partial transversals of two families of sets. *Studies in Pure Mathematics. Festschrift for R. Rado* (L. Mirsky, ed.), Academic Press (1971), 107–113.
72 Antiblocking polyhedra. *J. Combinatorial Theory* **B12** (1972), 50–71.

GALE, D.
57 A theorem on flows in networks. *Pacific J. Math.* **7** (1957), 1073–1082.
68 Optimal assignments in an ordered set: an application of matroid theory. *J. Combinatorial Theory,* **4** (1968), 176–180.

GALLANT, S. I.
70 The group of a matroid. *Tech. Report* 70–13 (1970), Stanford University.
71 Automorphism groups of matroids and clutters. *Tech. Report* 71–16 (1971), Stanford University.

GORN, S.
40   On incidence geometry. *Bull. Amer. Math. Soc.* **46** (1940), 158–167.
GRAHAM, R. L. and HARPER, L. H.
69   Some results on matching in bipartite graphs. *Siam. J. Appl. Math.* **17** (1969), 1017–1022.
GRAHAM. R. L. AND ROTHSCHILD. B.
71   Rota's geometric analogue to Ramsey's theorem. *Proc. Symp. Pure Math.* **19** (Amer. Math. Soc.) XI (*Combinatorics*) (1971), 101–104
GRAVER, J. E.
66   Lectures on the Theory of Matroids. University of Alberta (1966).
73   Boolean designs and self dual matroids (1973) (preprint).
GRAVES, W. H.
71   An algebra associated to a combinatorial geometry. *Bull. Amer. Math. Soc.* **77** (1971), 757–61.
71a A categorical approach to combinatorial geometry. *J. Combinatorial Theory*, **11** (1971), 222–232.
GREENE, C.
70   A rank inequality for finite geometric lattices. *J. Combinatorial Theory*, **9** (1970), 357–364.
71   An inequality for the Möbius function of a geometric lattice. *Proc. Conf. on Möbius Algebras* (Waterloo) (1971).
73   A multiple exchange property for bases. *Proc. Amer. Math. Soc.* **39** (1973). 45–50.
73a On the Möbius algebra of a partially ordered set. *Advances in Math.* **10** (1973), 177–187.
74   Weight enumeration and the geometry of linear codes (preprint) (1974).
74a Sperner families and partitions of a partially ordered set. *Combinatorics* (Math. Centre. Tract. **56**) (M. Hall and J. H. Van Lint, Eds) Amsterdam (1974) 91–106.
GREENE, C. and KLEITMAN, D. J
74   The structure of Sperner *k*-families (preprint) (1974).
GREENE, C., KLEITMAN, D. J. and MAGNANTI, T. L.
74   Complementary trees and independent matchings. *Studies in Appl. Math. LIII* (1974), 57–64.
GREENE, C. and MAGNANTI, T. L.
74   Some abstract pivot algorithms. *Siam J. Appl. Math.* (to appear) (preprint 1974).
GRÜNBAUM, B.
67   *Convex Polytopes*. Wiley-Interscience (N.Y.) (1967).
HALL, M. (Jr)
48   Distinct representatives of subsets. *Bull. Amer. Math. Soc.* **51** (1948), 922–926.
56   An algorithm for distinct representatives. *Amer. Math. Monthly*, **63** (1956), 716–717.
58   A survey of combinatorial analysis. "*Surveys in Applied Math. IV. Some Aspects of Analysis and Probability*". Wiley (1958) 37–104.
67   *Combinatorial Theory*. Blaisdell (Waltham) (1967).
HALL, P.
35   On representatives of subsets. *J. London Math. Soc.* **10** (1935), 26–30.
HANANI, H.

60 On quadruple systems. *Canad. J. Math.* **12** (1960), 145–157.
63 On some tactical configurations. *Canad. J. Math.* **15** (1963), 702–722.
75 Balanced incomplete block designs and related designs. *Discrete Math.* **11** (1975), 255–369.

HARARY, F.
69 *Graph Theory*. Addison Wesley (1969).

HARARY, F., PIFF, M. J. and WELSH, D. J. A.
72 On the automorphism group of a matroid. *Discrete Math.* **2** (1972), 163–171.

HARARY, F. and TUTTE, W. T.
65 A dual form of Kuratowski's theorem. *Canad. Math. Bull.* **8** (1965), 17–20.

HARARY, F. and WELSH, D. J. A.
69 Matroids versus graphs. *The Many Facets of Graph Theory*. Springer Lecture Notes **110** (1969), 155–170.

HARPER, L. H.
74 The morphology of partially ordered sets. *J. Combinatorial Theory* **17A** (1974), 44–59.

HARPER, L. H. and ROTA, G. C.
71 Matching theory: an introduction. *Advances in Probability,* **1** (1971), 169–213.

HARPER, L. H. and SAVAGE, J. E.
72 On the complexity of the marriage problem. *Advances in Math.* **9** (1972), 299–312.

HARTMANIS, J,
56 Two embedding theorems for finite lattices. *Proc. Amer. Math. Soc.* **7** (1956), 571–577.
57 A note on the lattice of geometries. *Proc. Amer. Math. Soc.* **8** (1957), 560–562.
59 Lattice theory of generalized partitions. *Canad. J. Math.* **11** (1959), 97–106.
61 Generalized partitions and lattice embedding theorems. *Proc. Symp. Pure Math.* **2** *(Lattice Theory)* *Amer. Math. Soc.* (1961), 22–30.

HAUTUS, M. L. J.
70 Stabilisation, controllability and observability of linear autonomous systems. *Nederl. Akad. Wetensch. Proc. Ser.* **A73** (1970), 448–55.

HELGASON, T.
70 On geometric hypergraphs. *Proc. Conf. N. Carolina* (1970), 276–284.
74 Aspects of the theory of hypermatroids. *Hypergraph Seminar,* Springer Lecture Notes , **411** (1974), 191–214.

HERON, A. P.
69 Transversal matroids: (Dissertation for Diploma in Math.) (Oxford) (1969).
72 Matroid polynomials. *Combinatorics* (Institute of Math. & Appl.) D. J. A. Welsh and D. R. Woodall, eds) (1972), 164–203.
72a Some topics in matroid theory (D.Phil. Thesis). Oxford (1972).
73 A property of the hyperplanes of a matroid and an extension of Dilworth's theorem. *J. Math. Analysis Appl.* **42** (1973), 119–132.

HIGGINS, P. J.
59 Disjoint transversals of subsets. *Canad. J. Math.* **11** (1959), 280–285.

HIGGS, D. A.
66 Maps of geometries. *J. London Math. Soc.* **41** (1966), 612–618.
68 Strong maps of geometries. *J. Combinatorial Theory,* **5** (1968), 185–191.

412 BIBLIOGRAPHY

69 Infinite graphs and matroids. *Recent Progress in Combinatorics,* Academic Press (1969), 245–253.
69a Matroids and duality. *Colloq. Math. XX* (1969), 215–220.
70 Equicardinality of bases in B-matroids. *Canad. Math. Bull.* (1970), 861–862.

HOFFMANN, A. J. and KUHN, H. W.
56 On systems of distinct representatives. *Linear Inequalities and Related Systems. Ann. Math. Studies* **38** (Princeton) (1956), 199–206.
63 Systems of distinct representatives and linear programming. *Amer. Math. Monthly,* **63** (1956), 455–460.

HOGGAR, S. G.
74 Chromatic polynomials and logarithmic concavity. *J. Combinatorial Theory* **16B** (1974), 248–255.

HORN, A.
55 A characterisation of unions of linearly independent sets. *J. London Math. Soc.* **30** (1955), 494–496.

HUGHES, N. J. S.
63 Steinitz' exchange theorem for infinite bases. *Compositio Math.* **15** (1963), 113–118.
66 Steinitz' exchange theorem for infinite bases II. *Compositio Math.* **17** (1966), 152–155.

INGLETON, A. W.
59 A note on independence functions and rank. *J. London Math. Soc.* **34** (1959), 49–56.
71 Representation of matroids. *Combinatorial Mathematics and its Applications* (D. J. A. Welsh, ed.) Academic Press (London & New York) (1971), 149–169.
71a A geometrical characterization of transversal independence structures. *Bull. London Math. Soc.* **3** (1971), 47–51.
71b Conditions for representability and transversality of matroids. *Proc. Fr. Br. Conf. (1970),* Springer Lecture Notes **211** (1971), 62–67.
75 Non base-orderable matroids. *Proc. Fifth British Conference on Combinatorics* (C. St. J. A. Nash-Williams and J. Sheehan, eds) *Utilitas* (1975) (to appear).

INGLETON, A. W. and MAIN, R. A.
75 Non-algebraic matroids exist, *Bull. London Math. Soc.* **7** (1975), 144–146.

INGLETON, A. W. and PIFF, M. J.
73 Gammoids and transversal matroids. *J. Combinatorial Theory,* **15** (1973), 51–68.

IRI, M.
68 A critical review of the matroid-theoretical and the algebraic-topological theory of networks. *RAAG Memoirs* **4A** (1968), 39–46.
69 The maximum-rank minimum-term-rank theorem for the pivotal transforms of a matrix. *Linear Alg. and its Applications,* **2** (1969), 427–446.

KANTOR, W. M.
69 Characterisations of finite projective and affine spaces. *Canad. J. Math.* **21** (1969), 64–75.
74 2-Transitive designs. *Combinatorics* (M. Hall and J. H. Van Lint, eds) (Math. Cent. Amsterdam) (1974), 44–98
74a Dimension and embedding theorems for geometric lattices. *J. Combinatorial Theory,* **A17** (1974), 173–196.
75 Envelopes of geometric lattices. *J. Combinatorial Theory,* **A18** (1975), 12–27.

KARP, R. M.

72 Reducibility among combinatorial problems. *Complexity of Computer Computations*, Plenum Press (New York) (1972), 85–105.

KELLY, D. G. and ROTA, G. C.

73 Some problems in combinatorial geometry. *A Survey of Combinatorial Theory*, North Holland (1973), 309–313.

KENNEDY, D.

73 Factorisation and majors of strong maps. *Lecture Note Series Univ. North Carolina* (1973).

KERTESZ, A.

60 On independent sets of elements in algebra. *Acta Sci. Math. Szeged*, **21** (1960), 260–269.

KLEE, V.

71 The greedy algorithm for finitary and cofinitary matroids. *Combinatorics* Amer. Math. Soc. Publ. XIX (1971), 137–152.

KLEITMAN, D.

66 Families of non-disjoint subsets. *J. Combinatorial Theory* (1966), 153–155.

KNUTH, D. E.

73 Matroid partitioning. (Preprint) Stanford University STAN-CS-73-342 (1973), 1–12.

74 The asymptotic number of geometries. *J. Combinatorial Theory* **A17** (1974), 398–401.

74a Random matroids. (Preprint) Stanford University STAN-CS-74-453 (1974), 1–26.

KÖNIG, D.

31 Graphen és matrixok. *Mat. Fiz. Lapok*, **38** (1931), 116–119.

36 *Theorie der endlichen und unendlich en Graphen*. (Leipzig) (1936).

KRUSKAL, J. B.

56 On the shortest spanning subgraph of a graph and the travelling salesman problem. *Proc. Amer. Math. Soc.* **7** (1956), 48–49.

KUNDU, S. and LAWLER, E. L.

73 A matroid generalisation of a theorem of Mendelsohn and Dulmage. *Discrete Math.* **4**, (1973), 159–163.

KURATOWSKI, K.

30 Sur le problème des courbes gauches en topologie. *Fund. Math.* **15**(1930), 271–283.

KURTZ, D. C.

72 A note on concavity properties of triangular arrays of numbers. *J. Combinatorial Theory*, **13** (1972), 135–139.

LANG, S.

64 *Introduction to Algebraic Geometry*, Interscience Tracts in Pure and Applied Mathematics, No. 5 (Wiley–Interscience) (1964).

65 *Algebra*, Addison Wesley (1965).

LAS VERGNAS, M.

70 Sur les systèmes de représents distincts d'une famille d'ensembles. *C. R. Acad. Sci.* Paris, **270** (1970), 501–503.

70a Sur un théoréme de Rado. *C. R. Acad. Sci.* (Paris), **270** (1970), 733–735.

70b Sur la dualité en théorie des matroides. *C. R. Acad. Sci.* (Paris), **13** (1970), 804–806.

70c Sur la dualité en théories des matroides. *Proc. Fr. Br. Conf.* (1970), Springer Lecture Notes **211**, 67–86.

73 Degree constrained subgraphs and matroid theory. *Colloq. Math. Societatis Janos Bolyai 10* (Infinite and Finite Sets), Keszthely (Hungary) (1973) 1473–1502.

75 Matroides orientables. *C. R. Acad. Sci.* (Paris), **280A** (1975), 61–64.

75a Sur les extensions principales d'un matroide. *C. R. Acad. Sci.* (Paris), **280A** (1975), 187–190.

75b Extensions normales d'un matroide, polynôme de Tutte d'un morphisme. *C.R. Acad. Sci.* (Paris), **280** (1975), 1479–1482.

75c On certain constructions concerning combinatorial geometries. *Proc. Fifth British Conference on Combinatorics* (C. St. J. A. Nash-Williams and J. Sheehan, eds) *Utilitas* (1975) (to appear).

LAWLER, E. L.

66 Covering problems: duality relations and a new method of solution. *SIAM Journal,* **14** (1966), 1115–1133.

70 Optimal matroid intersections. *Combinatorial Structures and their Applications.* Gordon & Breach (New York) (1970), 233–235.

71 Matroids with parity conditions: a new class of combinatorial optimization problems. *Electronics Research Laboratory, Berkeley Mem. No.* ERL-M334 (1971).

75 *Combinatorial Optimisation; Networks and Matroids* (to be published by Holt, Rinehart & Winston, New York) (1975).

75a Matroid intersection algorithms. *Mathematical Programming* (1975) (to appear).

LAZARSON, T.

57 Independence functions in algebra. (Thesis) University of London (1957).

58 The representation problem for independence functions. *J. London Math. Soc.* **33** (1958), 21–25.

LEHMAN, A.

64 A solution of the Shannon switching game. *J. Soc. Indust. Appl. Math.* **12** (1964), 687–725.

LINDSTROM, B.

73 On the vector representation of induced matroids. *Bull. Lond. Math. Soc.* **5** (1973), 85–90.

LOVÁSZ, L.

71 A brief survey of matroid theory. *Mathematikai Lapok.* **22** (1971), 249–267.

LOVASZ, L. and RECSKI, A.

73 On the sum of matroids. *Acta. Math. Acad. Sci. Hung.* **24** (1973), 329–333.

LUBELL, D.

66 A short proof of Sperner's lemma. *J. Combinatorial Theory,* **1** (1966), 299.

MACLANE, S.

36 Some interpretations of abstract linear dependence in terms of projective geometry. *Amer. J. Math.* **58** (1936), 236–240.

37 A combinatorial condition for planar graphs. *Fund. Math.* **28** (1937), 22–32.

38 A lattice formulation for transcendence degrees and *p*-bases. *Duke Math. J.* **4** (1938), 455–468.

MCCARTHY, P. J.

75 Matchings in graphs II. *Discrete Math.* **11** (1975), 141–147.

McDiarmid. C. J. H.

72 Strict gammoids and rank functions. *Bull. London Math. Soc.* 4 (1972), 196–198.

73 Independence structures and submodular functions. *Bull. London Math. Soc.* 5 (1973), 18–20.

73a An application of a reduction principle of Rado to the study of common transversals. *Mathematika, 20* (1973), 83–86.

73b A note on a theorem of R. Rado on independent transversals. *Bull. London Math. Soc.* 5 (1973), 315–316.

74 Path-partition structures of graphs and digraphs. *Proc. London Math. Soc.* 29 (1974), 750–768.

74a Independence structures and linking in graphs. (D.Phil. Thesis) Oxford (1974).

75 Extensions of Menger's theorem. *Quart. J. Math.* (Oxford) (1975), (to appear).

75a An exchange theorem for independence structures. *Proc. Amer. Math. Soc.* 47 (1975), 513–514.

75b On the number of systems of distinct representatives in an independence structure. *J. Math. Analysis Appl.* (1975) (to appear).

75c Rado's theorem for polymatroids. *Proc. Cambridge Phil. Soc.* (1975) (to appear).

MacWilliams, F. J.

63 A theorem on the distribution of weights in a systematic code. *Bell System Tech. J.* 42 (1963), 79–94.

Maeda, F.

52 Matroid lattices of infinite length. *J. Sci. Hiroshima Univ.* (A) 15 (1952), 177–182.

63 Parallel mappings and comparability theorems in affine matroid lattices. *J. Sci. Hiroshima Univ.* (A) 27 (1963), 85–96.

64 Perspectivity of points in matroid lattices. *J. Sci. Hiroshima Univ.* (A) 28 (1964), 10–32.

Maeda, F. and Maeda, S.

70 *Theory of Symmetric Lattices.* Springer-Verlag (1970).

Magnanti, T. L.

74 Complementary bases of a matroid. *Discrete Math.* 8 (1974), 355–361.

Main, R. A.

73 Matroids and block designs. (M.Sc. thesis) Oxford (1973).

Main, R. A. and Welsh, D. J. A.

75 A note on matroids and block designs. *J. Aust. Math. Soc.* (1975) (to appear).

Mann, H. B. and Ryser, H. J.

53 Systems of distinct representatives. *Amer. Math. Monthly* 60 (1953), 387–401.

Mason, J. H.

70 A characterization of transversal independence spaces. (Ph.D. Thesis) Univ. of Wisconsin (1970).

70a A characterisation of transversal independence spaces. *Proc. Fr. Br. Conf.* Springer Lecture Notes, 211 (1970), 86–95.

70b A characterisation of transversal independence spaces. *Combinatorial Structures and Their Applications.* Gordon & Breach (1970), 257–261.

71 Geometrical realization of combinatorial geometries. *Proc. Amer. Math. Soc.* 30 (1) (1971), 15–21.

72 On a class of matroids arising from paths in graphs. *Proc. London Math. Soc.* (3) 25 (1972), 55–74.

72a Matroids: unimodal conjectures and Motzkin's theorem. *Combinatorics*

(Institute of Math. & Appl.) (D. J. A. Welsh and D. R. Woodall, eds) (1972), 207–221.

73 Maximal families of pairwise disjoint maximal proper chains in a geometric lattice. *J. London Math. Soc.* **6** (1973), 539–542.

MAURER, S. B.

73 Matroid basis graphs 1. *J. Combinatorial Theory B* (1973) **14**, 216–240.

73a Matroid basis graphs 2. *J. Combinatorial Theory B* (1973) **15**. 121–145.

73b Basis graphs of pregeometries. *Bull. Amer. Math. Soc.* **79** (1973), 783–786.

75 Intervals in matroid basis graphs. *Discrete Math.* **11** (1975), 147–161.

75a A maximum-rank minimum term-rank theorem for matroids. *Lin. Alg. & Appl.* (1975) (to appear).

MENDELSOHN, N. S.

71 Intersection numbers of t-designs. *Studies in Pure Math.* Academic Press (1971).

MENDELSOHN, N. S. and DULMAGE, A. L.

58 Some generalisations of the problem of distinct representatives. *Canad. J. Math.* **10** (1958), 230–242.

MENGER, K.

27 Zur allgemeinen Kurventheorie. *Fund. Math.* **16** (1927), 96–115.

MILNER, E. C.

74 The matching matroid (unpublished) (1974).

75 Independent transversals for countable set systems. (to be published) (1975) *J. Combinatorial Theory.*

MINTY, G. J.

66 On the axiomatic foundations of the theories of directed linear graphs, electrical networks and network programming. *Journ. Math. Mech.* **15** (1966), 485–520.

MIRSKY, L.

66 Transversals of subsets. *Quart. J. Math* (Oxford) (2) **17** (1966). 58–60.

67 Systems of representatives with repetition. *Proc. Cambridge Phil. Soc.* **63** (1967), 1135–1140.

67a A hierarchy of structures. *Some Aspects of Transversal Theory* (Seminar), Chapter 6 Univ. of Sheffield (1967).

68 A theorem on common transversals. *Math. Annalen,* **177** (1968), 49–53.

68a Combinatorial theorems and integral matrices. *J. Combinatorial Theory,* **5** (1968), 30–44.

69 Hall's criterion as a 'self-refining' result. *Monatsh. für Math.* **73** (1969), 139–146.

69a Transversal theory and the study of abstract independence. *J. Math. Anal. Appl.* **25** (1969), 209–217.

69b Pure and applied combinatorics. *Bull. Inst. Math. Appl.* **5** (1969), 2–4.

71 *Transversal Theory.* Academic Press (London) 1971).

71a A proof of Rado's theorem on independent transversals. *Studies in Pure Mathematics. Festschrift for R. Rado* (L. Mirsky, ed.) Academic Press (London) (1971), 151–156

73 The rank formula of Nash-Williams as a source of covering and packing theorems. *J. Math. Analysis Appl.* **4** (1973), 328–347.

MIRSKY, L. and PERFECT; H.

66 Systems of representatives. *J. Math. Analysis Appl.* **15** (1966), 520–568.

67 Applications of the notion of independence to combinatorial analysis. *J. Combinatorial Theory,* **2** (1967), 327–357.

68 Comments on certain combinatorial theorems of Ford and Fulkerson. *Archiv. der Math.* **19** (1968). 413–416.

MISARE, R.
68 On the representability of matroids over finite fields (unpublished) (1968).

MOTZKIN, T.
51 The lines and planes connecting the points of a finite set. *Trans. Amer. Math. Soc.* **70** (1951), 451–464.

MURTY, U. S. R.
69 Sylvester matroids. *Recent Progress in Combinatorics.* Wiley (1969), 283–286.
70 Equicardinal matroids and finite geometries. *Combinatorial Structures and their Applications.* Gordon and Breach (1970), 289–293.
70a Matroids with Sylvester property. *Aequationes Math.* **4** (1970), 44–50.
71 On the number of bases of a matroid. *Proc. Second Louisiana Conf. on Combinatorics.* Louisiana State Univ., Baton Rouge (1971), 387–410.
71a Equicardinal matroids. *J. Combinatorial Theory*, **11** (1971), 120–126.
74 Extremal critically connected matroids. *Discrete Math.* **8** (1974), 49–59.

NARAYANAN, H.
74 Theory of matroids and network analysis. (Doctoral Thesis) Indian Institute of Technology Bombay (1974).

NARAYANAN, H. and VARTAK, M. N.
72 A partition for matroids (1972) (preprint).
73 Gammoids, base orderable matroids and series-parallel networks (1973) (preprint).

NASH-WILLIAMS, C. ST. J. A.
61 Edge-disjoint spanning trees of finite graphs. *J. London Math. Soc.* **36** (1961), 445–450.
64 Decomposition of finite graphs into forests. *J. London Math. Soc.* **39** (1964), 12.
66 An application of matroids to graph theory. *Theory of Graphs International Symposium* (Rome). Dunod (Paris) (1966). 263–265.

ORE, O.
42 Theory of equivalence relations. *Duke Math. J.* **9** (1942), 573–627.
55 Graphs and matching theorems. *Duke Math. J.* **22** (1955), 625–639.
62 *Theory of Graphs. Amer. Math. Soc. Colloq. Publ.* **38**, Providence (1962).
67 *The Four Colour Problem.* Academic Press (1967).

OSTRAND, P. A.
70 Systems of distinct representatives, 11. *J. Math. Analysis Appl.* **32** (1970), 1–4.

PERFECT, H.
66 Symmetrized form of P. Hall's theorem on distinct representatives. *Quart. J. Math.* (Oxford) (2) **17** (1966), 303–306.
68 Applications of Menger's graph theorem. *J. Math. Analysis Appl.* **22** (1968), 96–111.
69 Independence spaces and combinatorial problems. *Proc. London Math. Soc.* **19** (1969), 17–30.
69a Remark on a criterion for common transversals. *Glasgow Math. J.* **10** (1969), 66–67.
69b A generalisation of Rado's theorem on independent transversals. *Proc. Cambridge Phil. Soc.* **66** (1969), 513–515.
71 Marginal elements in transversal theory. *Studies in Pure Mathematics. Festschrift for R. Rado* (L. Mirsky, ed.). Academic Press (London) (1971), 203–220.

418   BIBLIOGRAPHY

72   Independent transversals with constraints. *J. London Math. Soc.* **5** (1972), 385–386.
73   Associated transversal structures. *J. London Math. Soc.* **6** (1973), 626–628.

      PERFECT, H. and PYM, J. S.

66   An extension of Banach's mapping theorem. with applications to problems concerning common representatives. *Proc. Cambridge Phil. Soc.* **62** (1966), 187–192.

      PIFF. M. J.

69   The representability of matroids. (*Dissertation for Diploma in Advanced Mathematics*) Oxford (1969).
71   Properties of finite character of independence spaces. *Mathematika*, **18** (1971), 201–208.
72   Some problems in combinatorial theory. (D.Phil. thesis) Oxford (1972).
73   An upper bound for the number of matroids. *J. Combinatorial Theory*, **13** (1973), 241–245.

      PIFF, M. J. and WELSH, D. J. A.

70   On the vector representation of matroids. *J. London Math. Soc.* **2** (1970), 284–288.

      PIFF, M. J. and WELSH, D. J. A.

71   The number of combinatorial geometries. *Bull. London Math. Soc.* **3** (1971), 55–56.

      PYM, J. S.

69   A proof of Menger's theorem. *Monatsh. für Math.* **73** (1969), 81–83.
69a   A proof of the linkage theorem. *J. Math. Analysis Appl.* **27** (1969), 636–638.
69b   The linking of sets in graphs. *J. London Math. Soc.* **44** (1969), 542–550.

      PYM, J. S. and PERFECT, H.

70   Submodular functions and independence structures. *J. Math. Analysis Appl.* **30** (1970), 1–31.

      RADO, R.

33   Bemerkungen zur Kombinatorik im Anschluss an Untersuchungen von Herr D. König. *Sitzber. Berliner Math. Ges.* **32** (1933), 60–75.
38   A theorem on general measure functions. *Proc. London Math. Soc.* (2) **44** (1938), 61–91.
42   A theorem on independence relations. *Quart. J. Math.* (Oxford), **13** (1942), 83–89.
43   Theorems on linear combinatorial topology and general measure. *Annals of Math.* (2) **44** (1943), 228–276.
49   Axiomatic treatment of rank in infinite sets. *Canad. J. Math.* **1** (1949), 337–343.
49a   Factorization of even graphs. *Quart. J. Math.* (Oxford), **20** (1949), 95–104.
57   Note on independence functions. *Proc. London Math. Soc.* **7** (1957), 300–320.
62   A combinatorial theorem on vector spaces. *J. London Math. Soc.* **37** (1962), 351–353.
66   Abstract linear dependence. *Colloq. Math.* **14** (1966), 258–264.
67   On the number of systems of distinct representatives of sets. *J. London Math. Soc.* **42** (1967), 107–109.
67a   Note on the transfinite case of Hall's theorem on representatives. *J. London Math. Soc.* **42** (1967), 321–324.

RAMSEY, F. P.

30 On a problem of formal logic. *Proc. London Math. Soc.* **30** (1930), 264–286.

RANKIN, R. A.

66 Common transversals. *Proc. Edinburgh Math. Soc.* (2) **15** (1966–7), 147–154.

READ, R. C.

68 An introduction to chromatic polynomials. *J. Combinatorial Theory,* **4** (1968), 52–71.

RECSKI, A.

72 On the sum of matroids with applications in electric network theory. (Doctoral Dissertation) Budapest (1972).

73 On partitional matroids with applications: *Colloq. Math. Societatis János 10* (Infinite and Finite Sets), Keszthely (Hungary) (1973), 1169–1179.

73a Matroids in network analysis (in Hungarian). *Proc. Res. Inst. Telecomm.* (1973), 89–99.

74 Enumerating partitional matroids. *Studia Sci. Math. Hung.* **9** (1974), 247–249.

75 On the sum of matroids II. *Proc. Fifth British Conference on Combinatorics* (C. St. J. A. Nash-Williams and J. Sheehan, eds) *Utilitas* (1975) (to appear).

REID, R.

73 Obstructions to representation of combinatorial geometries (unpublished) (1973).

ROBERT, P.

68 Sur l'axiomatique des systèmes générateurs, des rangs. Thèse Rennes 1966. *Bull. Soc. Math. France,* **14** (1968).

ROBERTSON, A. P. and WESTON, J. D.

58 A general basis theorem. *Proc. Edinburgh Math. Soc.* (2) **11** (1958–9), 139–141.

ROCKAFELLAR, R. T.

69 The elementary vectors of a subspace of R^N. *Combinatorial Math. and its Applications.* (R. C. Bose and T. A. Dowling, eds). Chapel Hill, N. C. (1969), 104–127.

ROTA, G. C.

64 On the foundations of combinatorial theory I. *Z. Wahrsch,* **2** (1964), 340–368.

70 Combinatorial theory, old and new. *Int. Cong. Math.* (Nice) (1970) **3,** 229–233.

71 On the combinatorics of the Euler characteristic. *Studies in Pure Maths. Festschrift for R. Rado* (L. Mirsky, ed.) Academic Press (1971), 221–234.

RYSER, H. J.

57 Combinatorial properties of matrices of zeros and ones. *Canad. J. Math.* **9** (1957), 371–377.

60 Matrices of zeros and ones. *Bull. Amer. Math. Soc.* **66** (1960), 442–464.

63 *Combinatorial Mathematics.* Carus Mathematical Monographs, **14,** *Math. Assoc. of America* (1963).

SABIDUSSI, G.

57 Graphs with given group and given graph-theoretical properties. *Canad. J. Math.* **9** (1957), 515–525.

SACHS, D.

61 Partition and modulated lattices. *Pac. J. Math.* **11** (1) (1961), 325–345.

61a Identities in finite partition lattices. *Proc. Amer. Math. Soc.* **6** (1961), 944.

66 Reciprocity in matroid lattices. *Rend. Sem. Mat. Univ. di Padova,* **36** (1966), 66–79.

70 Graphs, matroids and geometric lattices. *J. Combinatorial Theory,* **9** (1970), 192–199.

71 Geometric mappings on geometric lattices. *Canad. J. Math.* **23** (1971), 22–35.

SASAKI, U. and FUJIWARA, S.
52 The decomposition of matroid lattices. *J. Sci. Hiroshima Univ.* **A15** (1952), 183–188.
52a The characterisation of partition lattices. *J. Sci. Hiroshima Univ.* **A15** (1952), 189–201.

SCHRIJVER, A.
74 Linking systems. *Math. Centre* (Amsterdam) (1974) (preprint).
74a Linking systems II. *Math. Centre* (Amsterdam) (1975) (preprint).
75 Linking systems, matroids, and bipartite graphs. *Proc. Fifth British Conference on Combinatorics* (C. St. J. A. Nash-Williams and J. Sheehan, eds) *Utilitas* (1975) (to appear).

SEYMOUR, P. D.
75 An extension of the max-flow min-cut theorem. (1975) (to appear).
75a The matroids with the Menger property (1975) (to appear). *J. Combinatorial Theory*.
75b Matroids, hypergraphs and the max-flow min-cut theorem. (D.Phil. thesis) Oxford (1975).
75c The max-flow min-cut property in matroids. *Proc. Fifth British Conference on Combinatorics* (C. St. J. A. Nash-Williams and J. Sheehan, eds) *Utilitas* (1975) (to appear).
75d Matroid representation over GF(3) (1975) *J. Combinatorial Theory* (to appear).

SEYMOUR, P. D. and WELSH, D. J. A.
75 Combinatorial applications of an inequality from statistical mechanics. *Math. Proc. Cambridge Phil. Soc.* **77** (1975), 485–497.

SIERPINSKI, W.
52 *General Topology*. Univ. of Toronto. Mathematical Expositions, **7**, Toronto (1952).

SIMÕES-PEREIRA, J. M. S.
72 On subgraphs as matroid cells. *Math. Zeit.* **127** (1972), 315–322.
73 On matroids on edge sets of graphs with connected subgraphs as circuits. *Proc. Amer. Math. Soc.* **38** (1973), 503–6.

SLEPIAN, D.
60 Some further theory of group codes. *Bell System Tech. J.* **39** (1960), 1219–1252.

SMITH, C. A. B.
72 Electric currents in regular matroids. *Combinatorics* (Institute of Math. & Appl.) (D. J. A. Welsh & D. R. Woodall. eds) (1972), 262–285.
74 Patroids. *J. Combinatorial Theory*, **16** (1974), 64–76.

SMOLIAR, S. W.
75 On the free matrix representations of transversal geometries. *J. Combinatorial Theory* **A**   (1975), 60–71.

SPERNER, E.
28 Ein Satz über Untermengen einer endlichen Menge. *Math. Zeitschrift*, **27** (1928), 544–548.

STANLEY, R.
71 Modular elements of geometric lattices. *Algebra Universalis*, **1** (1971), 214–217.

STEINITZ, E.
30 *Algebraische Theorie der Korper*. (Berlin) (1930).

STOER, J. and WITZGALL, C.
70 *Convexity and Optimisation in Finite Dimensions I.* Springer-Verlag (Berlin) (1970).

STONESIFER, R.
75 Logarithmic concavity for a class of geometric lattices. *J. Combinatorial Theory,* **A18** (1975), 216–219.

SZASZ, G.
63 *Introduction to Lattice Theory,* Academic Press (New York) (1963).

TODD, J. A.
59 On representations of the Mathieu groups as collineation groups. *J. London Math. Soc.* **34** (1959), 406–416.

TUCKER, A. W.
63 Combinatorial theory underlying linear programmes. *Recent Advances in Mathematical Programming.* (R. E. Graves and P. Wolfe, eds), McGraw-Hill (New York) (1963), 1–16.

TURAN, P.
41 Eine Extremalaufgabe aus der Graphen theorie. *Mat. Fiz. Lapok,* **48** (1941), 436–452.

TUTTE, W. T.
47 A ring in graph theory. *Proc. Cambridge Phil. Soc.* **43** (1947), 26–40.
47a The factorization of linear graphs. *J. London Math. Soc.* **22** (1947), 107–111.
54 A contribution to the theory of chromatic polynomials. *Canad. J. Math.* **6** (1954), 80–91.
54a A short proof of the factor theorem for finite graphs. *Canad. J. Math.* **6** (1954), 346–352.
56 A class of Abelian groups. *Canad. J. Math.* **8** (1956), 13–28.
58 A homotopy theorem for matroids I and II. *Trans. Amer. Math. Soc.* **88** (1958), 144–174.
59 Matroids and graphs. *Trans. Amer. Math. Soc.* **90** (1959), 527–552.
60 An algorithm for determining whether a given binary matroid is graphic. *Proc. Amer. Math. Soc.* **11** (1960), 905–917.
61 On the problem of decomposing a graph into n connected factors. *J. London Math. Soc.* **36** (1961), 221–230.
61a A theory of 3-connected graphs. *Indag. Math.* **23** (1961), 441–455.
64 From matrices to graphs. *Canad. J. Math.* **16** (1964), 108–127.
65 Lectures on matroids. *J. Res. Nat. Bur. Stand.* **69B** (1965), 1–48.
65a Menger's theorem for matroids. *J. Res. Nat. Bur. Stand.* **69B** (1965), 49–53.
66 Connectivity in matroids. *Canad. J. Math.* **18** (1966), 1301–1324.
66a On the algebraic theory of graph colorings. *J. Combinatorial Theory,* **1** (1966), 15–50.
67 On even matroids. *J. Res. Nat. Bur. Stand.* **71B** (1967), 213–214.
69 Projective geometry and the 4-color problem. *Recent Progress in Combinatorics* (W. T. Tutte, ed.). Academic Press (1969), 199–207.
70 *Introduction to the Theory of Matroids.* American Elsevier (New York) (1970).
74 Codichromatic graphs. *J. Combinatorial Theory,* **16** (1974), 168–175.
74a A problem of spanning trees. *Quart. J. Math.* (Oxford), **25** (1974), 253–255.

ULLTANG, Ø.
72 Systems of independent representatives. *J. London Math. Soc.* **4** (1972), 745–752.

VAMOS, P.
68 On the representation of independence structures (unpublished).
71 A necessary and sufficient condition for a matroid to be linear (unpublished).

VAN DER WAERDEN, B. L.
27 Ein Satz über Klasseneinteilungen von endlichen Mengen. *Abl. Math. Sem. Hamburg Univ.* 5 (1927), 185–188.
37 *Moderne Algebra* (2nd edn). Springer (Berlin) (1937).

VAN LINT, J. H.
71 *Coding Theory.* Springer Lecture Notes, 201 (1971).

VEBLEN, O.
12 An application of modular equations in analysis situs. *Ann. of Math.* 14 (1912), 86–94.

WAGNER, K.
37 Uber eine Eigenschaft der ebenen Komplexe. *Math. Ann.* 114 (1937), 570–590.

WEINBERG, L.
72 Planar graphs and matroids. *Graph Theory and Applications.* Springer Lecture Notes (1972), 313–329.

WEISNER. L.
35 Abstract theory of inversion of finite series. *Trans. Amer. Math. Soc.* 38 (1935), 474–484.

WELSH, D. J. A.
67 On dependence in matroids. *Canad. Math. Bull.* 10 (1967), 599–603.
68 Some applications of a theorem of Rado. *Mathematika,* 15 (1968), 199–203.
68a Kruskal's theorem for matroids. *Proc. Cambridge Phil. Soc.* 64 (1968), 3–4.
69 A bound for the number of matroids. *J. Combinatorial Theory,* 6 (1969), 313–316.
69a Euler and bipartite matroids. *J. Combinatorial Theory,* 6 (1969), 375–377.
69b Transversal theory and matroids. *Canad. J. Math.* 21 (1969), 1323–1330.
69c On the hyperplanes of a matroid. *Proc. Cambridge Phil. Soc.* 65 (1969), 11–18.
70 On matroid theorems of Edmonds and Rado. *J. London Math. Soc.* 2 (1970), 251–256.
71 Generalised versions of Hall's theorem. *J. Combinatorial Theory 10* 2B (1971), 95–101.
71a Related classes of set functions. *Studies in Pure Mathematics.* Festschrift for R. Rado (L. Mirsky, ed.). Academic Press (1971), 261–274.
71b Combinatorial problems in matroid theory. *Combinatorial Mathematics and its Applications.* Academic Press (1971). 291–307.
71c Matroids and block designs. *Proc. Renc. Fr. Br.* (1970), Springer Lecture Notes, 211 (1971), 95–106.

WHITE, N. L.
70 Coordinatisation of combinatorial geometries. *Pros. Conf. North Carolina,* (1970), 484–486.
74 A basis extension property. *J. London Math. Soc.* 7 (1974), 662–665.
75 The bracket ring of a combinatorial geometry I. *Trans. Amer. Math. Soc.* 202 (1975). 79–95.

WHITNEY. H.
31. A theorem on graphs. *Annals. of Math.* (2) 32 (1931), 378–390.
32 Non-separable and planar graphs. *Trans. Amer. Math. Soc.* 34 (1932), 339–362.

32a Congruent graphs and the connectivity of graphs. *Amer. J. Math.* **54** (1932), 150–168.

32b A logical expansion in mathematics. *Bull. Amer. Math. Soc.* **38** (1932), 572–579.

32c The coloring of graphs. *Annals of Math.* **33** (1932), 688–718.

33 A set of topological invariants for graphs. *Amer. J. Math.* **55** (1933), 221–235.

33a On the classification of graphs. *Amer. J. Math.* **55** (1933), 236–244.

33b 2-isomorphic graphs. *Amer. J. Math.* **55** (1933), 245–254.

33c Planar graphs. *Fund. Math.* **21** (1933), 73–84.

35 On the abstract properties of linear dependence. *Amer. J. Math.* **57** (1935), 509–533.

WILDE, P. J.

75 The Euler circuit theorem for binary matroids. *J. Combinatorial Theory,* **18** (1975), 260–265.

75a A partial ordering of matroids. *Proc. Fifth British Conference on Combinatorics* (C. St. J. A. Nash-Williams and J. Sheehan, eds) *Utilitas* (1975) (to appear).

WILSON, R. J.

73 An introduction to matroid theory. *Amer. Math. Monthly,* **80** (1973), 500–525.

WITT, E.

38 Uber Steinersche Systeme. *Abh. Math. Sem. Univ. Hamburg,* **12** (1938), 265–275.

WOODALL, D. R.

73 Applications of polymatroids and linear programming to transversals and graphs. *Combinatorics.* Cambridge University Press (1973), 195–201.

74 An exchange theorem for bases of matroids. *J. Combinatorial Theory,* (*B*) **16** (1974). 227–229.

75 The induction of matroids by graphs. *J. London Math. Soc.* (1975) (to appear).

YOUNG, P. and EDMONDS, J.

72 Matroid designs. *J. Res. Nat. Bur. Stan.* **72B** (1972), 15–44.

YOUNG, P., MURTY, U. S. R. and EDMONDS, J.

70 Equicardinal matroids and matroid designs. *Combinatorial Mathematics and its Applications* (Chapel Hill) (1970), 498–542.

ZASLAVSKY, T.

75 Facing up to arrangements: face count formulas for partitions of space by hyperplanes. *Memoirs Amer. Math. Soc.* **154** (1975).

Index of Symbols

Throughout M denotes a matroid

$A(J)$	union of sets A_i. $i \in J$.
$A(M)$	automorphism group of M
$AG(n, q)$	affine geometry of rank $n + 1$ over $GF(q)$
$A_p(G)$	point (vertex) automorphism group of the graph G
$\mathscr{A}$	a family of sets
$b(M)$	number of bases of M
$\mathbb{B}(G)$	bond matroid of the infinite graph G
$\mathscr{B}(M)$	set of bases of M
$c(M)$	number of circuits of M
$c(M; q)$	critical number of M with respect to field $GF(q)$
$\text{co}\,(X)$	convex hull of the set X
C_n	circuit with n elements
$C(M)$	characteristic set of M
$\mathscr{C}(M)$	set of circuits of M
$C(x, B)$	fundamental circuit of element x in the base B
$\mathscr{D}_5$	forbidden sublattice for a modular lattice to be distributive
$D(M), D^*(M)$	circuit and cocircuit matrices of the binary matroid M
$D(b, v, r, k, \lambda)$	balanced incomplete block design with parameters b, v, r, k, λ
$E(G)$	edge set of graph G
$G \backslash U$	graph obtained by deleting the elements of the set $U \subseteq V(G) \cup E(G)$ from the graph G
$G \vert T$	restriction of graph G to edge set $T \subseteq E(G)$
$G . T$	contraction of graph G to edge set $T \subseteq E(G)$
G'_e, G''_e	graph G with the edge e respectively deleted or contracted
$GF(q)$	Galois field with q elements
$h(M)$	number of hyperplanes of M
$H(A)$	hereditary closure of the set $A \subseteq \mathbb{R}^n$
$\mathscr{H}(M)$	collection of hyperplanes of M
$i(M)$	number of independent sets of M
$i_k(M)$	number of independent sets of cardinality k of M

$R(M; x, y)$	rank generating function of M
S	in Chapters 1–19, S is a finite set; in Chapter 20 we allow it to be infinite
$\mathscr{S}(M)$	collection of spanning sets of M
$S(A, B)$	collection of separating sets of A and B
$T(M; x, y)$	Tutte polynomial of M
$U_{k,n}$	uniform matroid of rank k on n elements
$V(G)$	vertex set of graph G
$w(e)$	width of element e
$W_k = W_k(M)$	kth Whitney number of M
$\mathscr{W}_n$	wheel of n vertices
$\mathscr{W}^n$	whirl on n vertices
$\mathbb{Z}$	integers
$\mathbb{Z}^+$	non-negative integers

Greek Symbols

$\beta(M)$	matroid invariant see Section 15.4
$\delta(A)$	set of vertices of graph adjacent with a member of A
$\delta(.,.)$	identity function of incidence algebra
$\zeta(.,.)$	zeta function of incidence algebra
$\kappa(G)$	(vertex) connectivity of graph G
$\lambda(G)$	T-connectivity of G
$\lambda(M)$	connectivity of M
$\lambda^{(n)}$	$\lambda(\lambda - 1)\ldots(\lambda - n + 1)$
$\lambda(e)$	length of element e in a matroid
μ	general submodular set function
$\mu(.,.)$	Möbius function of partially ordered set
ρ	rank function of matroid, and groundset rank function of polymatroid
ρ^*	corank function
ρ^T	rank function of contraction of M to $T \subseteq S$
σ	closure operator of matroid
σ_T, σ^T	closure operators of $M \mid T$, $M.T$ respectively
$\chi(G), \chi(M)$	chromatic number of graph G, or respectively matroid M
$\Delta(S, T; E)$	bipartite graph with edges joining vertex sets S and T
$\Delta(\mathscr{A})$	bipartite graph determined by family of sets $\mathscr{A}$
$\Delta(M)$	matroid induced from M by directed bipartite graph Δ
$\Delta(\mathscr{I})$	collection of sets linked in Δ to a member of $\mathscr{I}$
Ω	derivation operator

Other Symbols

∂A	set of vertices adjacent to A but not belonging to A
$\lvert x \rvert$	modulus of vector x
$\lVert f \rVert$	domain of chain f
$\bigvee_{i \in I} a_i$	supremum of set of elements $\{a_i : i \in I\}$ in a lattice
$\bigwedge_{i \in I} a_i$	infimum of set of elements $\{a_i : i \in I\}$ in a lattice
xy'	(scalar) product of vectors x, y
A'	transpose of matrix A
$M_1 \vee M_2$	union of matroids M_1 and M_2
$M_1 + M_2$	direct sum of matroids M_1 and M_2
$\mathcal{L}_1 \times \mathcal{L}_2$	direct product of lattices $\mathcal{L}_1$ and $\mathcal{L}_2$
$G \prec H$	graph G is homeomorphic from H
$M_1 \to M_2$	the identity map is strong between the matroids M_1 and M_2
$x \sim A$	element x depends on the set A
$X \subset\subset A$	X a finite subset of A.

Subject Index

428